T0178110

Lecture Notes in Computer Science 13497

More information about this series at https://link.springer.com/bookseries/558

Stefano Berretti · Guan-Ming Su (Eds.)

Smart Multimedia

Third International Conference, ICSM 2022
Marseille, France, August 25–27, 2022
Revised Selected Papers

 Springer

Editors
Stefano Berretti 🆔
University of Florence
Florence, Italy

Guan-Ming Su
Dolby Labs
California, CA, USA

ISSN 0302-9743 ISSN 1611-3349 (electronic)
Lecture Notes in Computer Science
ISBN 978-3-031-22060-9 ISBN 978-3-031-22061-6 (eBook)
https://doi.org/10.1007/978-3-031-22061-6

This Springer imprint is published by the registered company Springer Nature Switzerland AG
The registered company address is: Gewerbestrasse 11, 6330 Cham, Switzerland

Preface

Welcome to the proceedings of the third International Conference on Smart Multimedia. In this conference, we continue our mission to build a focused forum to enable exchange of the latest advances in multimedia technologies, systems, and applications from the research, development, and industrial perspectives. The conference is organized in a single-track format to maximize the goals. We believe this approach has started to bring the desired fruitful results and persist in assisting new interdisciplinary collaborations and accelerating projects that need expertise in multiple disciplines. The ultimate goals will be going beyond the existing data-driven approach to arrive at efficient and effective smart solutions around the multimedia areas.

In our third conference, we received around 70 submissions. The review process was single blind. Every submission was assigned to at least three members of the Technical Program Committee to review and discuss. Only around 40% could be accepted in the regular tracks due to limited space. Papers in the conference were organized into 9 sessions plus invited talks and tutorials, and the Springer LNCS proceedings containing the papers are arranged following these sessions into 9 topics. The topics in the proceedings include Machine Learning for Multimedia, Image Processing, Multimedia Applications, Multimedia for Medicine and Healthcare, Smart Homes, Multimedia Environments and Metaverse, Deep Learning on Video and Music, Haptic, and Industrial. These areas cover a broad range of disciplines in the wider field of Smart Multimedia.

We thank several donors to the conference on Smart Multimedia whose gifts have not only assisted in covering the cost of organizing the conference, but have also made the variety of social events possible. We also would like to acknowledge the excellent efforts from all organization committee members, and leadership from General Co-chair, Sylvain Bouchigny (CEA-List Paris) and Arnaud Leleve (INSA, Lyon), Program Co-chair, Mohamed Daoudi (IMT, Lille), and Special Sessions Chair, Troy McDaniel (Arizona State University). In addition, we appreciate the contribution from authors and feedback from reviewers with many constructive suggestions to ensure the paper quality for this conference.

December 2022

Stefano Berretti
Program Chair
Guan-Ming Su
Industrial/Short Chair
Anup Basu
General Co-chair

Organization

General Chairs

S. Bouchigny   CEA-List Paris, France
A. Leleve   INSA Lyon, France
A. Basu   University of Alberta, Canada

Program Chair

S. Berretti   University of Florence, Italy

Program Co-chairs

M. Daoudi   IMT Lille, France
I. Curcio   Nokia, Finland

Area-Chair

A. K. Singh   NIT Patna, India

Industrial/Short Chair

G.-M. Su   Dolby, USA

Industrial/Short Co-chairs

H. Azari   Microsoft, USA
T. Wang   Stats Perform, USA
F. Zhai   Huawei, Hong Kong

Special Sessions Chair

T. McDaniel   Arizona State University, USA

Co-chair

H. Venkateswara   Arizona State University, USA

Registration Chairs

Y.-P. Huang	Taipei Tech, Taiwan
X. Sun	University of Alberta, Canada

Publicity Chairs

L. Gu	Riken, Japan
F. Xiong	Nanjing, China

Publicity Co-chairs

Z. Zhou	University of Alberta, Canada

Finance Chair

L. Ying	Together, USA

Local Co-chairs

S. Panëels	CEA-List Paris, France
L. Pantera	CEA-List Paris, France

Submissions Co-chairs

C. Zhao	University of Alberta, Canada
G. Dong	University of Alberta, Canada

Web Co-chairs

Y. Ma	University of Alberta, Canada
X. Wu	University of Alberta, Canada

Advisors

E. Moreau	SeaTech, France
I. Cheng	University of Alberta, Canada
A. El-Saddik	University of Ottawa, Canada

ICSM 2022 Program Committee

K. Abed-Meraim	Université d'Orléans, France
D. Aouada	University of Luxembourg, Luxembourg
S. Beauchemin	University of Western Ontario, Canada

C. Bhatt	FxPal, USA
F. Bouchara	University of Toulon, France
M. Brubaker	University of Toronto, Canada
V. Charvillat	The National Institute of Electrical Engineering, Electronics, Computer Science, Fluid Mechanics & Telecommunications and Networks, France
C. T. Y. Chet	Institute for Infocomm Research, Singapore
F. Denis	Aix-Marseille Université, France
C. Ferrari	University of Florence, Italy
L. Fillatre	UNS, CNRS, France
D. Fofi	University of Bourgogne Franche-Comté (UBFC), France
O. Garcia-Panella	Ramon Llull University, Spain
H. Ghennioui	Université Sidi Mohamed Ben Abdellah, Morocco
C. Grana	University of Modena and Reggio Emilia, Italy
A. Jepson	University of Toronto, Canada
S. Ji	Washington State University, USA
P. Karczmarek	The John Paul II Catholic University of Lublin, Poland
C. Leng	University of Alberta, Canada
H. Li	The Australian National University, Australia
Z. Li	Xidian University, China
A. W.-C. Liew	Griffith University, Australia
G. Lisanti	University of Pavia, Italy
V. Loia	University of Salerno, Italy
J. Martinet	University of Lille, France
I. Masi	University of Southern California, USA
F. J. S. Matta	CENATAV, Cuba
A. Mian	The University of Western Australia, Australia
A. Minghelli	University of Toulon, France
N. Mitton	Inria, France
M. Mohanty	The University of Auckland, New Zealand
E. Moreau	University of Toulon, France
G. Morin	The National Institute of Electrical Engineering, Electronics, Computer Science, Fluid Mechanics & Telecommunications and Networks, France
D. Mukherjee	Epson Canada Ltd., Canada
T. Nguyen	University of Alberta, Canada
P. Pala	University of Florence, Italy
M. Paul	Charles Sturt University, Australia
M. T. Pham	INSA Lyon, France

Contents

Multimedia for Medicine and Health-Care

Smart Homes

Multimedia Environments and Metaverse

Deep Learning on Video and Music

Haptic

Machine Learning for Multimedia

Normalizing Flow Based Defect Detection with Motion Detection

Zijian Kuang[1(✉)], Lihang Ying[2], Xinran Tie[1], and Shi Jin[2]

[1] Department of Computing Science, University of Alberta, Edmonton, Canada
{kuang,xtie}@ualberta.ca
[2] Zerobox Inc., Edmonton, Canada
{leo,shi}@zerobox.ai

Abstract. Visual defect detection is critical to ensure the quality of most products. However, the majority of small and medium-sized manufacturing enterprises still rely on tedious and error-prone human manual inspection. The main reasons include: 1) the existing automated visual defect detection systems require altering production assembly lines, which is time consuming and expensive 2) the existing systems require manually collecting defective samples and labeling them for a comparison-based algorithm or training a machine learning model. This introduces a heavy burden for small and medium-sized manufacturing enterprises as defects do not happen often and are difficult and time-consuming to collect. Furthermore, we cannot exhaustively collect or define all defect types as any new deviation from acceptable products are defects. In this paper, we overcome these challenges and design a three-stage plug-and-play fully automated unsupervised 360° defect detection system. In our system, products are freely placed on an unaltered assembly line and receive 360° visual inspection with multiple cameras from different angles. As such, the images collected from real-world product assembly lines contain lots of background noise. The products face different angles. The product sizes vary due to the distance to cameras. All these make defect detection much more difficult. Our system use object detection, background subtraction and unsupervised normalizing flow-based defect detection techniques to tackle these difficulties. Experiments show our system can achieve 0.90 AUROC in a real-world non-altered drinkware production assembly line.

Keywords: Visual defect detection · Normalizing flow · Object detection · Background subtraction · Computer vision

1 Introduction

Visual defects have a significant impact on the quality of industrial products. Small defects need to be carefully and reliably detected during the process of quality assurance [1,2]. It is important to ensure the defective products are

---

Z. Kuang and L. Ying—Equal contributions.

identified at earlier stages, which prevents a negative impact on a company's waste, reputation and additional financial loss. In recent research, visual defect detection has been increasingly studied again with deep learning approaches and has improved quality control in the industrial field [3,4]. However, visual defect detection is still challenging due to 1) collecting defective samples and manually labeling for training is time-consuming; 2) the defects' characteristics are difficult to define as new types of defects can happen any time; 3) and the product videos or images collected from SME's non-altered assembly lines usually contain lots of background noise as shown in Fig. 1, since a well designed production lines that can ensure high quality product videos or images can be prohibitively costly for SMEs. The results of defect detection become less reliable because of these factors.

Fig. 1. Examples of video collected from a real-world bottle manufacturer. It demonstrates the complexity and unpredictability of image background noise that could happen in a small to medium sized factory.

Most existing defect datasets [5] are either for one scenario (e.g. concrete, textile, etc.) or lack of defect richness and data scale. The popular anomaly defection dataset [5] is too "perfect" (e.g. all products are perfectly aligned in the center of the image, with clean and simple background) which cannot represent the realistic setup in SME factories or requires challenging perfect pre-processing (e.g. background removal, re-lighting, etc.). Specifically, the dataset is limited to a few categories of products and a smaller number of samples [1,2,6]. To ensure our experiments' realism and applicability, we introduce a new dataset collected from a commercially operating bottle manufacturer located in China. This dataset includes 21 video clips (with 1634 frames) consisting of multiple types of bottle products with both good samples with perfect surface and defective samples with less detectable scratches. An example of different types of collected bottles is shown in Fig. 2. These videos are provided by ZeroBox.

Since specialized cameras and well-designed turing assembling lines are too expensive for SME factories, it is highly desirable to have a fully automated defect detection system with minimal cost that can be plug-and-play added to the existing production lines. In this paper, we propose a three-stage deep learning powered, fully automated defect detection system based on object detection, background subtraction and normalizing flow-based defect detection. The system we proposed uses three novel strategies:

1. a novel object detection is used to narrow down the searching window and realign the product from each input video frames,
2. a novel video matting based background subtraction method is used to remove the background of the detected image so that the defect detection model can focus on the product,
3. and a semi-supervised normalizing flow-based model is used to perform product defect detection.

Extensive experiments are conducted on a new dataset collected from the real-world factory production line. We demonstrate that our proposed system can learn on a small number of defect-free samples of single product type. The dataset will also be made public to encourage further studies and research in visual defect detection.

Fig. 2. Samples of the ZeroBox bottle product dataset. Defective parts are labeled in the red squares. (Color figure online)

2 Related Work

Since this paper focus on an end to end three stage network for product defect detection, in this section, we will focus on the three areas of object detection, background subtraction and visual defect detection.

2.1 Object Detection

Object detection refers to the operation of locating the presence of objects with bounding boxes [7,8]. The types or classes of the located objects in an image are classified by the model with respect to the background. Currently, deep learning-based models are state-of-the-art on the problem of object detection. Top detection frameworks include systems such as deformable parts models, Faster R-CNN, and YOLO.

Deformable part models (DPM) [9] use a disjoint pipeline with a sliding window approach to detect objects. The system is disparate and only the static features are extracted. Faster R-CNN [10] and its variants utilize region proposals to find objects. The pipeline of Faster R-CNN consists of a convolutional neural network, an SVM, and a linear model. However, each of the stages needs to be finetuned precisely and independently. It can not be applied to real-time situations due to the slowness of the overall system.

In 2016, J. Redmon et al. introduced a unified real-time object detection model called "You only look once" (YOLO). Unlike DPM and Faster R-CNN, YOLO replaces disparate parts to a single convolutional neural network. It reframes object detection as a regression problem that separates bounding boxes spatially and associates them with their class probabilities [11]. YOLO is extremely fast, reasons globally, and learns a more generalized representation of the objects. It achieves efficient performance in both fetching images from the camera and displaying the detections. However, YOLO struggles with small items that appear in groups under strong spatial constraints. It also struggles to identify objects in new or unusual configurations from data it has not seen during the training [11]. Still, YOLO is so far the best objection detection algorithm.

2.2   Background Subtraction

Background subtraction is a technique that is widely used for detecting moving objects in videos from static cameras and eliminating the background from an image. A foreground mask is generated as the output, which is a binary image containing the pixels belonging to the moving objects [12,13]. The methods of background subtraction for videos include video segmentation and video matting.

In video segmentation, pixels are clustered into two visual layers of foreground and background. In 2015, U-Net [14] was proposed for solving the problem of biomedical image segmentation. The architecture of this network is in the shape of a letter "U", which contains a contracting path and an expansive path. A usual contracting layer is supplemented with successive layers and max-pooling layers. The other path is a symmetric expanding path that is used to assemble more precise localization. However, excessive data argumentation needs to be applied to retain a considerable size of features if there is a small amount of available training data.

Video matting, as another method of background subtraction, separates the video into two or more layers such as foreground, background and alpha mattes. Unlike video segmentation which generates a binary image by labelling the foreground and background pixels, the matting method also handles those pixels that may belong to both the foreground and background, called the mixed pixel

[12,13]. Recently, Background Matting V2 (BGM V2) has achieved the state-of-art performance to replace the background in a real-time manner [15]. The first version of Background Matting (BGM) was initially proposed to create a matte which is the per-pixel foreground colour and alpha of a person in 2020 [16]. It only requires an additional photo of the background that is taken without the human subject. Later, Background Matting V2 (BGM V2) is released to achieve real-time, high-resolution background replacement for video conferencing. However, in the final matting results, there is still some residual from the original background shown in the close-ups of users' hairs and glasses. However, in real industry environment, since the foreground object in each frame can be different, the generated composite mask cannot always fully remove all the background, therefore, the performance is not promising for unseen foreground object.

2.3 Defect Detection

In recent years, convolutional neural networks began to be applied more often to visual-defect classification problems in industrial and medical image processing. The segmentation approach plays a significant role in visualized data's anomaly detection and localization since it can not only detect defective products but also identify the anomaly area.

Autoencoder has become a popular approach for unsupervised defect segmentation of images. In 2019, P. Bergmann et al. proposed a model to utilize the structural similarity (SSIM) metric with an autoencoder to capture the inter-dependencies between local regions of an image. This model is trained exclusively with defect-free images and able to segment defective regions in an image after training [17].

Although segmentation-based methods are very intuitive and interpretable, their performance is limited by the fact that Autoencoder can not always yield good reconstruction results for anomalous images. In comparison, the density estimation-based methods can perform anomaly detection with more promising results.

The objective of density estimation is to learn the underlying probability density from a set of independent and identically distributed sample data [18]. In 2020, M. Rudolph et al. [19] proposed a normalizing flow-based model called DifferNet, which utilizes a latent space of normalizing flow to represent normal samples' feature distribution. Unlike other generative models such as variational autoencoder (VAE) and GANs, the flow-based generator assigns the bijective mapping between feature space and latent space to a likelihood. Thus a scoring function can be derived to decide if an image contains an anomaly or not. As a result, most common samples will have a high likelihood, while uncommon images will have a lower likelihood. Since DifferNet only requires good product images as the training dataset, defects are not present during training. Therefore, the defective products will be assigned to a lower likelihood, which the scoring function can easily detect the anomalies [19]. However, the normalizing flow based method cannot perform end to end defect detection task.

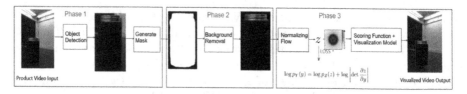

Fig. 3. Overview of our proposed system. Phase 1: our system first takes video clips as input and utilizes YOLO to detect and draw bounding boxes on each product in each frame. Phase 2: after YOLO detection, a pre-trained background matting model is applied along with our novel background subtraction algorithm to remove the background noises surrounding the product within the bounding box. Phase 3: the processed product images are further passed into the flow-based defect detection model to generate a normal distribution. After training the model, a scoring function is used to calculate likelihoods against the good sample's distribution to classify the input sample as defective or normal. We also created a visualization model to generate a video output with the bounding box and predicted label on each frame.

3   Our Approach

In this paper, we propose a low cost plug-and-play fully automated 360° deep learning defect detection system. Without requiring any major re-design of the existing production line, the system is a simple add-on "box" to the existing process. It utilizes multiple low-cost cameras to capture the product images from different angles to ensure all important visual areas are covered at least once. Then the captured images are used as the input in our proposed deep learning based system to perform defect detection. The overview of the proposed system's pipeline is shown in Fig. 3.

The general stages and tasks within our proposed product defect detection system can be divided into three main components, which are the object detection (Sect. 3.1), the background subtraction (Sect. 3.2) and the defect detection (Sect. 3.3).

3.1   Novel Object Detection Based on Deep Learning and Traditional Computer Vision Algorithms

Our system takes videos of products captured by four cameras installed on the assembling line as input. These cameras are arranged 90° apart around a center spot where all products will pass through along the assembling line. The 4 camera inputs are fed into the system independently so there is no complication of synchronizing all cameras.

In the video input, the product is moving on the convey belt viewed by a static camera. Therefore the position of the product in each frame is different. In our defect detection model, we focus on the product, eliminate the unnecessary information from each frame (such as background) and thus we decided to adopt a pre-trained YOLOv5 [11] object detection model to narrow down the defect detection searching window on input images collected from each cameras. The pre-trained YOLOv5 model was further fine-tuned with the ZeroBox dataset.

Even though YOLOv5 is able to detect product position for each frame of the video input, it is computationally too slow to continuously use YOLOv5 for all videos frames from all 4 cameras on a modest computer without GPU. In order to reduce the computational workload, a traditional computer vision based motion detection algorithm is utilized to first identify when a product has moved into the center of each camera view on the conveyor belt and then YOLOv5 is utilized only once per object instead of on all frames of the video stream. The traditional computer vision based motion detection algorithm we designed contains two parts, first, we calculate the absolute difference between each two frames, and then we dilute the image to make the differences of moving object more clear. Second, we set up a 400 by 400 pixels square region of interest set in the middle of the camera view, so it will start triggering YOLO only when the moving object entirely appear within the region of interest.

At the end of the object detection stage, the product will be realigned into the center of the bounding box, and around 80% background information will be eliminated from the original input frames. The output will be further normalized to a 512 by 512 pixels image, and we use O_i^n refers to the original images taken for $i - th$ object from $n - th$ view, after object detection, we got detected object D_i^n for each object and each view respectively.

$$f_{detection}(\mathbf{O}_i^n) = \mathbf{D}_i^n \tag{1}$$

3.2 Novel Background Subtraction Based on Video Matting and Traditional Computer Vision Algorithms

At the end of the object detection stage, most of the background has been removed by the pre-trained YOLOv5 algorithm. However, as depicted by Fig. 3, YOLOv5 would still keep a small margin around the product itself. The background in the margin makes it difficult for defect detection algorithms since background can vary significantly from image to image and is often mistreated as defective. Since YOLOv5 is only able to identify objects by rectangular boxes, this problem is particularly challenging for products that don't fit snugly in the bounding box such as some of the products shown in Fig. 2 (for example the cone shaped bottles and those with smaller necks).

To overcome this problem, an image background subtraction model is further utilized to remove the background in each YOLOv5 bounding box. After object detection and background subtraction, the processed images will be 100% of the product itself and then suitable to be passed on to defect detection phase.

We use the background matting technique from pre-trained BGMv2 [15] to draw a mask to remove the background. However, the matting performance is not very reliable. The mask generated in each frame is slightly different from the mask generated in other frames. To overcome this issue, we propose to linearly add the masks that are generated in all bounding boxes from sequential video

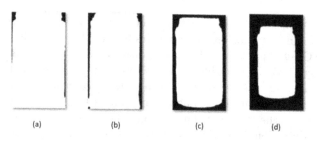

Fig. 4. Examples of mask generated in different stages. (a) Mask generated using BGMv2 on first frame of the black bottle product. (b) Composite first 10 masks generated using BGMv2 on first 10 frames. (c) The composite mask generated using the entire video dataset. (d) The 10% shrink and padding-resize of composite mask to minimize the background from each frame.

frames as a composite mask. In other words, in the single product video input, we will generate one single mask to segment the product and background in every single frame. Then we use the composite mask to remove the background from each bounding box generated by pre-trained YOLOv5 in each frame. We use M_i^n refers to the mask generated for $i-th$ object on the $n-th$ view from the differentiable matting function:

$$f_{matting}(\mathbf{D}_i^n) = \mathbf{M}_i^n \qquad (2)$$

Since the foreground object in each frame can be different can have multiple products on the conveyor, the generated composite mask cannot always fully remove all the background, therefore, the performance is not promising for unseen foreground object. As shown in Fig. 5(b), the bottom of the image still include some conveyor belt portion which is considered as the background noise. To solve this problem, we further shrink the final mask by 10%, and the mask is then padded to the original size as shown in Fig. 4(d).

The re-scaled mask can ensure all the background are removed in every frame. The areas near the boundaries of the product can also be masked, so our defect detection model might miss the defects in these boundary regions. However, this problem is compensated by the fact multiple cameras are employed in the system: most defects missed in one camera near the product boundary is fully visible close to the center view of another camera.

$$Y_{in} = concatenate(\mathbf{D}_i^n - f_{re}(\mathbf{M}_i^n)) \qquad (3)$$

where f_{re} is the shrink function and the final result for each object after background removal will be further concatenated as y_{in} which is used as the input feature in the next stage.

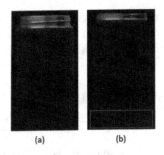

(a)   (b)

Fig. 5. Examples of results after background removal using composite masks. (a) Ideal background subtraction using composite mask (b) Some cases that the composite mask cannot remove all the background due to the product might not be in the center of the image.

3.3   Defect Detection Using Normalizing Flow Based Model

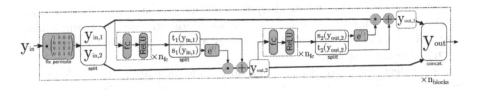

Fig. 6. Architecture of one block of normalizing flow based model proposed in Real-NVP [20].

After object detection and background subtraction, the processed images are further resized to the size of 448 by 448 pixels that only contains the product information excluding any background noise. Then the processed images are fed into a normalizing flow model to output a normal distribution by maximum likelihood training.

We adopted the design of coupling layers as proposed in Real-NVP [20], each individual layer of the normalizing flow based model is shown in Fig. 6. y_{in} which is the concatenated pre-processed images from four different angles, is split into $y_{in,1}$ and $y_{in,2}$ that manipulate each other by applying scale ($\odot$ as the element-wise product) and shift operations are described by:

$$y_{out,2} = y_{in,2} \odot e^{s_1(y_{in,1})} + t_1(y_{in,1})$$
$$y_{out,1} = y_{in,1} \odot e^{s_2(y_{out,2})} + t_2(y_{out,2})$$

$$(4)$$

where s and t are differentiable functions implemented as a fully connected network with a soft-clamping as our activation for better convergence:

$$\sigma_\alpha(h) = \frac{2\alpha}{\pi} \arctan \frac{h}{\alpha}$$

$$(5)$$

The goal of the normalizing flow based model is to maximize likelihoods for input features y which are quantifiable in the latent space Z, which is equivalent to maximizing the log-likelihood, which is also equivalent to minimizing the negative log-likelihood loss $L(y)$:

$$\log p_Y(y) = \log p_Z(z) + \log \left| \det \frac{\partial z}{\partial y} \right|$$
$$\mathcal{L}(y) = \frac{\|z\|_2^2}{2} - \log \left| \det \frac{\partial z}{\partial y} \right| \tag{6}$$

To classify if an input image is anomalous or not, our model uses a scoring function $T(x)$ that calculates the average of the negative log-likelihoods using multiple transformations of an image.

$$\mathcal{A}(x) = \begin{cases} 1 & \text{for } \tau(x) \geq \theta \\ 0 & \text{for } \tau(x) < \theta \end{cases} \tag{7}$$

The result will compare with the threshold value θ which is learned during training and validation process and is later applied to detect if the image contains an anomaly or not, where $A(x) = 1$ indicates an anomaly. Based on our experiments, we found 55.11 is the best threshold for our cases. More implementation details and threshold selection strategy along with experiment results are shown in the next section.

After defect detection, the information include anomaly prediction and predicted bounding box will be plot onto the original product video input as our visualized video output. The example frame of output result can be found in Fig. 3. Since we have four cameras to capture the 360° images of the product, the product will be classified as defective if any of the cameras detects a defect.

4   Experiments and Results

In this section, we evaluate the proposed system based on real-world videos provided by a factory in China. First, we briefly introduce the dataset that is used in the following experiments. Then, the results of several representative experiments are studied along with visual statistics. Since the complexity of experiments primarily stems from the noisy background in the video inputs, our experiments will solely concentrate on logo-free products and group them into single and multiple product categories for experiment purposes.

4.1   Dataset

In this paper, we evaluate our defect detection system based on videos recorded from real life. For fair and reliable experiment results, ZeroBox Inc. has created a brand new dataset collected from an industrial production line monitoring system. This dataset includes 21 video clips in total which consists of 13 types of products with both good and defective samples. Some of the product samples are shown in Fig. 2. In addition, there are 1381 good product images and

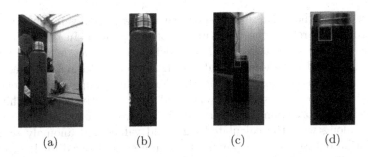

(a) (b) (c) (d)

Fig. 7. Example images from the ZeroBox dataset of products from a real-world bottle factory. Figure 7(a) and Fig. 7(b) show examples of the original image and cropped images of a good product. Figure 7(c) and Fig. 7(d) show examples of the original and cropped images of a defective product (the defect is labeled in the red square). (Color figure online)

253 defective product images generated through YOLO detection and cropping. Examples of defective and defective-free samples are presented in Fig. 7.

Since our normalizing flow-based defect detection model utilizes semi-supervised learning, it only requires approximately 150 good sample images to efficiently learn how to use a simple normal distribution to represent the complex distribution of a group of good samples. Within our experiments on the product of a white jar and a black jar, a total of 150 good sample images are used for training. Another 121 sample images are used for validation, and the rest of 47 sample images are used for testing purposes. For both validation and testing, a mixture of good and defective samples is used to evaluate the proposed system.

4.2 Implementation Details

For all experiments, we train our system for 10 meta epochs. Each meta epoch contains 8 sub epochs which result in a sum of 80 epochs. Additional transformation settings are applied to manipulate and adjust brightness, contrast and saturation for the input images. During our experiments, we manipulated the contrast, brightness and saturation of the product images with a uniformly distributed random factor in the interval of $[0.5, 1.5]$ to pre-process input video frames. Results in later sections also compare each experiment outcome while including or excluding this process from the experiment. Although the model DifferNet proposed in the paper [19] does not need defective samples during training, this process is still necessary for validation. In fact, the validation process plays a critical role in determining the threshold of the anomaly score. Within the evaluation stage, our proposed system will be validated once at the end of each epoch based on the anomaly score calculated from the current training stage.

During the testing stage, the threshold for detection is chosen based on the corresponding true-positive rate and false-positive rate of the trained model and a given true-positive rate as the target used for training purposes. More specifically, the threshold value has the true-positive rate greater than our target-true positive rate but the smallest false-positive rate will finally be chosen in the testing process to reflect the performance of our system. In order to predict if a given input sample has a good or a defective product in it, we will use the aforementioned threshold to evaluate. If the predicted anomaly score provided by the model is less than the threshold, this sample will be classified as good. Otherwise, the sample is classified as a defective one. Within each meta epoch, the corresponding Area Under Receiver Operator Characteristics (AUROC) along with the calculated threshold values and anomaly scores on our validation dataset for the system are computed and saved for later evaluation stages. The metric AUROC is calculated using build-in roc_curve function imported from scikit-learn library [21]. In the last meta epoch, the system's aforementioned parameters are saved into the system as well for evaluation. After training and evaluation, the test accuracy is calculated based on the percentage of the test dataset that is correctly classified. Moreover, a ROC curve is plotted at the end of the training and testing process and is saved locally for further analysis and exploration.

4.3   Evaluation

Our proposed system is tested on two products; one is jars with black visuals and the other is jars with white visuals to compare and report the performance. The two jars are in a shape that is very similar to a cone with no logo on their surfaces. Moreover, we have compared the performance of the system trained on two different types of input images along with the performance of the original differNet [19]. The two types of input images that our model was trained on are product images with cropping and product images with a mask used for background removal. From our experiments, the best performance occurs on the input images using the strategy of mask for background removal with a target true positive rate set to 0.85 in training. Since many of the defects happen far from the edges of the product from the input frames, an extra 10% mask extension can further enhance the performance and achieve a promising test accuracy. As a result, the accuracy of defect detection is increased by 20% with a final test accuracy above 80% as our best performance from the proposed system. Later experiments also show that the effect of background factors in each frame can be further reduced by extending the mask. Detailed experiment results are displayed in the following sections.

Experiment results are displayed in detection accuracy and its calculated threshold for each trial and are compared for each of the aforementioned strategies in the tables below. Table 1 presents the detailed performance of detection in terms of test accuracy and its corresponding anomaly threshold on the black jar product. Experiments with and without the image transformation process (manipulation of contrast, brightness and saturation settings of input images) are performed for comparison. Two experiment settings are applied which include using images with cropping technique or using images with a mask. From Table 1 and Table 2, the original model of differNet [19] achieves the same test accuracy before and after adding the transformation process. With a 10% cropping on each side of the image, our proposed system can obtain a better result on defect detection with cropping performed during training. Finally, with an adaptive mask applied to the input images, our proposed system can obtain the best result of 87.00% as the test accuracy in all the experiments and a value of 55.11 as the corresponding anomaly threshold on defect detection while the mask helps sufficiently eliminate other factors that potentially affect the accuracy in performance.

Table 1. Performance of detection on images of black jar represented in Accuracy/Threshold.

Accuracy/Threshold	Without transformations	With transformations
differNet	24.19%/4.55	24.19%/15.88
(Ours)with 10% Cropping	61.29%/84.5	75.81%/13.91
(Ours)with Mask	67.42%/12.79	**87.00%/55.11**

Table 2 presents the detailed performance of detection in test accuracy and its corresponding threshold on the product of the white jar under the same test setup. Test results are compared with and without the image transformation process for each proposed strategy. In this case, the proposed system has achieved a generally higher test accuracy on images under image transformation settings. The best performance still happens in the case which utilizes the adaptive mask to eliminate the impact from the background.

Table 2. Performance of detection on images of white jar represented in Accuracy/Threshold.

Accuracy/Threshold	Without transformations	With transformations
differNet	24.87%/1.71	24.87%/1.85
(Ours)10% Cropping	65.88%/35.76	74.11%/3.79
(Ours)with Mask	67.27%/8.89	**83.63%/8.76**

5 Conclusion

In this paper, we introduce a new dataset for product visual defect detection. This dataset has several challenges regarding defect types, background noise, and dataset sizes. We have proposed a three-stage defect detection system that is based on the techniques of object detection, background subtraction and normalizing flow-based defect detection. Finally, extensive experiments show that the proposed approach is robust for the detection of visual defects on real-world product videos. In the future, we plan to work on using background and foreground segmentation with an end-to-end trained mask to eliminate the background noise in images identified by YOLO. Also, more data samples will be collected for training, validation and testing.

References

1. Patel, N., Mukherjee, S., Ying, L.: EREL-Net: a remedy for industrial bottle defect detection. In: Basu, A., Berretti, S. (eds.) ICSM 2018. LNCS, vol. 11010, pp. 448–456. Springer, Cham (2018). https://doi.org/10.1007/978-3-030-04375-9_39
2. Ma, N., Zhu, X.: Computational stoning method for surface defect detection (2013)
3. Varun Chandola, A.B., Kumar, V.: Anomaly detection: a survey, no. 15 August 2009. https://dl.acm.org/doi/abs/10.1145/1541880.1541882
4. Chalapathy, R., Chawla, S.: Deep learning for anomaly detection: a survey (2019)
5. Bergmann, P., Fauser, M., Sattlegger, D., Steger, C.: MVTec AD - a comprehensive real-world dataset for unsupervised anomaly detection. In: 2019 IEEE/CVF Conference on Computer Vision and Pattern Recognition (CVPR) (2019)
6. Zhang, C.: Surface defect detection, segmentation and quantification for concrete bridge assessment using deep learning and 3d reconstruction (2020)
7. Vijayalakshmi, M.N., Senthilvadivu, M.: Performance evaluation of object detection techniques for object detection. In: 2016 International Conference on Inventive Computation Technologies (ICICT) (2016)
8. Lu, Y., Zhang, L., Xie, W.: Yolo-compact: an efficient yolo network for single category real-time object detection. In: 2020 Chinese Control and Decision Conference (CCDC) (2020)
9. Felzenszwalb, P.F., Girshick, R.B., McAllester, D., Ramanan, D.: Object detection with discriminatively trained part-based models. IEEE Trans. Pattern Anal. Mach. Intell. **32**(9), 1627–1645 (2010)
10. Girshick, R.: Fast R-CNN. In: Proceedings of the IEEE International Conference on Computer Vision (ICCV), December 2015
11. Jocher, G., et al.: ultralytics/yolov5: v5.0 - YOLOv5-P6 1280 models, AWS, Supervise.ly and YouTube integrations (2021). https://doi.org/10.5281/zenodo.4679653
12. Bouwmans, B.H.T., Porikli, F., Vacavant, A.: Statistical models for background subtraction. In: Background Modeling and Foreground Detection for Video Surveillance, pp. 153–172 (2014)
13. Rashid, M., Thomas, V.: A background foreground competitive model for background subtraction in dynamic background. Procedia Technol. **25**, 536–543 (2016)
14. Ronneberger, O., Fischer, P., Brox, T.: U-net: convolutional networks for biomedical image segmentation (2015)
15. Lin, S., Ryabtsev, A., Sengupta, S., Curless, B., Seitz, S., Kemelmacher-Shlizerman, I.: Real-time high-resolution background matting (2020)
16. Sengupta, S., Jayaram, V., Curless, B., Seitz, S., Kemelmacher-Shlizerman, I.: Background matting: the world is your green screen (2020)
17. Bergmann, P., Löwe, S., Fauser, M., Sattlegger, D., Steger, C.: Improving unsupervised defect segmentation by applying structural similarity to autoencoders (2019). http://dx.doi.org/10.5220/0007364503720380
18. Nachman, B., Shih, D.: Anomaly detection with density estimation. Phys. Rev. D **101**, 075042 (2020). https://link.aps.org/doi/10.1103/PhysRevD.101.075042

19. Rudolph, M., Wandt, B., Rosenhahn, B.: Same same but DifferNet: semi-supervised defect detection with normalizing flows (2020)
20. Dinh, L., Sohl-Dickstein, J., Bengio, S.: Density estimation using real NVP (2017)
21. Trappenberg, T.P.: Machine learning with sklearn. In: Fundamentals of Machine Learning, pp. 38–65 (2019)

FUNet: Flow Based Conference Video Background Subtraction

Zijian Kuang[1]([⊠]), Xinran Tie[1], Xuanyi Wu[1], and Lihang Ying[2]

[1] Department of Computing Science, University of Alberta, Edmonton, Canada
{kuang,xtie,xuanyi}@ualberta.ca
[2] Zerobox Inc., Edmonton, Canada
leo@zerobox.ai

Abstract. Video background subtraction has applied widely in video conferencing. However, it still cannot handle well for motion blurring, e.g., shaking the head or waving hands. To overcome the motion blur problem in background subtraction, we propose a novel optical flow-based encoder-decoder network (FUNet) that combines both traditional Horn-Schunck optical-flow estimation technique and autoencoder neural networks to perform robust real-time video background subtraction. We concatenate the optical flow motion feature and original image's appearance feature, and pass the concatenated value into an encoder-decoder structure to perform video background subtraction. We also introduce a video and image subtraction dataset: Conference Video Segmentation Dataset. Code and pre-trained models are available on our GitHub repository: https://github.com/kuangzijian/Flow-Based-Video-Matting.

Keywords: Background subtraction · Optical flow · Convolutional neural network · Autoencoder

1 Introduction

Video subtraction is widely used in movie editing, virtual reality applications, and video conferencing applications like Zoom, Google meet, and Microsoft Teams [1]. Besides the entertainment purpose, such as the post-production of movies or virtual reality scenes, the video subtraction can also protect people's privacy during video conferencing by hiding the location and environment behind the users [1].

The traditional Gaussian Mixture model-based background subtraction algorithms such as GMG, KNN, MOG and MOG2 [2–5] assume a moving object in the foreground to a static background [6,7]. However, if the user stops moving in the video, then the traditional background subtraction methods will fail. On the other hand, most deep learning-based video subtraction requires a huge amount of training data and some models even require a clean background image as input before performing prediction [1,8]. The video subtraction with motion blur is also challenging that even the state-of-the-art learning-based model cannot handle appropriately [9]; for example, video subtraction is not accurate when people are shaking the head or waving their hands in most of the conferencing applications as show in Fig. 1.

In this paper, a novel flow-based encoder-decoder network(FUNet) is proposed to detect a human head and shoulders from a video and remove the background to create elegant media for video conferencing and virtual reality applications. The proposed model combines both traditional Horn-Schunck optical-flow-based technique [10] and convolutional neural networks to perform robust real-time video subtraction. In our model, an optical-flow-based model is utilized to extract motion features between every two frames. Then we combine both the motion feature and the appearance feature from the original frame and utilize an encoder-decoder network to learn and predict a mask output for human head and shoulders segmentation from the background. The convolutional neural networks are implemented to speed up the video subtraction process to enable processing video frames in a real-time manner.

Fig. 1. Zoom's background replacement application cannot detect waving hand accurately.

2   Related Work

Background replacement aims to extract the foreground from the input video and then combine it with a new background. The methodologies of background replacement can be divided into matting, or subtraction [1,11–13]. Image or video matting can perform visually detailed composites but requires very detailed manual annotations. While the background subtraction focuses segmenting the object by labeling the pixels between the foreground objects and background, its performance is much faster and more efficient [1]. In this section, we discuss related works regarding background replacement methods and related methods for potential improvements on subtraction methods.

2.1   Deep Learning Based Background Subtraction

Image and video matting estimates the foreground objects accurately. Unlike image or video subtraction generates a binary image by labeling the foreground and background pixels, the matting method can handle those pixels that may belong to the foreground and background, called the mixed pixels.

Background Matting V2 (BGM V2) is one state of the art network introduced to overcome the challenge that users do not typically have access to green screens to replace the background during a video conferencing. Consequently, there is residual of the original background shown in the close-ups of users' hairs and glasses. In the paper [1], S. Lin et al. proposed to employ two neural networks to achieve real-time, fully automated, and high-resolution background matting. One neural network in the architecture is worked as the base network by providing an extra background image. The other neural network is served as the refinement network. The base network is responsible for predicting the alpha matte and foreground layer at a lower resolution. After that, an error prediction map specifies the areas that need refinement to be passed to the refinement network. Then, the refinement network takes the low-resolution result and the original image to refine the regions with significant predicted errors. Lastly, the alpha and foreground residuals are produced at the original resolution [1].

Another state-of-the-art image and video matting algorithm is called MODNet [9] introduced by Z. Ke et al. in 2020. The proposed MODNet takes only one RGB image as the input and then decomposes the matting process into three correlated sub-tasks to learn and predict simultaneous. The three sub-tasks include predicting human semantics s_p, boundary details d_p, and final alpha matte α_p through three separate pipelines S, D, and F [9]. To predict the coarse semantic sp, the model uses L2 loss under supervised learning using low-resolution representation S(I) against a thumbnail of the ground truth matte α_g. The equation is formed as below:

$$\mathcal{L}_s = \frac{1}{2} \left\| s_p - G\left(\alpha_g\right) \right\|_2 . \tag{1}$$

The detailed boundary feature is calculated using D(I, S(I)) and L1 loss function as below:

$$\mathcal{L}_d = m_d \left\| d_p - \alpha_g \right\|_1 , \tag{2}$$

where m_d is a generated binary mask through dilation and erosion to let $\mathcal{L}_d$ focus on the boundaries [9].

Finally, the model combines both semantics and details by concatenating S(I) and D(I, S(I)) to predict the final alpha matte α_p using:

$$\mathcal{L}_\alpha = \left\| \alpha_p - \alpha_g \right\|_1 + \mathcal{L}_c . \tag{3}$$

2.2 Traditional Gaussian Mixture Model-Based Background Segmentation

The traditional Gaussian Mixture model-based background subtraction algorithms such as GMG, KNN, MOG and MOG2 [2–5] assume a moving object in the foreground to a static background [6,7]. However, if the person stops moving in the video, then the traditional background subtraction methods will fail.

A. Basu, I. Cheng et al. introduced a methodology for airways segmentation using Cone Beam CT data and measuring airways' volume [14]. Since many of the researches focus on the segmentation and reconstruction of lower airway trees, a new strategy for detecting the boundary of slices of the upper airway and tracking the airway's contour using Gradient Vector Flow (GVF) snakes [15,16] is introduced.

The traditional GVF snakes [15,16] are not performing well for airway segmentation when applied directly to CT images. Therefore, this paper's new method has modified the GVF algorithm with edge detection and sneak-shifting steps. By applying edge detection before the GVF snakes and using snake shifting techniques, the prior knowledge of airway CT slices is utilized, and the model works more robustly. The previous knowledge of the shape of the airway can automatically detect the airway in the first slice. The detected contour will then be used as the second slice's sneak initialization and so on [14]. A heuristic is also applied to differentiate bones from the airway by the color to make sure the snake converges correctly. Following this, the airway volume is estimated based on the 3D model constructed with automatically detected contours. However, the GVF based model has the drawbacks of sensitivity to noise and adaptability of the parameters, which can decrease the performance of image/video segmentation.

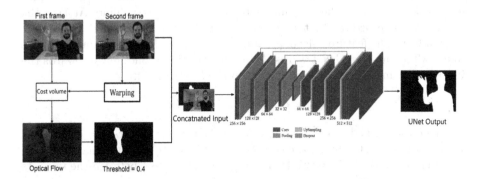

Fig. 2. Overview of our proposed network. It first estimates optical flows for every two frames, and extract the magnitude information as motion feature $[M_1, M_2, \ldots, M_N]$ for each frame (Sect. 3.1). Then we fuse the motion feature and the appearance feature from the original frame and feed it into an encoder-decoder architecture to perform video background subtraction (Sect. 3.2).

3   Our Approach

In this section, we elaborate on the architecture of FUNet and discuss more details of the network design, as shown in Fig. 2. Our goal is to detect a human head and shoulders from a video and remove the background to create elegant media for video conferencing and virtual reality applications. We proposed a network to predict a binary mask for video background subtraction: given a sequence of video frames $[I_1, I_2, \ldots, I_N]$, our model can label the pixel as foreground or background in each frame, and predict a sequence of binary masks $[O_1, O_2, \ldots, O_N]$.

We first estimate optical flows for every two frames, and extract the magnitude information as motion feature $[M_1, M_2, \ldots, M_N]$ for each frame (Sect. 3.1). Then we fuse the motion feature and the appearance feature from the original frame and feed it into an encoder-decoder architecture that inspired from UNet [17] to perform video background subtraction (Sect. 3.2).

3.1   Motion Feature Extraction

Optical flow is a pattern of apparent motion of image objects between two consecutive frames caused by the movement. The pattern is represented in the 2D vector field.

PWC-Net [18] is a CNN model that is proposed to use the principles of pyramidal processing, warping, and cost volumes for optical flow. This model is 17 times smaller in size, two times faster in inference, much easier to train, and outperforms all the published optical flow models on two of the benchmarks. In the paper, PWC-Net firstly uses the optical flow estimate to warp the features extracted from the second image. Then, it utilizes both of the warped features and the features from the first image to construct a cost volume. At last, the cost volume is processed by the network to predict the optical flow. In the first stage, a learnable feature L-level pyramid is constructed with raw images as the input. After that, PWC-Net warps the features of the second image towards the first image using the upsampled flow from the $l + 1$th level:

$$c_w^l(\mathbf{x}) = c_2^l \left(\mathbf{x} + \mathrm{up}_2 \left(\mathbf{w}^{l+1}\right)(\mathbf{x})\right). \tag{4}$$

Then, the warped features along with the features of the first image are used to compute the cost volume. A cost volume is utilized to store the matching cost of associating two pixels in two frames. The matching cost between the two sets of features is defined as

$$\mathbf{cv}^l(\mathbf{x}_1, \mathbf{x}_2) = \frac{1}{N} \left(\mathbf{c}_1^l(\mathbf{x}_1)\right)^\top \mathbf{c}_w^l(\mathbf{x}_2), \tag{5}$$

where $\top$ is a transpose operator and N is the length of the column vector $\mathbf{c}_1^l(\mathbf{x}_1)$. Next, the cost volume and original image features, and upsampled features are processed by the multi-layer optical flow estimator. At last, a context network is used to post-process the flow.

Since the optical flow is a pattern of apparent motion of image objects between two consecutive frames caused by the movement, inspired by the PWC-Net [10], we use the same learnable feature pyramids and layer-wise warping operation along with cost volume calculation to estimate optical flow.

For two images input I_1 and I_2, the model generates L-level pyramids of extracted features from each frame. The model warps I_2's features toward I_1's features at each lth level. Then the model constructs a cost volume to store the matching costs between each pixel in the current frame with its corresponding pixels in the previous frame.

With given sequence of video frames $[I_1, I_2, \ldots, I_N]$, the motion feature extraction process can output a sequence of optical flow vectors $[v_1, v_2, \ldots, v_N]$ for each frame, we further formed a indicator function $\mathbb{I}(x)$ to separate the optical flow vector value into either foreground or background using a threshold α:

$$\mathcal{M}_n(x) = \begin{cases} 1 & \text{for } v_n(x) \geq \alpha \\ 0 & \text{for } v_n(x) < \alpha \end{cases}, \tag{6}$$

where $v_n(\mathbf{x})$ denotes the optical flow magnitude for pixel x in nth frame. The M_n is the extracted motion feature mask for nth frame.

3.2   Motion and Appearance Fusion

After retrieving the sequence of extracted motion feature masks $[M_1, M_2, \ldots, M_N]$ from the motion feature extraction process (Sect. 3.1), we further concatenate the motion feature M and original image's appearance feature I, and pass the concatenated value into an encoder-decoder structure to perform video background subtraction.

U-Net [17] is a fully convolutional network proposed for the problem of biomedical segmentation problem. The architecture of this network is modified to work with few training samples and produce more precise segmentation. A usual contracting layer is supplemented with successive layers, and the upsampling operators are used instead in the network. Features from the contracting path are combined with upsampled output to boost the accuracy of the localization better. In order to assemble a more precise output, a successive convolution layer is used for learning. In our proposed model, we use U-Net to learn video background subtraction from fused motion and appearance feature.

In our model, many feature channels are used in the upsampling part for propagating context information to high-resolution layers. Moreover, excessive data argumentation is applied due to a small amount of available training data. During training, both input images and the corresponding segmentation images are used as input. The soft-max is defined as

$$p_k(\mathbf{x}) = \exp\left(a_k(\mathbf{x})\right) / \left(\sum_{k'=1}^{K} \exp\left(a_{k'}(\mathbf{x})\right)\right), \tag{7}$$

where $a_k(\mathbf{x})$ represents the activation in feature channel k at the pixel position x. To penalize the deviation of $p_{\ell(\mathbf{x})}(\mathbf{x})$ from 1 for each position, the cross-entropy is defined as

$$E = \sum_{\mathbf{x} \in \Omega} w(\mathbf{x}) \log \left(p_{\ell(\mathbf{x})}(\mathbf{x}) \right), \tag{8}$$

where ℓ is the ground truth for each pixel and w is a weight map. After computing the separation border, the weight map is computed as

$$w(\mathbf{x}) = w_c(\mathbf{x}) + w_0 \cdot \exp \left(-\frac{(d_1(\mathbf{x}) + d_2(\mathbf{x}))^2}{2\sigma^2} \right), \tag{9}$$

where w_c is the weight map, d_1 is the distance to the border of the nearest cell, and d_2 is the distance to the border of the second nearest cell.

In our proposed model, the encoder part is built with the repeated application of two 3×3 convolution layers, each followed by a ReLU and a 2×2 max pooling operation with downsampling using stride 2. The decoder part consists of an upsampling along with a 2×2 convolutional layer. The model also includes a concatenation (shown as light blue lines on top of the model in Fig. 2) from the corresponding cropped feature map from the encoder side and two 3×3 convolutions, each followed by a ReLU.

We use binary cross-entropy loss with a pixel-wise softmax (BCE With logits Loss) as our loss function:

$$L = -\frac{1}{n} \sum \left(O_n \times \ln C_n + (1 - O_n) \times \ln (1 - C_n) \right), \tag{10}$$

where C_n denotes the predicted probability of the nth pixel, and O_n denotes the ground-truth value of the nth pixel.

4   Implementation Details

4.1   Training

Our model is trained on eight video sequences that contain 2600 frames. During training, optical flows between every two frames are firstly estimated. Since there is no previous frame for the first frame, the first optical flow generated using the first and second frames is duplicated to keep the number of optical flow images consistent with the number of training images. Next, the estimated optical flow magnitude is computed and is later split into the foreground and background under a threshold, which resulted in a binary mask as our motion feature layer. We found the threshold $\alpha = 0.4$ works best to separate foreground and background during our experiments. Then, we concatenate the motion feature mask as the fourth channel into the original image and feed it to the convolutional neural network.

The model is implemented with the BCE With Logits Loss and the Root Mean Square Propagation (RMSProp) optimizer. RMSProp is a gradient-based optimization technique that uses an adaptive learning rate and a moving average of squared gradients to normalize the gradient. During training, the learning rate is set to 0.0001 with a weight decay of 1e-8, and the momentum equals 0.9. BCE With Logits Loss combines a plain sigmoid after the Binary Cross Entropy between the target and the output in a single class, promoting numerical stability for training.

4.2   Dataset

We created our own video segmentation dataset. The source data includes ten online conference-style green screen videos. We extracted 3600 frames from the videos and generated the ground truth masks for each character in the video, and then we applied virtual background to the frames as our training/testing dataset. The examples of our dataset are shown in Fig. 3. The dataset is available for download under our GitHub repository[1].

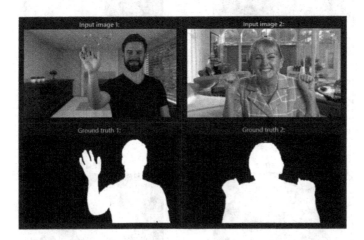

Fig. 3. Examples of video frames and ground truth masks of our Conference Video Segmentation Dataset.

5   Experiments Results

Input images and their corresponding segmentation masks are used as input during training. The model is implemented based on the PyTorch framework and is trained on a piece of RTX 2080 Ti GPU. A total of 10 epochs is trained for the model with a learning rate of 0.0001 and a batch size of 1. Original images of the dataset are used for training with an available down scaling factor that could be adjusted based on different needs. BCE With Logits Loss is used as the loss function for the model to minimize the error between the predicted mask and its ground truth.

---

[1] https://github.com/kuangzijian/Flow-Based-Video-Matting.

5.1   Evaluation Method

During validation and testing, the Dice coefficient (F1 score) is used to evaluate the performance of the proposed network. Dice coefficient calculates the size of the overlap of the two segmentation masks divided by the total size of two masks resulted in a value range from 0 to 1. A higher Dice coefficient indicates a better performance which detects a greater similarity between the prediction and the ground truth. In our experiments, 60% of the dataset is used for training, and 20% is used for validation. The rest 20% of the dataset is used for testing.

Fig. 4. Evaluation on the Conference Video Segmentation Dataset with an average Dice coefficient (F1 score) of 0.96 as the result.

5.2   Experiment - Evaluate the Video Background Subtraction on the Conference Video Segmentation Dataset

This experiment evaluated the proposed model on the Conference Video Segmentation Dataset, which contains two video sequences consisting of 720 frames. In each video sequence of the Conference Video Segmentation Dataset, the background image and foreground human are unique and different from our training dataset. The overall performance on the video background subtraction is very promising as the averaging Dice coefficient (F1 score) on these testing video sequences is 0.96. The example frames of our evaluation results are shown as follow:

6   Conclusion

In conclusion, we proposed a novel flow-based encoder-decoder network to detect a human head and shoulders from a video and remove the background to create elegant media for videoconferencing and virtual reality applications. We also created our own conference video style segmentation dataset called Conference Video Segmentation Dataset for further studies and researches.

Acknowledgment. The authors would like to thank our mentor Xuanyi Wu, for her guidance and feedback throughout the research and study. We would also thank our advisor Dr. Anup Basu and Dr. Lihang Ying for their motivation and support to bring out the novelty in our research.

References

1. Lin, S., Ryabtsev, A., Sengupta, S., Curless, B., Seitz, S., Kemelmacher-Shlizerman, I.: Real-time high-resolution background matting (2020)
2. Godbehere, A. ., Matsukawa, A., Goldberg, K.: Visual tracking of human visitors under variable-lighting conditions for a responsive audio art installation. In: 2012 American Control Conference (ACC) (2012)
3. Zivkovic, Z., Heijden, F.V.D.: Efficient adaptive density estimation per image pixel for the task of background subtraction. Pattern Recogn. Lett. **27**(7), 773–780 (2006)
4. KaewTraKulPong, P., Bowden, R.: An improved adaptive background mixture model for real-time tracking with shadow detection. In: Remagnino, P., Jones, G.A., Paragios, N., Regazzoni, C.S. (eds) Video-Based Surveillance Systems, Springer (2002). https://doi.org/10.1007/978-1-4615-0913-4_11
5. Zivkovic, Z.: Improved adaptive gaussian mixture model for background subtraction. In: Proceedings of the 17th International Conference on Pattern Recognition. ICPR (2004)
6. Zhou, D., Zhang, H., Ray, N..: Texture based background subtraction. In: 2008 International Conference on Information and Automation (2008)
7. Qin, X., Zhang, Z., Huang, C., Dehghan, M., Zaiane, O.R., Jagersand, M.: U2-Net: Going deeper with nested u-structure for salient object detection. Pattern Recogn. vol. 106, pp. 107404 (2020). https://doi.org/10.1016%2Fj.patcog.2020.107404

8. Zhao, X., Chen, Y., Tang, M., Wang, J.: Joint background reconstruction and foreground segmentation via a two-stage convolutional neural network. In: 2017 IEEE International Conference on Multimedia and Expo (ICME) (2017)

9. Ke, Z., et al.: Is a green screen really necessary for real-time portrait matting? https://arxiv.org/abs/2109.15130 (2020)

10. Niklaus, S.: A reimplementation of PWC-Net using PyTorch. https://github.com/sniklaus/pytorch-pwc (2018)

11. Ding, S.: Motion-aware contrastive video representation learning via foreground-background merging. https://arxiv.org/abs/2109.15130 (2021)

12. Tezcan, M.O., Ishwar, P., Konrad, J.: BSUV-Net: a fully-convolutional neural network for background subtraction of unseen videos. https://arxiv.org/abs/1907.11371 (2019)

13. Yang, Z., Wei, Y., Yang, Y.: Collaborative video object segmentation by foreground-background integration. https://arxiv.org/abs/2003.08333 (2020)

14. Cheng, I., Nilufar, S., Flores-Mir, C., Basu, A. : Airway segmentation and measurement in CT images. In: 2007 29th Annual International Conference of the IEEE Engineering in Medicine and Biology Society (2007)

15. Zhao, F., Zhao, J., Zhao, W., Qu, F.: Guide filter-based gradient vector flow module for infrared image segmentation. **54**(33), 9807–17 (2015). https://opg.optica.org/ao/abstract.cfm?uri=ao-54-33-9809

16. Xu, C., Prince, J.L.: Snakes, shapes, and gradient vector flow. https://jhu.pure.elsevier.com/en/publications/snakes-shapes-and-gradient-vector-flow-4

17. Ronneberger, O., Fischer, P., Brox, T.: U-Net: convolutional networks for biomedical image segmentation. In: Navab, N., Hornegger, J., Wells, W.M., Frangi, A.F. (eds.) MICCAI 2015. LNCS, vol. 9351, pp. 234–241. Springer, Cham (2015). https://doi.org/10.1007/978-3-319-24574-4_28

18. Sun, D., Yang, X., Liu, M.-Y., Kautz, J.: PWC-Net: CNNs for optical flow using pyramid, warping, and cost volume (2018)

IARG: Improved Actor Relation Graph Based Group Activity Recognition

Zijian Kuang[(✉)] and Xinran Tie

Department of Computing Science, University of Alberta, Edmonton, Canada
{kuang,xtie}@ualberta.ca

Abstract. Group Activity Recognition is to recognize and classify different actions or activities appearing in the video. The detailed description of human actions and group activities is essential information, which can be used in real-time CCTV video surveillance, health care, sports video analysis, etc. The existing methods, such as pose estimation based and graph network based group activity recognition can perform reasonable group activity understanding, however those models have bad performance on video with extreme brightness and contrast condition. This study proposes an improved actor relation graph based model (IARG) that mainly focused on group activity recognition by learning the pairwise actor appearance similarity and actor positions. We propose to use Normalized cross-correlation (NCC) and the sum of absolute differences (SAD) to calculate the pair-wise appearance similarity and build the actor relationship graph to allow the graph convolution network to learn how to classify group activities. We demonstrate that our approach significantly outperforms existing state-of-the-art techniques on the public group activity recognition datasets called collective activity dataset and augmented dataset. Visualized results (sample frames can be found in Appendix) can further demonstrate each input video frame with predicted bounding boxes on each human object and both predicted individual action and collective activities.

Keywords: Group activity recognition · Actor relation graphs · Video understanding

1 Introduction

Group Activity Recognition is an extensively studied topic widely used in the video content analysis area [1]. Traditional video captioning techniques such as LSTM-YT [2] and S2VT [3] use recurrent neural networks, specifically LSTMs [4], to train the models with video-sentence pairs [1–3,5]. The models can learn the association between video frames' sequence and the sequence of sentences to generate a description of videos [3]. Krishna et al. indicated that those video captioning approaches only works for a short video with only one major event [5]. Therefore, they introduced a new captioning module that uses contextual information from the timeline to describe all the events during a video clip [5]. However, it is still very limited in video captioning approaches to generate a detailed description of human actions, and their interactions [6]. Recent studies in pose estimation, human action recognition, and group activity recognition areas show the capability to describe more detailed human actions, and human group activities [7,8].

Human action recognition and group activity recognition are an important problem in video understanding [9,10]. The action and activity recognition techniques have been widely applied in different areas such as social behavior understanding, sports video analysis, and video surveillance. To better understanding a video scene that includes multiple persons, it is essential to understand both each individual's action and their collective activity. Actor Relation Graph (ARG) based group activity recognition is the state-of-the-art model that focuses on capturing the appearance and position relationship between each actor in the scene and performing the action and group activity recognition [9].

In this paper, we propose several approaches to improve the functionality and the performance of the Actor Relation Graph-based model, which called the IARG model to perform a better group activity recognition under extreme lighting conditions. To enhance human action and group activity recognition performance, we apply MobileNet [11] in the CNN layer and use Normalized cross-correlation (NCC) and the sum of absolute differences (SAD) to calculate the pair-wise appearance similarity to build the Actor Relation Graph. We also introduce a visualization model that plots each input video frame with predicted bounding boxes on each human object and predicted individual action and group activity. The output examples are shown in Fig. 2.

2   Related Work

2.1   Video Captioning Based Group Activity Recognition

One important study area of group activity understanding is video captioning. In 2015, S. Venugopalan et al. proposed an end-to-end sequence-to-sequence model which exploited recurrent neural network, specifically Long Short-Term Memory (LSTM [4]) networks as trained on video-sentence pairs and learned to associate a sequence of frames in a video to sequential words to generate the descriptions of the event in the video as captions [3]. A stack of two LSTMs was used to learn the frames' sequence's temporal structure and the sequence model of the generated sentences. In this approach, the entire video sequence needs to be encoded using the LSTM network at the beginning. Long video sequences could lead to vanishing gradients and prevent the model from being trained successfully [5].

In 2017, R. Krishna et al. introduced a Dense-Captioning Events (DCE) model that can detect multiple events and generate a description for each event using the contextual information from past, concurrent, and future in a single pass of the video [5]. In this paper, the process is divided into two steps: event detection and description of detected events. The DCE model leverages a multi-scale variant of the deep action proposal model to localize temporal proposals of interest in short and long video sequences. In addition, a captioning LSTM model is introduced to exploit the context from the past and future with an attention mechanism.

In 2018, X. P. Li et al. introduced a novel attention-based framework called Residual attention-based LSTM (Res-ATT [12]). This new model benefits from the existing attention mechanism and further integrates the residual mapping into a two-layer LSTM network to avoid losing previously generated words information. The residual attention-based decoder model is designed with five separate parts: a sentence encoder, temporal attention, a visual and sentence feature fusion layer, a residual layer, and an MLP [12]. The sentence encoder is an LSTM layer that explores important syntactic information from a sentence. The temporal attention is designed to identify the importance of each frame. The visual and sentence feature fusion LSTM layer is working on mixing natural language information with image features, and the residual layer is proposed to reduce the transmission loss. The MLP layer is used to predict the word to generate a description in natural language [12]. However all the video captioning approaches above have good performance on individual activity recognition, but not group activity recognition.

2.2 Pose Estimation Based Group Activity Recognition

To better understanding a video scene that includes multiple persons, it is essential to understand both each individual's action. OpenPose is an open-source real-time system which is used for 2D multi-person pose detection [8]. Nowadays, it is also widely used in body and facial landmark points detection in video frames [7,13,14]. It produces a spatial encoding of pairwise relationships between body parts for a variable number of people, followed by a greedy bipartite graph matching to output the 2D keypoints for all people in the image. In this approach, both prediction of part affinity fields (PAFs) and detection of confidence maps are refined at each stage [8,15,16]. By doing this, the real-time performance is improved while it maintains the accuracy of each component separately. The online OpenPose library supports jointly detect the human body, hand, and facial keypoints on a single image, which provides 2D human pose estimation for our proposed system.

Recurrent Neural Networks (RNNs) with Long Short-Term Memory (LSTM) cells are widely used for human action recognition with the emerging accessible human activity recognition methods. In this paper, F. M. Noori et al. proposes an approach that first extracts anatomical keypoints from RGB images using the OpenPose library and then obtains extra-temporal motions features after considering the movements in consecutive video frames, and lastly classifies the features into associated activities using RNN with LSTM [7]. Improved performance is shown as organizing activities performed by several different subjects from various camera angles. However, the pose estimation based human action recognition has poor performance on multi-person interactions and group activity recognition.

2.3 Learning Actor Relation Graphs for Group Activity Recognition

Human action recognition and group activity recognition are an important problem in group activity understanding [9]. In 2019, J. Wu et al. proposed to use Actor Relational Graph (ARG) to model relationships between actors and recognize group activity with multiple persons involved [9]. Using the ARG in a multi-person scene, the relation between actors from respect to appearance similarity and the relative location is inferred and captured. Compared with using a CNN to extract person-level features and later aggregate the features into a scene-level feature or using RNN to capture temporal information in the densely sampled frames, learning with ARG is less computationally expensive and more flexible while dealing with variation in the group activity. Given a video sequence with bounding boxes and ground truth labels of action for actors in the scene, the trained network can recognize individual actions and group activity in a multi-person scene. For long-range video clips, ARG's efficiency is improved by forcing a relational connection only in a local neighborhood and randomly dropping several frames while maintaining the training samples' diversity and reducing the risk of overfitting. At the beginning of the training process, the actors' features are first extracted by CNN and RoIAlign model [17] using the provided bounding boxes. After obtaining the feature vectors for actors in the scene, multiple actor relation graphs are built to represent the diverse information for the same set of actors' features. Finally, Graph Convolutional Network (GCN) is applied to perform learning and inference to recognize individual actions and group activity based on ARG. Two classifiers used for individual actions and group activity recognition are applied respectively to the pooled ARG. Scene-level representation is generated by max-pooling individual actor representations, which later uses for group activity classification. However the ARG based model has bad performance on video with extreme brightness and contrast condition.

3   Our Approach

We propose using an improved Actor relation graph-based model (IARG) that focused on group activity recognition without restrictions of extreme brightness and contrast condition. The overview of the Improved ARG-based model is shown in Fig. 1.

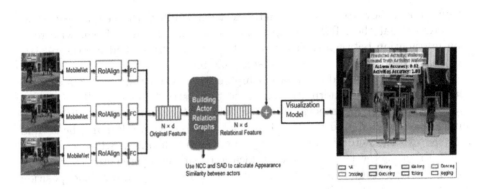

Fig. 1. The improved ARG-based human actions and group activity recognition model.

The model first extracts actor features from sampled video frames with manually labeled bounding boxes using CNN, and RoIAlign [17]. Next, it builds an N by d dimensional feature matrix, using a d-dimension vector to represent each actor's bounding box and using N to present the total number of bounding boxes in video frames. The actor relation graphs are then built to capture each actor's appearance and position relationship in the scene. Afterward, the model uses Graph Convolutional Networks (GCN) to analyze each actor's relationship from the ARG. Finally, the original and relational features are aggregated and used by two separate classifiers to perform actions, and group activity recognition [9]. Since the study is mainly focused on group activity recognition, the individual action recognition is not very accurate because the model only uses the region of interest and CNN to perform action recognition.

To improve the overall accuracy, we propose to apply MobileNet [11] in the CNN layer to extract image feature maps and use Normalized cross-correlation (NCC) and the sum of absolute differences (SAD) to calculate the pair-wise appearance similarity to build the Actor Relation Graph. More details of our proposed methodology are mentioned in section V.

To make our model have a more visualized result, we also introduce a visual-ization model that plots each input video frame with predicted bounding boxes on each human object and predicted individual action and group activity as the output. The output examples are shown in Fig. 2.

4   Methodology

In our model, how to build this Actor relation graph is the key. J. Wu et al. has proved that the ARG can represent the graph structure of pair-wise rela-tion information between each pair of actors in each frame and use the related information for group activity understanding [9].

To better understand the relationship between two actors, both appearance features and position information are used to construct the ARG. The relation value is defined as a composite function below, which function fa indicates the appearance relation, and function fs indicates the position relation. The x_i^a and x_j^a refers to the actor i's and actor j's appearance features, while x_i^s and x_j^s refers to the actor i's and actor j's location features (the center coordinates of each actor's bounding box). The function h fuses appearance and position relation to a scalar weight [9]:

$$G_{ij} = h(f_a(x_i^a, x_j^a), f_s(x_i^s, x_j^s)). \tag{1}$$

The normalization is further adopted on each actor node with SoftMax func-tion so that the sum of all the corresponding values of each actor node will always equal to one [9]:

$$\mathbf{G}_{ij} = \frac{f_s\left(\mathbf{x}_i^s, \mathbf{x}_j^s\right) \exp\left(f_a\left(\mathbf{x}_i^a, \mathbf{x}_j^a\right)\right)}{\sum_{j=1}^{N} f_s\left(\mathbf{x}_i^s, \mathbf{x}_j^s\right) \exp\left(f_a\left(\mathbf{x}_i^a, \mathbf{x}_j^a\right)\right)}. \tag{2}$$

4.1   Appearance Relation

In J. Wu's paper, the Embedded Dot-Product is implemented to compute the similarity between the two actors' appearance features (the image feature inside each actor's bounding box) in embedding space [9]:. The corresponding function is written in the way below:

$$f_a\left(\mathbf{x}_i^a, \mathbf{x}_j^a\right) = \frac{\theta\left(\mathbf{x}_i^a\right)^{\mathrm{T}} \phi\left(\mathbf{x}_j^a\right)}{\sqrt{d_k}}. \tag{3}$$

The θ and ϕ are two functions using Wx + b, in which W and b are learnable weight. The learnable transformations of original appearance features can better understand the relation between two actors in a subspace.

Our model evaluates two other methods for the appearance relation calculation: the Normalized cross-correlation (NCC) and the sum of absolute differences (SAD).

Normalized cross-correlation (NCC) is a method to evaluate the degree of similarity between two compared images. The brightness of the compared images can vary due to lighting and exposure conditions, so the images can be first normalized to get a more accurate similarity score by applying NCC. The advantage of the normalized cross-correlation is that it is less sensitive to linear changes in the amplitude of illumination in the two compared images, and the corresponding function can be written in the following way [18]:

$$\varphi'_{x_i^a x_j^a}(t) = \frac{\varphi_{x_i^a x_j^a}(t)}{\sqrt{\varphi_{x_i^a x_i^a}(0)\varphi_{x_j^a x_j^a}(0)}}. \tag{4}$$

The normalized quantity $\varphi'_{x_i^a x_j^a}(t)$ will be vary between -1 and 1, while value 1 indicates exactly matching between two images, and value of 0 indicates no matching between two images. The NCC value can help us better understand the appearance relation between each pair of actors.

Sum of absolute differences (SAD) is another method we evaluate to calculate the appearance relation when building the ARG. SAD calculates the distance between two matrices by computing the sum of absolute difference of the components of the matrices as the formula:

$$\text{SAD}\left(x_i^a, x_j^a\right) = \sum_k^n \left|x_{ik}^a - x_{jk}^a\right|. \tag{5}$$

Since SAD is more resistant to extreme values in the data, it is more robust when comparing appearance features and better captures the appearance relation.

4.2 Position Relation

Besides, spatial structural information is considered to capture the position relation between actors better. A distance mask has been applied to obtain signals from entities that are not distantly apart. Since relation in a local scope is more crucial compared with global relation for group activity understanding, a measure of Euclidean distance $\mathbf{G}_{ij}$ between two actors is computed as:

$$f_s\left(\mathbf{x}_i^s, \mathbf{x}_j^s\right) = \mathbb{I}\left(d\left(\mathbf{x}_i^s, \mathbf{x}_j^s\right) \leq \mu\right), \tag{6}$$

where $\mathbb{I}(\cdot)$ denotes the indicator function, $d(\mathbf{x}_i^s, \mathbf{x}_j^s)$ calculates the Euclidean distance between the center coordinates of two actors' bounding boxes, and μ is a distance threshold.

5   Experiments and Results

5.1   Datasets

In this paper, we used the public group activity recognition datasets called collective activity dataset and augmented dataset [19] to train and test our model. This dataset has 74 video scene that includes multiple persons in each scene. The manually defined bounding boxes on each person and the ground truth of their actions and the group activity are also labeled in each frame.

5.2   Implementation Details and Results

We use a minibatch size of 16 with a learning rate of 0.0001 and train our network in 100 epochs. The individual action loss weight *lambda* = 1 is used. The GCN parameters are set as dk = 256, ds = 32, and 1/5 of the image width is adopted to be the distance mask threshold μ. The default backbone CNN network for feature extraction is set as Inception-v3 [20], and the default of the appearance relation function is set as embedded dot-product. Our implementation is based on the PyTorch framework and two pieces of RTX 2080 Ti GPUs.

Evaluation 1 - Evaluate with Different Backbone Networks. In this subsection, we conduct detailed studies on the Collective Activity dataset to understand the proposed backbone networks' relation modeling using group activity prediction accuracy as the assessment metric. The results of the experiments are shown in Table 1.

In our 2-stage training, we first finetune the ImageNet pre-trained backbone network with the randomly selected frame from each of the training samples. Then, the weights of the feature extraction part of the backbone network are fixed in stage 2. We further train the network with GCN and calculate appearance relation using embedded dot-product. We begin our experiments with Inception-v3 [20] as the backbone network. The first stage of training takes approximately 2.6 h, while the second stage takes a longer time, about 3.3 h. With Inception-v3 [20], the group activity recognition accuracy after the training of stage 1 achieves 90.91%. With the additional training of GCN in stage 2, our model yields a higher recognition accuracy of 92.71%.

We further adopt MobileNet [11] as the backbone network to boost the speed of our model. MobileNet [11] is a deep convolutional neural network that is lightweight but efficient. With MobileNet [11], the training time of stage 1 is 1.8 h, which reduces the time spent at stage 1 by 32.7%. The training time of stage 2 is 2.4 h, which is 26% less than the training time of Inception-v3 at the same stage. In summary, the training speed of our model is increased by 35%. However, the activity recognition accuracy is slightly dropped from 92.71% to 91.44%.

Table 1. Accuracy (%) of group activity prediction from two backbone networks

Backbone network	Stage	Time cost (seconds)	Best group activity prediction accuracy
ARG	1	9569.8	90.91%
ARG	2	11784.4	92.71%
IARG(our model)	1	**6443.8**	89.37%
IARG(our model)	2	**8719.7**	91.44%

Evaluation 2 - Evaluate with Different Appearance Relation Functions. In this experiment, we evaluate the group activity recognition performance with different appearance relation functions. We first train and validate the group activity recognition performance based on the default Inception-v3 [20] backbone and the embedded dot-product for appearance relation calculation. The best result we get is 92.71%. Then we update the appearance relation function with Normalized cross-correlation (NCC) to draw the actor relation graph, and the best result we achieve is 93.50%. We further evaluate the sum of absolute distance (SAD) function to calculate the appearance similarity, and the best score we achieve is 93.98%.

Table 2. Accuracy (%) of group activity prediction on backbone network inception-v3 [20] with different appearance relation functions

Method	Best group activity prediction accuracy
ARG	92.71%
IARG(our model)	**93.50%**
IARG(our model)	**93.98%**

In this experiment 2, we prove that our proposed model with either NCC or SAD as the appearance relation calculation function will achieve better group activity prediction accuracy as expected. The results of the experiments are shown in Table 2.

6   Conclusion

This paper utilizes the actor relation graph (ARG) based model with novel improvements for group activity recognition. To enhance our model's performance, we learn ARG to perform appearance relation reasoning on graphs using normalized cross-correlation (NCC) and the sum of absolute difference (SAD). Besides, to improve the computational speed, we introduce MobileNet [11] as the backbone network into our proposed model. Furthermore, extensive experiments demonstrate that the proposed methods are robust and effective for enhancing both accuracy and speed on the Collective Activity dataset. Since our project is mainly focused on group activity recognition, the individual action recognition is not very accurate because we only use the region of interest and CNN to perform action recognition.

Acknowledgment. The authors would like to thank our mentor Dr. Nasim Hajari, Postdoctoral Fellow, Department of Computing Science, University of Alberta, for her guidance and feedback throughout the research and study. We would also thank our advisor Dr. Anup Basu for their motivation and support to bring out the novelty in our research. Finally, We would like to thank the researchers of the previous work, which is the inspiration and starting point for our research.

Appendix

Fig. 2. Visualization of results on 6 test video clips

References

1. Gao, L., Guo, Z., Zhang, H., Xu, X., Shen, H.T.: Video captioning with attention-based LSTM and semantic consistency. IEEE Trans. Multimedia **19**(9), 2045–2055 (2017)
2. Venugopalan, S., Xu, H., Donahue, J., Rohrbach, M., Mooney, R., Saenko, K.: Translating videos to natural language using deep recurrent neural networks. In: Proceedings of the 2015 Conference of the North American Chapter of the Association for Computational Linguistics: Human Language Technologies (2015)
3. Venugopalan, S., Rohrbach, M., Donahue, J., Mooney, R., Darrell, T., Saenko, K.: Sequence to sequence - video to text. In: 2015 IEEE International Conference on Computer Vision (ICCV) (2015)
4. Sherstinsky, A.: Fundamentals of recurrent neural network (RNN) and long short-term memory (LSTM) network. Phys. D Nonlinear Phenomena **404**, 132306 (2020). https://www.sciencedirect.com/science/article/pii/S0167278919305974
5. Krishna, R. Hata, K., Ren, F., Fei-Fei, L., Niebles, J. C.: Dense-captioning events in videos. In: 2017 IEEE International Conference on Computer Vision (ICCV) (2017)
6. Fernando, B., Chet, C.T.Y., Bilen, H.: Weakly supervised gaussian networks for action detection. In: 2020 IEEE Winter Conference on Applications of Computer Vision (WACV) (2020)
7. Noori, F.M., Wallace, B., Uddin, M.Z., Torresen, J.: A robust human activity recognition approach using openpose, motion features, and deep recurrent neural network. In: Felsberg, M., Forssén, P.-E., Sintorn, I.-M., Unger, J. (eds.) SCIA 2019. LNCS, vol. 11482, pp. 299–310. Springer, Cham (2019). https://doi.org/10.1007/978-3-030-20205-7_25
8. Cao, Z., Martinez, G.H., Simon, T., Wei, S.-E., Sheikh, Y.A.: Openpose: realtime multi-person 2D pose estimation using part affinity fields. IEEE Trans. Pattern Anal. Mach. Intell. **PP**, 1 (2019)
9. Wu, J., Wang, L., Wang, L., Guo, J., Wu, G.: Learning actor relation graphs for group activity recognition. In: 2019 IEEE/CVF Conference on Computer Vision and Pattern Recognition (CVPR) 2019
10. Ibrahim, M.S., Mori, G.: Hierarchical relational networks for group activity recognition and retrieval. In: Ferrari, V., Hebert, M., Sminchisescu, C., Weiss, Y. (eds.) ECCV 2018. LNCS, vol. 11207, pp. 742–758. Springer, Cham (2018). https://doi.org/10.1007/978-3-030-01219-9_44
11. Howard, A.G., et al: MobileNets: efficient convolutional neural networks for mobile vision applications (2017).https://arxiv.org/abs/1704.04861
12. Li, X., Zhou, Z., Chen, L., Gao, L.: Residual attention-based LSTM for video captioning. World Wide Web **22**(2), 621–636 (2018). https://doi.org/10.1007/s11280-018-0531-z
13. Heath, C.D.C., Heath, T., McDaniel, T., Venkateswara, H., Panchanathan, S.: Using participatory design to create a user interface for analyzing pivotal response treatment video probes. In: McDaniel, T., Berretti, S., Curcio, I.D.D., Basu, A. (eds.) ICSM 2019. LNCS, vol. 12015, pp. 183–198. Springer, Cham (2020). https://doi.org/10.1007/978-3-030-54407-2_16
14. Raaj, Y., Idrees, H., Hidalgo, G., Sheikh, Y.: Efficient online multi-person 2D pose tracking with recurrent spatio-temporal affinity fields. In: 2019 IEEE/CVF Conference on Computer Vision and Pattern Recognition (CVPR) (2019)

15. Yuan, H., Ni, D., Wang, M.: Spatio-temporal dynamic inference network for group activity recognition. In: 2021 IEEE/CVF International Conference on Computer Vision (ICCV) (2021)
16. Li, et al.: GroupFormer: group activity recognition with clustered spatial-temporal transformer. In: 2021 IEEE/CVF International Conference on Computer Vision (ICCV) (2021)
17. He, K., Gkioxari, G., Dollár, P., Girshick, R.: Mask R-CNN. In: IEEE International Conference on Computer Vision (ICCV), vol. 2017, pp. 2980–2988 (2017)
18. Raghavender Rao, Y.: Application of normalized cross correlation to image registration. Int. J. Res. Eng. Technol. **03**(17), 12–16 (2014)
19. Choi, W., Shahid, K., Savarese, S.: What are they doing?: collective activity classification using spatio-temporal relationship among people. In: 2009 IEEE 12th International Conference on Computer Vision Workshops, ICCV Workshops (2009)
20. Szegedy, C., Vanhoucke, V., Ioffe, S., Shlens, J., Wojna, Z.: Rethinking the inception architecture for computer vision, pp. 2818-2826 (2015). https://arxiv.org/abs/1512.00567

SPGNet: Spatial Projection Guided 3D Human Pose Estimation in Low Dimensional Space

Zihan Wang$^{(\boxtimes)}$ ⓘ, Ruimin Chenⓘ, Mengxuan Liuⓘ, Guanfang Dongⓘ, and Anup Basuⓘ

University of Alberta, Edmonton, Canada
{zihan6,ruimin6,mliu2}@ualberta.ca

Abstract. We propose a method SPGNet for 3D human pose estimation that mixes multi-dimensional re-projection into supervised learning. In this method, the 2D-to-3D-lifting network predicts the global position and coordinates of the 3D human pose. Then, we re-project the estimated 3D pose back to the 2D key points along with spatial adjustments. The loss functions compare the estimated 3D pose with the 3D pose ground truth, and re-projected 2D pose with the input 2D pose. In addition, we propose a kinematic constraint to restrict the predicted target with constant human bone length. Based on the estimation results for the dataset Human3.6M, our approach outperforms many state-of-the-art methods both qualitatively and quantitatively.

Keywords: Machine learning for multimedia · Pattern processing

1 Introduction

3D human shape and posture estimation from a single image or video is a fundamental topic in computer vision. Unfortunately, it is not easy to estimate 3D body shape and posture directly from monocular images without any 3D information. The problem of 3D human shape and posture estimation can be defined as giving images as the input, and generating a 3D skeleton as the output. A typical 3D skeleton consists of 3D points for 17 joints. Mathematically it can be written as a mapping function $f(M) = x$, where M is a fixed-sized matrix representing the image input and x is a matrix representing the 3D joints (size is $(17, 3)$ in our case). Generally, f is applied to each frame of the video.

Before deep learning was widely used, massive 3D labeled data and 3D parameters with prior knowledge were necessary to deal with this problem [1]. After introducing the deep learning strategy, some approaches extract the 3D human pose based on the input image directly without any intermediate stage. Some of the existing strategies rely on convolutional neural networks to learn visual representations successfully from a very large dataset [19,23]. However, recent research has shifted to two-stage approaches. The 2D key points detection algorithm is first used to acquire the 2D poses from images. Then, 2D-to-3D pose

© The Author(s), under exclusive license to Springer Nature Switzerland AG 2022
S. Berretti and G.-M. Su (Eds.): ICSM 2022, LNCS 13497, pp. 41–55, 2022.
https://doi.org/10.1007/978-3-031-22061-6_4

lifting is applied as the second stage [3,15,18]. Existing methods have focused on optimizing the loss functions or neural network structures [3,15,18,20].

Inspired by the cycle consistency in unsupervised learning, some approaches re-map the predicted 3D poses to 2D poses in a semi-supervised learning framework [20]. Like following the CycleGAN Loss, the network is designed incorporating two components. One is the mapping from 2D to 3D. Another one is re-mapping 3D to 2D and comparing the re-projected 2D poses with the 2D input. However, the results for unlabeled videos are relatively unremarkable and the training process is relatively slow. In our work, we also extract the intermediate 2D pose. However, we do not intuitively deploy the CycleGAN semi-supervised learning framework, which leads us to avoid the abovementioned drawbacks.

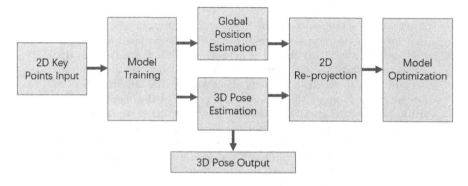

Fig. 1. General workflow of our SPGNet, take 2D key points as input, and estimate the 3D pose with its global position. The 2D re-projection combines both estimations and loss to optimize the model.

We design SPGNet, a novel neural network architecture that combines traditional supervised learning with a spatial projection that re-projects the predicted 3D poses to 2D poses. Instead of traditional loss functions based on the ground truth and 3D pose output, we introduce the 2D MPJPE loss (defined as Eq. 3). As shown in Fig. 1, SPGNet computes the 2D projection of 3D poses output with the estimated global position. The 2D MPJPE loss minimizes the re-projected 2D poses with the 2D poses input, which reuses the 2D poses input during the learning process. This approach increases the accuracy of 3D pose prediction. Overall, our main contributions are:

1. Introducing an adaptive supervised training framework for 3D human pose estimation under the category of 2D-to-3D lifting approach.
2. Exploiting the 2D pose input efficiently and improving robustness by presenting a re-projection loss, which is based on global pose estimation from 2D poses and the 2D poses themself.

Our model achieves 45.3 mm accuracy in Protocol 1 and 35.7 mm in Protocol 2 in the Human3.6M dataset, which are 0.6% and 1.4% relative improvements over previous approaches [28].

2   Related Work

Previous research on 3D Human Pose Estimation can be classified into two main categories. The first category extracts the 3D human pose based on the input image directly without any intermediate stage. Under this category, some recent approaches rely on convolutional neural networks to learn visual representations successfully from very large datasets. The predicted accuracy has increased considerably during the past few years [19,23]. The second category consists of two stages, namely 2D key points detection from video and 2D-to-3D pose lifting [3,15,18].

2.1   2D Key Joint Detection

Before estimating 3D human pose, many methods require labeled key joints. Inaccurate key joint labels may cause pose prediction to fail. Initially, 2D human pose was estimated based on Deep Neural Networks (DNNs), later Convolutional Neural Networks (CNNs) have shown more advantages in 2D key joint detection [16,24,25,27,29]. CNN-based models have outstanding performances in extreme test conditions (body occlusion or low-resolution images). A heatmap is generated to show the possibilities of a specific joint shown in an image and the estimate is refined to improve localization. Reusing the hidden layer in the CNN-based model, Tompson et al. proposed a method that uses heatmap, works as a regularizer to find their output, and increases the accuracy [24]. A similar method proposed by Yang et al. uses an end-to-end mixture of parts model [29]. In their method, the probability is calculated by the softmax function and a Max-sum algorithm is applied between pose parts. The pose machine uses multi-stage differentiable iterations to the joint on the heat-map to finally converge to one solution [16,27]. The pose machine combines the previous output and the updated prediction for the same input image in each subsequent stage.

2.2   Image to 3D Pose

Without estimating the 2D human pose, the 3D pose can be constructed directly through an image. This method minimizes the effects of error prediction during the 2D pose estimation, and in general, it can be more robust [9,21,26]. Additionally, it eliminates the limitation of the unlabeled image used for training [30]. One of the methods uses Pose Orientation Net (PONet) and generates heat maps: limb confidence maps and 3D orientation maps. With these heat maps, the model uses a fixed-length skeleton to match with the 3D orientation maps and complete the missing limbs in a 3D pose using sub-networks [26]. For a more specific pose, Ruiz et al. used three loss functions for each angle's rotation, with classification and regression [21]. Another approach proposed by Yang et al. uses the Generative Adversarial Networks (GANs) to directly predict the 3D pose using unlabeled images in outdoor scenarios [30]. Therefore, skipping 2D pose detection provides a solution for unpaired 2D-to-3D data training [9,12].

2.3   2D-to-3D Pose Lifting

3D pose estimation can be lifted from 2D human pose joints [3,15,18]. During the training, some approaches directly using 2D human pose ground truth as input [5,11]. This approach ensures the accuracy of the inputs and distraction is minimized. Moreover, 2D label on the image is easier to obtain. Some datasets, like Human3.6M and HumanEval, contain multiple 2D views, and EpipolarPose with another method estimate 3D pose from different directions at the same frame. The mixture of the multi-views provides consistency loss and recovers the pose via triangulation [11]. The model can be optimized by blending multiple 2D views into the same 3D pose. Depth prediction is important for some 2D-to-3D approaches. The simpler method uses binary ordinal depth relation prediction, while other have explicit depth prediction on each joint [3,17,18].

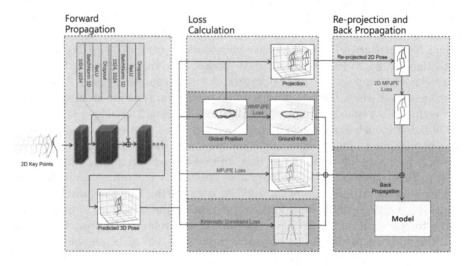

Fig. 2. Supervised training framework with a 3D pose model with predicted 2D pose sequence input. In addition to the kinematic constraint and MPJPE loss, we concatenate the global position and key points of the 3D poses to perform the projection. Finally, we compare the re-projected 2D pose and original input 2D poses to compute a 2D MPJPE loss and perform backpropagation.

3   Proposed Method

In this section, we briefly explain the details of the proposed SPGNet architecture, as shown in Fig. 2.

3.1   Problem Formulation

To better illustrate our method, we first formulate the 3D pose estimation problem as a 2D-to-3D lifting pipeline. Then, we assume the dataset defined as $\mathcal{D} = \{(x_i, y_i)\}_{i=1}^{N}$ consisting of N data, where each data x_i is associated with a corresponding label y_i. Here, $x_i \in \mathbb{R}^{M \times 2}$ represents the one frame input of 2D pose key points. Similarly, $y_i \in \mathbb{R}^{M \times 3}$ represent estimated 3D pose key points labels, where M denoted as number of joints for each human pose. In order to take advantage of temporal information between frames, we define the sequence of input for one frame as vector $\{\mathbf{x}^k \in \mathbb{R}^{1 \times M \times 2} | k = 1, 2, 3...J\}$, where J represents the number of frames in the input sequence. The goal is to optimize the prediction function (estimated 3D pose key points in three dimensions) f from the training data,

$$f(\mathbf{x}; W) = \arg\max_{y \in \mathcal{D}} F(\mathbf{x}, y; W) \tag{1}$$

where W is the weight of the neural network and F is the optimizing function.

3.2   Spatial Projection Guided Approach

In the training process, SPGNet contains three components: an encoder, a projector and loss functions. First, the encoder processes the input 2D key points, aiming to transform the data into precise 3D coordinates that represent the estimated human pose. Then, the projector transforms the estimated 3D human pose into a re-projected 2D pose for computing the similarity with the original input. Finally, multiple loss functions guide the backpropagation of the neural network in order to learn helpful representations in the latent space.

The Encoder. First, we encode our input data through an encoder and obtain the predicted 3D pose key points in three dimensions. Then, we decompose our output $\hat{t}_i$ for a certain frame into $\hat{t}_i^k$ and $\hat{t}_i^p$, where $\hat{t}_i^k$ is the predicted 3D pose key points coordinates in a certain 2D plane and $\hat{t}_i^p$ is the spatial information to confirm the global position of pose. In this case, $\hat{t}_i^k \in \mathbb{R}^{M \times 2}$, which is the (x, y) coordinates of each joint of a 3D pose in three dimensions. Similarly, $\hat{t}_i^p \in \mathbb{R}^M$ is just the (z) coordinate as predicted by the position. We do same decomposition for label y_i and gain y_i^k, y_i^p for further loss computation.

The Projector. A projector is utilized to map the estimated 3D human pose into the re-projected 2D pose. In order to minimize the impact of lens distortion on projection, we chose the nonlinear projector. The schematic diagram is shown in Fig. 3. This schematic diagram shows the forward projection onto the image plane that maps (x, y, z) coordinate into the (x, y) plane. The center is the camera center, and the focal length is the distance between the camera center and the image plane along the principal axis perpendicular to the image plane. These two parameters represent the spatial geometry relationship of the 3D pose in low dimensions. Another essential aspect to consider is the camera

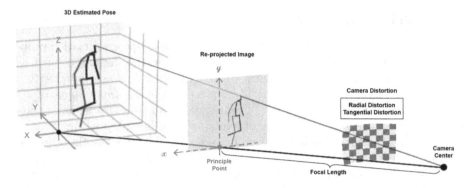

Fig. 3. Schematic diagram of the spatial dimension projection for SPGNet. The projector simulates the real camera to maintain consistency. The 3D estimated pose is projected to a 2D image with the same principle point, focal length, radial distortion and tangential distortion.

distortion. It is a kind of optical aberration that causes a straight line in the scene to not remain straight in an image. There are two common camera distortions. First, the radial distortion causes the magnification of the image to decrease or increase with distance from the principal axis. Second, tangential distortion occurs because the lens assembly is not centered and parallel to the image plane. Fixing lens distortion during projection can increase projection accuracy. Consequently, the projector has several intrinsic camera parameters as input to ensure the projection is at a right angle, specified by the particular dataset. Here, we define focal length, center, radial distortion, and tangential distortion as f_c, c_e, d_r, and d_t, respectively. The pseudocode of nonlinear projection is summarized in Algorithm 1.

Algorithm 1: Nonlinear projector mapping estimated 3D pose to reprojected 2D pose.

---

1 Projector $(\hat{t}_i^k, \hat{t}_i^p, f_c, c_e, d_r, d_t)$;

    Input   : Estimated 3D pose $\hat{t}_i^k$, estimated global position $\hat{t}_i^p$, intrinsic camera parameters (f_c, c_e, d_r, d_t)

    Output: Re-projected 2D pose $pose_{2d}$

2 $pose_{depth} = \hat{t}_i^k / \hat{t}_i^p$;

3 $pose_{cons} =$ Clamp $(pose_{depth})$ with $(min = -1, max = 1)$;

4 $r =$ Sum $(pose_{cons})$ value between (x, y) coordinates;

5 $concat =$ Concatenates the given sequence (r, r^2, r^3) ;

6 $R =$ Sum $(1 + (d_r \cdot concat))$ value between (x, y, z) coordinates;

7 $T =$ Sum $(d_t \cdot pose_{cons})$ value between (x, y) coordinates;

8 $pose_{trans} = pose_{cons} \cdot (R + T) + d_t \cdot r$;

9 $pose_{2d} = f_c \cdot pose_{trans} + c_e$;

10 **return** $pose_{2d}$

Loss Functions. Usually, deep learning models predict the target without restraint. Thus, we propose the kinematic constraint loss as a penalty to maintain consistency. This is under the assumption that the length of human bones is constant from beginning to end. We construct the output 3D pose pairs $(\hat{\alpha}, \hat{\beta})$, where $\hat{\alpha}$ is the previous predicted frame $\hat{t}$, and $\hat{\beta}$ is current predicted frame. Then, we define kinematic constraint loss as follows:

$$\mathcal{L}_{kc}(\hat{\alpha}; \hat{\beta}) = \frac{1}{2M} \sum_{i=1}^{M} \sum_{j=1}^{P} abs(\|\hat{\alpha}_i - \hat{\alpha}_j\|_2 - \|\hat{\beta}_i - \hat{\beta}_j\|_2) \qquad (2)$$

where, $\hat{\alpha}_j$ is the parent joint of the current joint $\hat{\alpha}_i$. The 3D human pose can be regarded as a tree-like structure. Each joint has at least one connection with the other joint (parent), and some of them have multiple connections. Thus, for a particular joint, the number of connections is determined by the P denoted as parents. The dummy variable P depends on a specific dataset that consists of an indeterminate number of joints. The constant 0.5 is a multiplier for model stability because the repeated bone length would be calculated twice. Note that the first predicted frame does not have $\hat{\alpha}$. Thus, we omit the computation of the loss function for the first predicted frame. This loss can be regarded as a penalty between consecutive frames and effectively maintaining the length of the bone, as shown in Sect. 4.5.

The loss function for estimated 3D human pose is simply the mean per joint position error (MPJPE), which is the Protocol 1 used in many existing work:

$$\mathcal{L}_{mpjpe}(\hat{t}_i^k; y^k) = \frac{1}{M} \sum_{i=1}^{M} \|\hat{t}_i^k - y_i^k\|_2 \qquad (3)$$

The MPJPE loss will calculate the euclidean distance between all the joints of the predicted 3D pose and the ground truth. During the backpropagation, the loss gradient provides information for optimizing the degree of the key point's accuracy. Notably, the ground truth is in the camera space, transformed by using the intrinsic and extrinsic camera parameters. Therefore, for the global position y^p, MPJPE cannot hold the depth information of the 3D pose. We use the weighted MPJPE loss function for estimated global position to retain the maximum spatial feature:

$$\mathcal{L}_{wmpjpe}(\hat{t}_i^p; y^p) = \frac{1}{M} \sum_{i=1}^{M} \frac{1}{y_i^k} \|\hat{t}_i^p - y_i^k\|_2 \qquad (4)$$

The inverse term $1/y_i^k$ is the regularization term compared with MPJPE loss to force the predicted 3D pose centered around the trim area. This is assuming that the pose object cannot move very far away from the camera position. The model learns the 3D pose characteristics of centralization in terms of results. In addition, the weighted MPJPE loss significantly increases the accuracy of the projected 2D pose and reduces the error caused by the abnormal predicted global position.

3.3 Design of Encoder

The architecture of the 2D-to-3D lifting neural network we designed is a temporal dilated convolutional model inspired by previous lifting approaches [15,20]. The neural network is fully implemented with the residual connections [6] in order to transform the sequence of input $\mathbf{x}$ defined in Sect. 3.1 through temporal convolutional layers. In detail, each residual blocks can be defined as

$$\mathbf{z} = D(\sigma(N(C(\mathbf{x})))), \tag{5}$$

where $\mathbf{z}$ is the extracted feature, C is the convolutional layer with 1024 input sizes except the input layer and 1024 output sizes, B is the batch normalization layer, σ is the ReLU activation layer, and finally, D is the dropout layer. Two residual blocks form a residual connection shown in Fig. 2. The number of residual blocks depends on the number of input frames. For example, input $\mathbf{x}$ with $J = 243$ needs 8 residual blocks for residual connections. Reminder, the input convolutional layer has $2 \cdot M$ input size to adapt the 2D key points, where M is defined in Sect. 3.1. Finally, the output layer is a convolutional layer with output size $3 \cdot M$, fitting with size of estimated 3D pose, defined as:

$$\hat{t}_i = C_{out}(\mathbf{z}), \tag{6}$$

where $\hat{t}_i$ is the output of encoder for one frame defined in Sect. 3.2.

Table 1. Detailed results under Protocol 1 (MPJPE)

Methods	Martinez [15]	Sun [22]	Yang [30]	Lee [13]	Pavllo [20]	Cai [2]	Xu [28]	Our
Direct.	51.8	52.8	51.5	40.2	45.1	44.6	**37.4**	<u>37.5</u>
Discuss.	56.2	54.8	58.9	49.2	47.4	47.4	**43.5**	<u>44.7</u>
Eat	58.1	54.2	50.4	47.8	<u>42.0</u>	45.6	42.7	**41.8**
Greet	59.0	54.3	57.0	52.6	46.0	48.8	<u>42.7</u>	**42.1**
Phone	69.5	61.8	62.1	50.1	49.1	50.8	<u>46.6</u>	**45.5**
Photo	78.4	67.2	65.4	75.0	56.7	<u>59.0</u>	59.7	**58.9**
Pose	55.2	53.1	49.8	50.2	44.5	47.2	**41.3**	<u>42.0</u>
Purch	58.1	53.6	52.7	43.0	44.4	**43.9**	<u>45.1</u>	46.7
Sit	74.0	71.7	69.2	55.8	57.2	57.9	**52.7**	<u>52.8</u>
SitD	94.6	86.7	85.2	73.9	66.1	61.9	<u>60.2</u>	**59.4**
Smoke	62.3	61.5	57.4	54.1	47.5	49.7	**45.8**	<u>46.7</u>
Wait	59.1	53.4	58.4	55.6	44.8	46.6	<u>43.1</u>	**42.8**
WalkD	65.1	61.6	**43.6**	58.2	49.2	51.3	47.7	<u>46.6</u>
Walk	49.5	47.1	60.1	43.3	32.6	37.1	**33.7**	<u>34.7</u>
WalkT	52.4	53.4	47.7	43.3	34.0	39.4	<u>37.1</u>	**36.8**
Avg	62.9	59.1	58.6	52.8	47.1	48.8	<u>45.6</u>	**45.3**

*The table reports the result with CPN 2D detection pose key points as input. The last line is the average of all 15 action results in millimeter. Best results in bold, second best underlined.

4 Experiments

4.1 Dataset and Metrics

We evaluate our method on the public dataset **Human3.6M**, widely used in other work [8]. The Human3.6M dataset is collected by a motion capture system. As one of the largest 3D human pose estimation datasets, 11 professional actors performed 15 scenarios consisting of 3.6 million video frames. There are four digital video cameras, one time-of-flight sensor, and ten motion cameras to capture the human pose. Our experiments follow the previous work [20] to adopt a standard 17-joints skeleton and split the dataset into a training set (S1, S5, S6, S7, S8) and a testing set (S9, S11).

We take two widely used protocols to evaluate our model: **Protocol 1** is the mean per joint position error (MPJPE) defined in Eq. 3. MPJPE calculates the Euclidean distance between the estimated positions of the joints and ground truth in millimeters. **Protocol 2** is the Procrustes mean per joint position error (P-MPJPE), which calculates the error after aligning the estimated 3D pose to the ground truth in a rigid transformation, such as translation, rotation, and scale.

4.2 Implementation Details

We train our model with input in camera space for consistency of other work through quaternion transformation. Here, we set our temporal convolutional model with 243 frames to benefit from the consecutive video stream. We choose Adam [10] as an optimizer and train our model with 100 epochs. The learning rate starts from 0.001 and decays exponentially every epoch. We adopt fine-tuned 2D pose detection key points through the Cascaded Pyramid Network [4] and ground truth 2D as our input. We apply the data augmentation method, pose flipping horizontally in the training set, with settings similar to [20].

4.3 Comparison with State-of-the-art Methods

This section reports our model's performance on 15 actions belonging to S9 and S11. First, we use the CPN network as the 2D pose detector to obtain the 2D key points as our input data. We compare them using Protocol 1 and Protocol 2, shown in Tables 1 and 2. Our model has a lower average error than all other approaches under both protocols and does not rely on additional data as many other approaches. Under Protocol 1 (Table 1), our model slightly outperforms the previous best result [28] by 0.3 mm on the average, corresponding to a 0.6% error reduction. To be more specific, we got the best 7 out of 15 actions and 7 s best actions. This indicates that our model's architecture has a better learning ability to extract features in the latent space in order to keep the spatial and temporal information. For Protocol 2, our model achieves the best results in terms of average P-MJPJE, with a 1.4% error reduction. However, for individual actions, most actions predicted by our model only achieve the second-best results.

Table 2. Detailed results under Protocol 2 (P-MPJPE)

Methods	Martinez [15]	Sun [22]	Yang [30]	Lee [13]	Pavllo [20]	Liu [14]	Xu [28]	Our
Direct.	39.5	42.1	**26.9**	34.9	34.2	32.5	31.0	<u>30.8</u>
Discuss.	43.2	44.3	**30.9**	35.2	36.8	35.3	36.8	<u>34.6</u>
Eat	46.4	45.0	36.3	43.2	**33.9**	34.3	34.7	<u>34.1</u>
Greet	47.0	45.4	39.9	42.6	37.5	36.2	**34.4**	<u>35.8</u>
Phone	51.0	51.5	43.9	46.2	37.1	37.8	<u>36.2</u>	**35.3**
Photo	56.0	53.0	47.4	55.0	43.2	**43.0**	43.9	<u>43.2</u>
Pose	41.4	43.2	**28.8**	37.6	34.4	33.0	<u>31.6</u>	31.9
Purch	40.6	41.3	**29.4**	38.8	33.5	<u>32.2</u>	33.5	32.5
Sit	56.5	59.3	**36.9**	50.9	45.3	45.7	42.3	<u>42.1</u>
SitD	69.4	73.3	58.4	67.3	52.7	51.8	**49.0**	<u>49.9</u>
Smoke	49.2	51.0	41.5	48.9	<u>37.7</u>	38.4	**37.1**	39.0
Wait	45.0	44.0	**30.5**	35.2	34.1	32.8	33.0	<u>32.6</u>
WalkD	49.5	48.0	**29.5**	50.7	38.0	37.5	39.1	<u>37.2</u>
Walk	38.0	38.3	42.5	31.0	**25.8**	**25.8**	26.9	<u>26.9</u>
WalkT	43.1	44.8	32.2	34.6	**27.7**	<u>28.9</u>	31.9	29.7
Avg	47.7	48.3	37.7	43.4	36.8	36.8	<u>36.2</u>	**35.7**

*The table reports the result with CPN 2D detection pose key points as input. The last line is the average of all 15 action results in millimeter. Best in bold, second best underlined.

Yang's [30] work reports a significant improvement in the actions with complex spatial relationships, such as sitting or direction, achieving the seven best results. Compared to results in Table 1, their method uses the feature of GANs but makes it hard to detect the global position of the 3D human pose, leading the better performance in Protocol 1.

To further study our method, we utilize the ground truth of 2D key points as our input to evaluate our model, with results shown in Table 3 under Protocol 1 and Protocol 2. By using 2D ground truth, the models generally get better performance than Table 2. Compared to the previous best result [14], our model outperforms by 1.3 mm on the average. Our method highlights the reuse of the 2D key points for the 2D MPJPE loss, so the model depends more on the 2D key points than other models that only use those data at the input stage. This is evident from the higher error deduction comparing our model to previously implemented methods between Tables 2 and 3. More discussion about the improvement in performance in this case is presented in a later section. We also report our model's performance in Protocol 2, resulting in better performance of 25.3 mm on the average, outperforming the state-of-the-art.

4.4   Ablation Study

Analysis of Re-Projected 2D Pose Loss. We further analyze the MPJPE loss between the re-projected 2D pose and the original input. We choose the encoder only consisting of MPJPE loss for the estimated 3D human pose as our

Table 3. Detailed results based on ground truth of 2D human pose.

| Methods | Protocol 1 | | | | Protocol 2 |
	Hossain [7]	Lee [13]	Liu [14]	Our	Our
Direct.	35.2	32.1	34.5	**29.5**	21.1
Discuss.	40.8	36.6	37.1	**33.6**	23.9
Eat	37.2	34.3	33.6	**32.8**	24.5
Greet	37.4	37.8	34.2	**32.5**	24.6
Phone	43.2	44.5	32.9	**32.5**	25.3
Photo	44.0	49.9	37.1	**33.6**	29.9
Pose	38.9	40.9	39.6	**37.5**	21.1
Purch.	35.6	36.2	35.8	**29.8**	23.7
Sit	42.3	44.1	40.7	**30.7**	30.4
SitD	44.6	45.6	41.4	**38.1**	38.9
Smoke	39.7	35.3	**33.0**	46.4	26.9
Wait	39.7	35.9	**33.8**	35.9	23.1
WalkD	40.2	30.3	33.0	**31.8**	27.7
Walk	32.8	37.6	26.6	**25.5**	18.8
WalkT	35.5	35.5	**26.9**	28.5	19.6
Avg	39.2	38.4	34.7	**33.4**	25.3

*Utilize the ground truth 2D human pose to predict the target. The table reports results of both Protocal 1 (left side) and Protocol 2 (right side). We label the best in bold.

encoder and add kinematic constraint as Baseline* to compare with SPGNet, as shown in Fig. 4a. The histogram indicates that our model dramatically benefits from the MPJPE loss of the re-projected 2D pose, leading to the re-projected 2D pose being closer to the ground truth 2D human pose. Furthermore, the kinematic constraint slightly improves the model's accuracy, by adding a bone length constraint.

Analysis of Size of Input Frames. The size of the input frames has a significant impact on the pose estimation performance. For example, the chart in Fig. 4b shows that the error in Protocol 1 decreases by 0.9 mm and 1.8 mm while the size of the input frames increases from 27 to 81 and 243, respectively. A similar error decrease trend is also reflected in Protocol 2. Consequently, we conclude that the larger size of sequential input provides more temporal information to allow our model to capture the movement of the 3D human pose between frames, similar to [20].

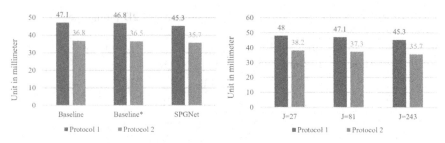

(a) Comparisons of different settings   (b) Comparisons of size of input frames

Fig. 4. Clustered column charts (a) and (b) illustrate the comparisons of different sittings and input frames, respectively. The error shows a decreasing trend as the model is applied from the Baseline to our SPGNet or the size of input frames increases. Notably, in Fig. 4a, Baseline is the model only learned from MPJPE loss, and Baseline* represents the Baseline adding kinematic constraints. In the chart 4b, J represents the number of frames in the input sequence. (a) Comparisons of different settings. (b) Comparisons of size of input frames.

4.5   Effect of Kinematic Constraints

In this section, we compare the sixteen bone lengths measured in different frames in one particular walking scenario, shown in Fig. 5. We can see that the difference in most small bones is almost negligible. Some large bones, such as leg bone or spine, have some relatively large variance compared to the small bone. However, the errors are all within 0.065 m, which is acceptable and may be caused by movement of the frame. As a consequence, this line chart indicates that our kinematic constraint is effective.

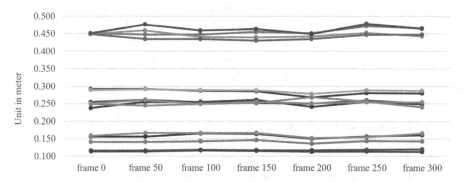

Fig. 5. Line chart of the measured bone lengths in seven different frames extracted from S11 walking scenarios. The different colors of lines represent the sixteen bones of a 3D human pose (tree-like structure).

4.6   Shielding Problem

We further analyze the limitation of our method from visualization results, shown in Fig. 6. We found that in the first row, which represents the action of sitting, the human limbs barely overlap in the camera's view. Thus, our results cannot be distinguished by the human eye if the consideration of global position is ignored. However, in the third row, the red arm of the human overlaps with the human body, which forms a shield masking the inner limbs. Consequently, the spatial relationship is not perfectly represented for both human pose in the third row, the variance of body tilt, and the difference of the arm's spatial positions, respectively.

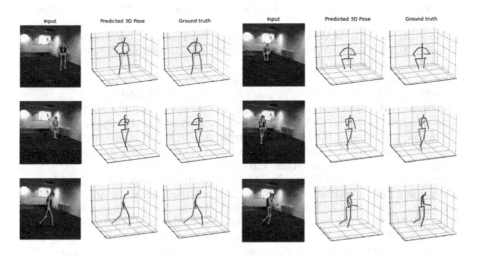

Fig. 6. Visualization of qualitative results on video clips.

5   Conclusion

We proposed SPGNet, a fully convolutional network based on supervised learning for human 3D pose estimation. To utilize our 2D-to-3d-lifting network, we used 2D key points in both input and re-projection stages and introduced kinematic constraints of human bone length and the corresponding loss function. Our model achieved more reliable estimates than state-of-the-art methods. SPGNet utilizes 2D labels in a more effective way, so the performance is expected to increase using image-to-2D methods with higher accuracies. Furthermore, besides the popular Human3.6M dataset, more datasets need to be tested for better analysis of our method.

References

1. Andriluka, M., Roth, S., Schiele, B.: Monocular 3D pose estimation and tracking by detection. In: 2010 IEEE Computer Society Conference on Computer Vision and Pattern Recognition, pp. 623–630 (2010). https://doi.org/10.1109/CVPR.2010.5540156
2. Cai, Y., et al.: Exploiting spatial-temporal relationships for 3D pose estimation via graph convolutional networks. In: 2019 IEEE/CVF International Conference on Computer Vision (ICCV), pp. 2272–2281 (2019). https://doi.org/10.1109/ICCV.2019.00236
3. Chen, C., Ramanan, D.: 3D human pose estimation = 2D pose estimation + matching. CoRR abs/1612.06524 (2016). https://arxiv.org/abs/1612.06524
4. Chen, Y., Wang, Z., Peng, Y., Zhang, Z., Yu, G., Sun, J.: Cascaded pyramid network for multi-person pose estimation. CoRR abs/1711.07319 (2017). https://arxiv.org/abs/1711.07319
5. Drover, D., Rohith, M.V., Chen, C., Agrawal, A., Tyagi, A., Huynh, C.P.: Can 3D pose be learned from 2D projections alone? CoRR abs/1808.07182 (2018). https://arxiv.org/abs/1808.07182
6. He, K., Zhang, X., Ren, S., Sun, J.: Deep residual learning for image recognition. CoRR abs/1512.03385 (2015). https://arxiv.org/abs/1512.03385
7. Hossain, M.R.I., Little, J.J.: Exploiting temporal information for 3D pose estimation. CoRR abs/1711.08585 (2017). https://arxiv.org/abs/1711.08585
8. Ionescu, C., Papava, D., Olaru, V., Sminchisescu, C.: Human3.6m: large scale datasets and predictive methods for 3D human sensing in natural environments. IEEE Trans. Pattern Anal. Mach. Intell. **36**(7), 1325–1339 (2014)
9. Kanazawa, A., Black, M.J., Jacobs, D.W., Malik, J.: End-to-end recovery of human shape and pose. CoRR abs/1712.06584 (2017). https://arxiv.org/abs/1712.06584
10. Kingma, D.P., Ba, J.: Adam: a method for stochastic optimization. CoRR abs/1412.6980 (2014). https://dblp.uni-trier.de/db/journals/corr/corr1412.html
11. Kocabas, M., Karagoz, S., Akbas, E.: Self-supervised learning of 3D human pose using multi-view geometry. CoRR abs/1903.02330 (2019). https://arxiv.org/abs/1903.02330
12. Lassner, C., Romero, J., Kiefel, M., Bogo, F., Black, M.J., Gehler, P.V.: Unite the people: closing the loop between 3D and 2D human representations. CoRR abs/1701.02468 (2017). https://arxiv.org/abs/1701.02468
13. Lee, K., Lee, I., Lee, S.: Propagating LSTM: 3D pose estimation based on joint interdependency. In: Ferrari, V., Hebert, M., Sminchisescu, C., Weiss, Y. (eds.) ECCV 2018. LNCS, vol. 11211, pp. 123–141. Springer, Cham (2018). https://doi.org/10.1007/978-3-030-01234-2_8
14. Liu, R., Shen, J., Wang, H., Chen, C., Cheung, S.C., Asari, V.: Attention mechanism exploits temporal contexts: real-time 3D human pose reconstruction. In: Proceedings of the IEEE/CVF Conference on Computer Vision and Pattern Recognition (CVPR), pp. 5064–5073 (2020)
15. Martinez, J., Hossain, R., Romero, J., Little, J.J.: A simple yet effective baseline for 3D human pose estimation. CoRR abs/1705.03098 (2017). https://arxiv.org/abs/1705.03098
16. Newell, A., Yang, K., Deng, J.: Stacked hourglass networks for human pose estimation. CoRR abs/1603.06937 (2016). https://arxiv.org/abs/1603.06937
17. Nie, X., Wei, P., Zhu, S.C.: Monocular 3D human pose estimation by predicting depth on joints. 2017 IEEE International Conference on Computer Vision (ICCV), pp. 3467–3475 (2017)

18. Pavlakos, G., Zhou, X., Daniilidis, K.: Ordinal depth supervision for 3D human pose estimation. In: 2018 IEEE/CVF Conference on Computer Vision and Pattern Recognition, pp. 7307–7316 (2018). https://doi.org/10.1109/CVPR.2018.00763
19. Pavlakos, G., Zhou, X., Derpanis, K.G., Daniilidis, K.: Coarse-to-fine volumetric prediction for single-image 3D human pose. CoRR abs/1611.07828 (2016). https://arxiv.org/abs/1611.07828
20. Pavllo, D., Feichtenhofer, C., Grangier, D., Auli, M.: 3D human pose estimation in video with temporal convolutions and semi-supervised training. CoRR abs/1811.11742 (2018). https://arxiv.org/abs/1811.11742
21. Ruiz, N., Chong, E., Rehg, J.M.: Fine-grained head pose estimation without keypoints. In: Proceedings of the IEEE Conference on Computer Vision and Pattern Recognition (CVPR) Workshops, pp. 2074–2083 (2018)
22. Sun, X., Shang, J., Liang, S., Wei, Y.: Compositional human pose regression. CoRR abs/1704.00159 (2017). https://arxiv.org/abs/1704.00159
23. Sun, X., Xiao, B., Liang, S., Wei, Y.: Integral human pose regression. CoRR abs/1711.08229 (2017). https://arxiv.org/abs/1711.08229
24. Tompson, J., Goroshin, R., Jain, A., LeCun, Y., Bregler, C.: Efficient object localization using convolutional networks. In: Proceedings of the IEEE Conference on Computer Vision and Pattern Recognition (CVPR), pp. 648–656 (2015)
25. Toshev, A., Szegedy, C.: DeepPose: human pose estimation via deep neural networks. CoRR abs/1312.4659 (2013). https://arxiv.org/abs/1312.4659
26. Wang, J., Huang, S., Wang, X., Tao, D.: PONet: robust 3D human pose estimation via learning orientations only. CoRR abs/2112.11153 (2021). https://arxiv.org/abs/2112.11153
27. Wei, S.E., Ramakrishna, V., Kanade, T., Sheikh, Y.: Convolutional pose machines. In: Proceedings of the IEEE Conference on Computer Vision and Pattern Recognition (CVPR), pp. 4724–4732 (2016)
28. Xu, J., Yu, Z., Ni, B., Yang, J., Yang, X., Zhang, W.: Deep kinematics analysis for monocular 3D human pose estimation. In: Proceedings of the IEEE/CVF Conference on Computer Vision and Pattern Recognition (CVPR), pp. 899–908 (2020)
29. Yang, W., Ouyang, W., Li, H., Wang, X.: End-to-end learning of deformable mixture of parts and deep convolutional neural networks for human pose estimation. In: 2016 IEEE Conference on Computer Vision and Pattern Recognition (CVPR), pp. 3073–3082 (2016). https://doi.org/10.1109/CVPR.2016.335
30. Yang, W., Ouyang, W., Wang, X., Ren, J.S.J., Li, H., Wang, X.: 3D human pose estimation in the wild by adversarial learning. CoRR abs/1803.09722 (2018). https://arxiv.org/abs/1803.09722

Image Processing

Unsupervised Face Frontalization GAN Driven by 3D Rotation and Symmetric Filling

Guanfang Dong$^{(\boxtimes)}$ and Anup Basu

University of Alberta,Edmonton, Canada
{guanfang,basu}@ualberta.ca

Abstract. Face frontalization has been partially solved by deep learning methods, such as Generative Adversarial Networks (GANs). However, due to the lack of paired training datasets, current generative models are limited to specific poses. Similarly, current unsupervised frameworks do not utilize properties of human faces, which burdens the neural network training. To improve and overcome current challenges, we design a novel self-supervised method that takes full advantage of human face modeling and facial properties. With our proposed method, single-view images collected in the wild can be utilized in training and testing. Also, the synthesized images are robust to input faces with large variations. We utilize the symmetric properties of human face to texture unseen parts in a human face model. Then, a GAN is used to fix undesired artifacts. Experiments show that our method outperforms many existing methods.

Keyword: Image processing

1 Introduction

Face frontalization is widely used in face recognition, face editing, and virtual and augmented reality. The problem of face frontalization is described as rotating various head poses into the frontal view of a face. As a fundamental challenge in computer vision, traditional methods do not perform well for this problem. This is because of the variations in facial details [1]. Most traditional methods struggle to find corresponding patterns and translate a face with various poses into the frontal view. However, the advent of the Generative Adversarial Network (GAN) brings a different aspect to solving the problem of face frontalization. Specifically, numerous input-output pairs are used for training to find the pattern of the face rotation for a generator and a discriminator. It is no coincidence that many large-scale face datasets [2,3,5–7] allow deep Convolutional Neural Networks to train for pose-invariant recognition. Compared to pose-invariant datasets, collecting pose-variant datasets (target person is photographed by different angles) is more expensive. Also, existing pose-variant datasets are not comparable with pose-invariant datasets in terms of the number of images [8,9]. For example, the

S. Berretti and G.-M. Su (Eds.): ICSM 2022, LNCS 13497, pp. 59–74, 2022.
https://doi.org/10.1007/978-3-031-22061-6_5

Multi-PIE face dataset [9] only contains images from 337 people, which is far less than a wild face dataset [6]. Also, these datasets are usually constrained to a controlled environment. The trained model only reflects the result from a specific domain. This kind of model lacks generalization ability. As a result, current methods which directly apply pose-variant datasets to GANs have difficulty in achieving promising results [10–14].

To avoid the problems caused by insufficient training datasets, we propose a novel unsupervised learning structure that allows face frontalization from single-view images. With this new unsupervised learning structure, all known face datasets can be used for training. This measure also vastly increases the generalization ability of our trained model. Compared to inpainting-based unsupervised face frontalization methods [16–18], our method further exploits the symmetric information of a face in a 3D space. This better preserves facial details and identity information for a GAN. We call our method URSF-GAN (Unsupervised Rotation and Symmetric Filling Driven GAN).

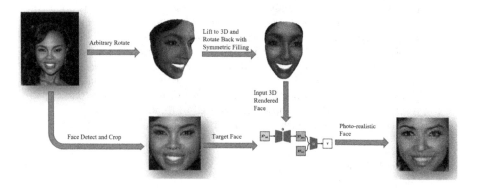

Fig. 1. The pipeline our URSF-GAN (Unsupervised Rotation and Symmetric Filling Driven GAN). Our method supports pose-invariant datasets. In-the-wild single-view facial images can be directly used in our method. First, an input face image is lifted to a 3D Morphable Model (3DMM). Then, the model is rotated by an arbitrary angle. We render the model and lift it back to 3D. A specially designed symmetric repair algorithm fixes missing textures. We rotate the model back to the front and use Pix2Pix GAN to eliminate artifacts.

As Fig. 1 shows, our method consists of two main parts. First, the single-view facial image is tested to see if they are front facing. For front faces, we lift faces from the 2D plane into 3D. Then, we rotate faces by arbitrary angles. The rotated faces are rendered into the 2D plane. For these rendered faces, we lift them into 3D again and rotate them back to the front. Clearly, some facial textures are hidden due to the arbitrary rotations, such as the opposite side of the nose bridge. We repair these missing information using the other side of the face. Until this step we do not use GAN. Nevertheless, the rendered faces are already good. The reason for using GAN is trying to eliminate the artifacts caused by rotation and repairs. The steps above can be described in the following mathematical way.

$$P_b = f_{\text{render}} \left(f_{\text{lift}} \left(P_a \right) \times R_d \right)$$

$$P_c = f_{\text{render}} \left(f_{\text{lift}} \left(P_b \right) \times R_{-d} \oplus P_{-b} \right)$$

Here, P_a means the original front face input. f_{lift} is the function that lifts the face into 3D. R_d is a rotation matrix with degree d. f_{render} means rendering a 3D scene into 2D. P_b is the rotated face, and P_c is the face rotated back. Following this order, $\{P_a, P_c\}$ becomes the training pair that is ready to fit a GAN for eliminating artifacts. As for the selection of a GAN, we modify Pix-to-Pix, which allows it to be more suitable for eliminating artifacts and enhance facial details. More details will be given in Sect. 3.

Remarkably, differing from previous supervised learning models, our proposed structure can fit any existing human face dataset. Our key contributions can be summarized as follows:

1. We propose an unsupervised framework that exploits the concept of 3D lifting and rotation. Single-view face images can be directly used in our framework. This brings unlimited training data and increases the models generalization ability.
2. We apply symmetric repair during rotation. It makes training-free facial images suitable for use. Our approach shifts the training focus of GAN from inpainting faces to eliminating artifacts, making our network easier to train. This structure benefits from GANs to generate more photo-realistic face frontalization results.

2 Related Work

2.1 Traditional Approaches

The problem of face frontalization aims to generate frontalized faces. The classical approach establishes a 3D model from facial images. Then, the 3D model can be rotated to the desired position. In a seminal work, 3D Morphable Model (3DMM) [19] sets a baseline for subsequent studies. 3DMM decomposes facial shape and appearance into PCA spaces. The decomposed information can be further analyzed in studies like face reconstruction or face landmark localization [20]. Based on 3DMM, Claudio et al. proposed an image patch localization and interpolation method by extracting feature descriptors [21]. However, limited by the number of feature descriptors, their method cannot keep enough facial details. Similarly, Zhu et al. proposed a Possion Editing based method to fill the area occluded by 3DMM rotation [22]. Taking advantage of 3DMM fitting, both methods can be used in constrained and unconstrained environments. However, the missing area usually shows more artifacts than deep learning methods.

2.2   Supervised Learning Approaches

The development of deep learning and GAN drew more activity for solving face frontalization problems [10–14, 16–18]. Tran et al. proposed a method leveraging a disentangled representation GAN model to learn frontalized face synthesis [10,11]. Also, the encoder-decoder structure is used to estimate variations in different faces. However, there are artifacts in their results due to the lack of effective interpretation from the encoder and decoder. Other than this, Huang et al. proposed a new GAN structure called TP-GAN to solve the problem of face frontalization [12]. This GAN structure involves perceiving and analysis of global structure and local details. They involve a landmark located patch network to assists encoder-decoder preserving local texture. Similarly, Hu et al. proposed a GAN structure that is driven by facial landmark heatmaps, which is called CAPG-GAN [13]. CAPG-GAN also introduces identity information preservation loss to maintain local textures. Besides this, Yin et al. designed a GAN structure that incorporates 3DMM [14]. To avoid undesired effects from occlusion, they also introduce mask symmetry loss to reproduce missing information. Cao et al. proposed a High Fidelity Pose Invariant Model to solve the problem of face frontalization [17]. They designed a new texture wrapping method to bind faces in 2D and 3D surfaces. However, most of the current methods require input-output pairs for training. As we mentioned before, obtaining paired data is very expensive. Thus, most of the current GAN methods are trained on the Multi-PIE dataset. However, the Multi-PIE dataset contains only 337 people, leading to a lack the generalization ability.

2.3   Unsupervised Learning Approaches

Recently, some research has looked into face frontalization for in-the-wild single view images. Deng et al. proposed a GAN structure to repair the missing areas of the facial UV map [16]. Finally, they perform a reconstruction on the repaired UV map. However, without real 3D rotations, an incomplete UV map cannot represent the real missing area, causing risks in robustness. Zhou el at. proposed a GAN inpainting network to fix the missing areas after rotation [18]. However, the convergence speed of their network is slow due to the lack of missing area pre-filling. Our approach both preserves facial details and improves training speed.

2.4   Image-to-Image Translation GAN

In recent years, many GAN structures try to solve the problem of image-to-image translation, which means the input images will be encoded in a low-dimensional space and stylized for the output [23–26]. Almost all such architectural solutions include decoders and encoders. However, l_2 loss often blurs output images during decoding. Thus, l_1 loss is usually used for the Image-to-Image translation GAN structure. Corresponding loss functions are adapted for specific problems. For classical network architectures, Pix2Pix focuses on mapping the training pairs by traditional l_2 loss [23]. It is applicable and easy to use for many image-to-image translation tasks. Also, collecting paired datasets is expensive. Unsupervised Image-to-Image GAN structures take advantage of cycle consistency to stylize the input images [26]. As a classical example, CycleGAN can translate images without feeding it paired data. Although we have the same goals, face frontalization requires the maximum preservation for identity information and facial details. Thus, we modify the Pix2Pix method to eliminate artifacts and enhance facial details.

3   Proposed Approach

3.1   Summary of Our Framework

Our approach can be conceptually divided into two parts. The first part aims to model, rotate and infer for single-view face images to generate a training dataset. The second part is to de-fake the inferred faces by an Image-to-Image translation GAN. We will then explain the implementation details and ideology behind each part.

3.2   Basic Concepts on Lifting Faces into 3D Space

Due to the needs of our framework, we model the face while preserving almost all of the facial textures. Currently, many deep learning-based studies are available to fit faces as 3D parameters. Like many models in 3D space, 3D face models are defined by their vertices. The mathematical representation is as follows:

$$V_{\text{face}} = [v_1, v_2, v_3, \ldots, v_n] \quad \text{while} \quad v_i = [x_i, y_i, z_i]^T$$

V_{face} is the collection of facial vertices. Based on this 3D feature, 3D Morphable Models (3DMMs) allow fitting 3D facial models with given 3D parameters [19]. Recall the fitting formula:

$$S = \bar{S} + A_{id}\alpha_{id} + A_{exp}\alpha_{exp}$$

$$T = \bar{T} + A_{tex}\alpha_{tex}$$

Here, S has a similar meaning as V, i.e., 3D vertex coordinates. $\bar{S}$ represents the mean facial shape from 200 young adults (100 male and 100 female). A_{id} denotes shape basis. A_{exp} means expression basis. Here, $\bar{S}$, A_{id} and A_{exp} are constant values. With given 3D parameters α_{id} and α_{exp}, the input 2D facial images can be fitted into a 3D model. Similarly, $\bar{T}$ means mean texture, and A_{tex} represents the texture basis. However, from the perspective of fitting accuracy, the results of texture fitting are poor. Thus, we restore the texture information by recording the vertex RGB information for the corresponding 2D face image. The process of obtaining vertex RGB information can also be called vertical projection. It can be mathematically expressed as:

$$\mathbb{C}_i = \mathcal{H}_{\tilde{x}_i, \tilde{y}_i} \left(\widetilde{\mathbb{C}}_i \right) \quad \text{while} \quad \mathbb{C}_i = [r_i, g_i, b_i]^T$$

Here $\mathcal{H}_{\tilde{x}_i, \tilde{y}_i}(\widetilde{\mathbb{C}}_i)$ are the vertical projections of vertices at 2D position $\tilde{x}_i, \tilde{y}_i$. Suppose we do not perform the rotation matrix operation after 3DMM modeling. In this case, we are able to ignore the depth information in 3D space and get all facial textures if the input image is a frontal face. In other words, if we are only fitting a face as 3D Morphable Model, the face's pitch, yaw, and roll will remain the same as the original 2D facial image. However, if the input image is not a frontalized face, the problems of occlusion and texture persist. We will discuss this issue in a later section (Figs. 2, 3, 4 and 5).

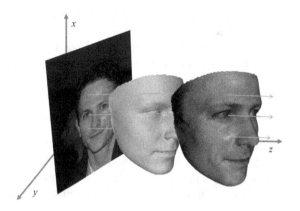

Fig. 2. The 3DMM fitting and texturing pipeline. The face is fed into 3DDFAv2 to estimate 3DMM parameters. Then, we lift face images to the 3DMM model. For frontal faces, we can use the vertical projection directly to obtain the facial texture.

Note that α_{id}, α_{exp} are two 3D parameters that we need to calculate for a unique face. We use 3D Dense Face Alignment Version 2 (3DDFAv2) [27], as one of the most efficient, stable and accurate methods, to model the facial images. The most crucial core of this method is the 3D-assisted short video synthesis method, which can simulate in-plane and out-of-plane face movement to convert a static image into a short video. Then, the model can be trained based on the short video. Thus, it makes 3D Dense Face Alignment possible for only one input static image, which matches our need.

Another important concept in our work is rendering. Rendering means projecting the 3D model into 2D space. Before rendering, the model can be rotated by a certain Euler angle to the desired position. Mathematically, the step of rendering can be represented as:

$$V_{2d}(\mathbf{p}) = f * \mathbf{Pr} * \mathbf{R} * \left(\bar{\mathbf{S}} + \mathbf{A}_{id}\alpha_{id} + \mathbf{A}_{exp}\alpha_{exp} \right) + \mathbf{t}_{2d}$$

Here, $V_{2d}(\mathbf{p})$ is the projection result. f is the scale factor. $\mathbf{Pr}$ represents the orthographic projection matrix. $\mathbf{R}$ is the rotation matrix which contains the Euler angles of pitch, yaw and roll. $\mathbf{t}_{2d}$ is a translation matrix. $(\bar{\mathbf{S}} + \mathbf{A}_{id}\alpha_{id} + \mathbf{A}_{exp}\alpha_{exp})$ is the 3DMM fitted model.

3.3 Unsupervised Training Strategy

Recall the overview in the introduction section. The purpose of lifting 2D face images to 3D space and rotating, filling, and rendering is to create effective training data pairs. The whole set of operations described here (lifting, adding texture, rotating and rendering) can be replaced by the mathematical notations $\{S_a, T_a, R_\theta, V\}$. In order to create the input and output datasets, we have to rotate the face model to the front face position after rotating it by an arbitrary angle. This means that the above process will be performed twice. In terms of notation sets, $\{S_a, T_a, R_\theta, V_b, S_b, R_{-\theta}, T_{-a}, V_c\}$ can represent the entire process. Among them, since the input face image must be a frontal face, the facial texture information would be completely obtained. However, when faces are rotated by an Euler angle, there will be some missing texture patches on the model. Thus, to solve this problem, we first need to find the location of missing patches.

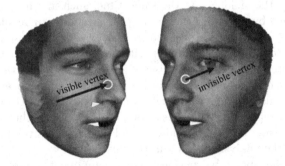

Fig. 3. For vertices that have different depths, only the vertex in the surface can obtain the correct texture when rendering a 3D model. In this example, the orange vertex is visible, and the gray vertex is invisible.

Recall the vertical projection $\mathbb{C}_i = \mathcal{H}_{\tilde{x}_i, \tilde{y}_i}(\tilde{\mathbb{C}}_i)$ *while* $\mathbb{C}_i = [r_i, g_i, b_i]^T$. Now, we consider the depth information z_i. Due to the properties of the projection, the texture information is valid only when it first crosses the surface of the object. So, within the same coordinate, only the uppermost z value will yield the correct texture information. The remaining vertices are invisible due to occlusion. Thus, we modify the original vertical projection formula and categorize the vertices into two classes (with and without visible texture).

$$v_i = \begin{cases} v_{visib} = \mathbb{C}_i = \mathcal{H}_{\tilde{x}_i \tilde{y}_i}\left(\tilde{\mathbb{C}}_i\right) \text{ while } \mathbb{C}_i = [r_i, g_i, b_i]^T \text{ and } i = \arg\max\left(x, y, \forall z_i\right) \\ v_{\text{invisib}} = \mathbb{C}_i = \{x, y, \forall z_i\} \setminus \arg\max\left(x, y, \forall z_i\right) \end{cases}$$

Here, $\mathcal{V}_{visib}$ can only be vertices that are located at the surface of the object. For the remaining vertices with lower depth, we consider them as invisible vertices and do not add texture on them.

As an intuitive idea to filling the $\mathcal{V}_{invisib}$, the perpendicular distance between an invisible vertex to the face midline can be calculated. Based on this distance, we can find the vertex that is perpendicular to the midline at the other half of the face. Thus, we can find a one-to-one correspondence for the left and right face vertices. Mathematically, it can be represented as:

$$d\left(v_{i\sim\text{midline}}\right) = |\ V_i \perp \text{ midline }|$$

$$V_{i'} = -d\left(v_{i\sim\text{midline}}\right) \perp \text{ midline}$$

Here, $V_{i'}$ is the corresponding vertex. $d\left(v_{i\sim\text{midline}}\right)$ is the Euclidean distance between the current vertex to the point perpendicular at midline. Since we compute the distance, we can find the corresponding vertex. Finally, the texture information (RGB value) for invisible vertices can copy and paste vertices from another half of the face.

However, this intuitive idea does not produce good results. There are two main reasons. First, during face photography, different parts of the face are exposed to different light levels. This also results in a small area of the face being affected by the shadow. At the same time, such an effect is difficult to completely remove by normalization and other methods. In regular modeling, the shaded part of the face will feel natural because of the smooth transition. However, if we force the corresponding positions of the left and right faces to be swapped, the vertex textures at the missing positions will make the overall look feel abrupt. Second, the human face is not perfectly symmetrical from left to right. If we do not use a landmark to determine the position of the face's organs, direct left-right swapping usually does not perfectly match the corresponding positions. Thus, using an intuitive idea would lead to massive artifacts, which inevitably affect the analysis of GAN in the next step.

Considering the above, we perform a pre-component before the symmetric copy. By landmark detection of the base 3DMM face model, we perform a regional division for the left and right faces. The region can be divided as eyes, nose, mouth and other parts for each left and right faces. For each classification, if there are vertices that are not visible in this classification, we swap and copy directly from the other half of the face that is visible. Furthermore, suppose more than two classifications have invisible vertices. This case indicates that we have a large missing range and we directly copy the other half of the face to the invisible part. The advantage of this approach is that only a small amount of artifacts appear in the marginal parts of each category. Also, these artifacts are patterned. This weakens the difficulty when analyzing input-output pairs in the next step of GAN.

$$\mathcal{P}_{\text{left}} = \mathcal{P}_{\text{right}} \text{ while } V_i \in \mathcal{P}_{\text{right} \vee \text{ left}} \tag{1}$$

Here, $\mathcal{P}$ represents the parts for eyes, nose, mouth, and other parts of one side of the face. The texture value can be found in the corresponding part if the current part has invisible vertices.

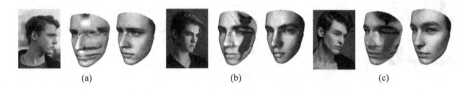

(a)   (b)   (c)

Fig. 4. Comparison of texturing with and without restoration. The left image is the original facial image. The middle image is texturing result without restoration. The right image is the texturing result with our restoration method. When there is no repair, the occluded vertices will also get texture information from the corresponding 2D position for the image. This causes the model texture to be normal on one side of the face and distorted on the other side. We added detection and repair for texture. The model obtained is almost free of artifacts. The results are already in a usable state even before processing by GAN.

3.4 Photo-Realization by URSF-GAN

In order to achieve photo-realistic results, we need to eliminate artifacts by applying an Image-to-Image translation network. Again, compared to other GAN-based methods for encoding 2D facial images or inpanting missing parts, we have less burden on the network. Thus, we choose Pix2Pix as the base network to translate the images. We have made small modifications to this network. We trained the network on RTX 2080*2. Eventually, after three hours of training, the network can generate realistic facial images.

URSF-GAN uses U-Net as its Generator. The reason is that our task needs to translate from one facial image to another. That is, the input and output need to be roughly aligned. Although the input and output differ in surface detail, they have the same underlying general structure. U-Net can be employed to capture both underlying structure and high-level semantic information.

To train our network, we apply the conditional GAN loss defined as:

$$\mathcal{L}_{cGAN}(G, D) = \mathbb{E}_{x,y}[\log D(x, y)] +$$
$$\mathbb{E}_{x,z}[\log(1 - D(x, G(x, z)))], \arg \min_G \max_D \mathcal{L}_{cGAN}(G, D).$$

Here, $D(x, y)$ is the probability of the target facial image being real. $(1 - D(x, G(x, z)))$ is the probability of the generated facial image being real. Thus, the generator tries to minimize the loss, and the discriminator tries to maximize

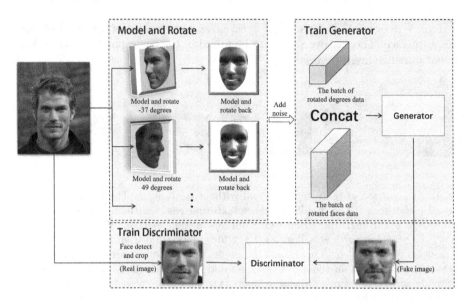

Fig. 5. The GAN structure of our URSF-GAN method. We have two differences from the original Pix2Pix. First, we embed the degrees when the model is rotated back. Second, we add random Gaussian noise to the input face image to avoid over-fitting.

the loss. Also, current research show that l_2 loss empirically leads to blurry output [28]. We involve the l_1 loss to avoid blurred images. This is expressed as:

$$\mathcal{L}_{L1}(G) = \mathbb{E}_{x,y,z}\left[\|y - G(x,z)\|_1\right]$$

Eventually, the final loss function for our URSF-GAN generator is:

$$G^* = \arg\min_G \max_D \mathcal{L}_{cGAN}(G,D) + \lambda\mathcal{L}_{L1}(G)$$

Before training, random Gaussian noise is added to the input facial images to avoid over-fitting. Then, to standardize the face size, we use a face detection algorithm on the image dataset to crop the images. Besides this, we also embed the 3D model rotation degree since we found that higher degree tends to have a wider face and lower light. Embedding the rotation degree helps the network analyze the repairing environment. The whole network setup is shown below.

Our model is tested differently than the way it is trained. For testing, the face image is lifted to a 3D model. The same method of obtaining textures requires the occluded texture to be repaired. Then, the model is rotated to a frontal face. Finally, the rendered model is used as the model input for the GAN. There is no need to add random noise during testing.

4   Experiments and Evaluations

4.1   Experimental Settings

We use the CelebA dataset for preprocessing and training. CelebA contains 202599 face images [29]. Among all facial images, we pick the first 60000 for training purpose, and use 5000 images (index from 60000 to 65000) for testing. Before training, we perform facial detection based on the Dlib C++ Library to standardize the face size. Then, each face image is modeled after rotation by 10 different arbitrary angles. After this, the face models are rotated back, filled, and rendered. The abovementioned steps allow us to create training input-output pairs. These data is run for 50 epochs. After that, the training data is discarded, and new training data needs to be generated by performing the abovementioned steps again.

Our model is trained in the Pytorch 1.6 environment with 2 * RTX 2080 settings. The size of input and output is $100 * 100 * 3$. During the preprocessing, we add Gaussian noise to input images with mean 0 and standard deviation 1. We choose the Adam optimizer to optimize the network with $\beta_1 = 0.5$ and $\beta_2 = 0.999$. The learning rate is 0.0001. The model is well-trained after 400 Epochs.

4.2   Evaluation

Within the field of face frontalization, most of the methods do not have publicly available source code. Thus, we directly use the demonstration images of various papers for comparison. Figure 6 shows the result of comparison between our URSF-GAN and other state-of-the-art methods. Since we cannot find the original facial images, we just show screenshots from other papers. Thus, the input resolution in our method is very low. However, although our method has low resolution, it still generates good results.

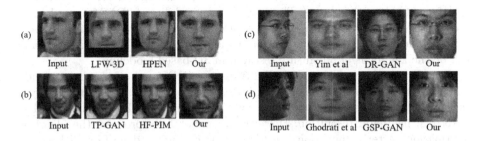

Fig. 6. Results of comparison with other popular methods.

4.3   Comparison with Other Methods

We analyze the results from different methods. Figure 6 contains four comparison pairs (a), (b), (c) and (d). For (a), as a critical issue, LFW-3D [15] and HPEN [22] do not rotate the face into an exact frontal position. For each method, LFW-3D uses a single, unmodified 3D reference for all input faces to generate frontal images. Apparently, a single reference cannot adapt well to all human faces, which leads to significant misalignment of the results. HPEN also employs a method that combines 3D modeling and interpolation. However, without further artifact elimination, the results have significant distortions at the facial boundary. Compared to other methods, our URSF-GAN generates a true frontal face and maximally preserves facial details.

For (b), since TP-GAN [12] takes a supervised method to train the network, it has chromatic aberration problems in the output. Specifically, their chromatic aberration is biased towards the chromatic aberration used in the MultiPIE dataset [9]. This also demonstrates the lack of generalization capability for supervised methods at the current stage. HF-PIM [17] decomposes the facial rotation problem into dense field estimation and facial texture map recovery. However, both TP-GAN and HG-PIM cannot fully rotate the face into frontal faces. Finally, our method has some problems in the nose tip area, where we recover some parts of the nose tip as bread. We believe this is because of the low input resolution. Our 3D facial model contains 38365 vertices, which is larger than the number of input pixels. Thus, when we fit the model, many vertices will share the same pixel value. This leads to misalignment when we recover the model. Besides this issue, our URSF-GAN can rotate the input faces quite well.

For (c), Yim et al. proposed a method that only utilizes DNN instead of GAN [31]. Their DNN can decode, rotate and reconstruct the final image. However, for image synthesis, DNN is inferior to GAN. Their result has issues in reconstructing glasses. DR-GAN [10] is deficient in terms of retaining identity information. Also, the synthetic human facial image has problems on the face contour and positions. Our method is better at preserving identity information. One small drawback is that our method does not have satisfactory restoration for the right side of the glasses. This is due to the inability of 3DMM in modeling accessories like glasses. Thus, accessories are directly captured as skin texture, which influences the accuracy of positions after rotation. Fixing this problem will be the key focus of our future research in 3D-based modeling.

For (d), Ghodrati et al. proposed a two-stage DNN method to generate and refine the frontal face [32]. Their result is excessively smooth and hardly conveys the effect of light and shade when a person is photographed. The result for GSP-GAN [33] has good resolution. However, it cannot fully preserve the identity information. After comparison, we found that our method performed best in terms of overall performance.

4.4   More Results for CelebA Dataset

Figure 7 shows more testing results for CelebA dataset. In this figure, we roughly divide the results into three categories (faces rotated in minor degree, moderate

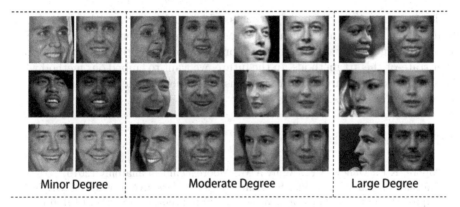

| Minor Degree | Moderate Degree | Large Degree |

Fig. 7. More results on CelebA dataset.

degree and large degree). The left is the input face and the right is the output. By comparison among three categories, our facial identity feature decays slowly as the rotation degree increases. Benefitting from the strategy of modeling rotation, this strategy also limits the facial details that are generated by GAN, which simplifies the inference processing and leads to the preservation of identity feature. Facial recognition results from Sect. 4.5 also proves this point.

4.5 Face Recognition Results

We also perform the face recognition test on Multi-PIE dataset with baseline of LightCNN [30]. The quantitative results show that our method has greater robustness when rotating faces with larger angles (Table 1).

Table 1. Face recognition correct rate by LightCNN on the Multi-PIE dataset.

Angle	30°	45°	60°	75°	90°
FF-GAN	92.5	89.7	85.2	77.2	61.2
TP-GAN	98.1	95.4	87.7	77.4	64.6
CAPG-GAN	**99.6**	**97.3**	**90.3**	76.4	66.1
URSF-GAN (Our)	91.4	90.7	88.6	**81.3**	**70.8**

5 Conclusion

We designed an unsupervised face frontalization model based on 3D rotation and symmetric filling. Then, image translation GAN was used to fix artifacts. Compared to other unsupervised GAN models, our model reduces the GAN's burden from inpainting to fixing artifacts, which makes our model easy to train. Results show that our model can generate photo-realistic frontal face images. For faces with large angles, our method can preserve more identity information than other state-of-the-art methods. In the future, we want to embed our method with the supervised structure to improve robustness when people wear accessories or are occluded by others.

References

1. Moore, K.L., Dalley, A.F., Agur, A.M.: Moore's clinical anatomy (2010)
2. Karras, T., Aila, T., Laine, S., Lehtinen, J.: Progressive growing of GANs for improved quality, stability, and variation. arXiv preprint arXiv:1710.10196 (2017)
3. Guo, Y., Zhang, L., Hu, Y., He, X., Gao, J.: MS-Celeb-1M: a dataset and benchmark for large-scale face recognition. In: Leibe, B., Matas, J., Sebe, N., Welling, M. (eds.) ECCV 2016. LNCS, vol. 9907, pp. 87–102. Springer, Cham (2016). https://doi.org/10.1007/978-3-319-46487-9_6
4. Huang, G.B., Mattar, M., Berg, T., Learned-Miller, E.: Labeled faces in the wild: a database forstudying face recognition in unconstrained environments. In: Workshop on faces in 'Real-Life' Images: Detection, Alignment, and Recognition (2008)
5. Klare, B.F., et al.: Pushing the frontiers of unconstrained face detection and recognition: IARPA Janus benchmark A. In: Proceedings of the IEEE Conference on Computer Vision and Pattern Recognition, pp. 1931–1939 (2015)
6. Nech, A., Kemelmacher-Shlizerman, I.: Level playing field for million scale face recognition. In: Proceedings of the IEEE Conference on Computer Vision and Pattern Recognition, pp. 7044–7053 (2017)
7. Liu, Z., Luo, P., Wang, X., Tang, X.: Deep learning face attributes in the wild. In: Proceedings of the IEEE International Conference on Computer Vision, pp. 3730–3738 (2015)
8. Zhu, X., Lei, Z., Liu, X., Shi, H., Li, S.Z.: Face alignment across large poses: a 3D solution. In: Proceedings of the IEEE Conference on Computer Vision and Pattern Recognition, pp. 146–155 (2016)
9. Gross, R., Matthews, I., Cohn, J., Kanade, T., Baker, S.: Multi-pie. Image vision comput. **28**(5), 807–813 (2010)
10. Tran, L., Yin, X., Liu, X.: Disentangled representation learning GAN for pose-invariant face recognition. In: Proceedings of the IEEE Conference on Computer Vision and Pattern Recognition, pp. 1415–1424 (2017)
11. Tran, L., Yin, X., Liu, X.: Representation learning by rotating your faces. IEEE Trans. Pattern Anal. Mach. Intell. **41**(12), 3007–3021 (2018)
12. Huang, R., Zhang, S., Li, T., He, R.: Beyond face rotation: global and local perception GAN for photorealistic and identity preserving frontal view synthesis. In: Proceedings of the IEEE International Conference on Computer Vision, pp. 2439–2448 (2017)
13. Hu, Y., Wu, X., Yu, B., He, R., Sun, Z.: Pose-guided photorealistic face rotation. In: Proceedings of the IEEE Conference on Computer Vision and Pattern Recognition, pp. 8398–8406 (2018)
14. Yin, X., Yu, X., Sohn, K., Liu, X., Chandraker, M.: Towards large-pose face frontalization in the wild. In: Proceedings of the IEEE International Conference on Computer Vision, pp. 3990–3999 (2017)

15. Hassner, T., Harel, S., Paz, E., Enbar, R.: Effective face frontalization in unconstrained images. In: Proceedings of the IEEE Conference on Computer Vision and Pattern recognition, pp. 4295–4304 (2015)

16. Deng, J., Cheng, S., Xue, N., Zhou, Y., Zafeiriou, S.: UV-GAN: adversarial facial UV map completion for pose-invariant face recognition. In: Proceedings of the IEEE Conference on Computer Vision and Pattern Recognition, pp. 7093–7102 (2018)

17. Cao, J., Hu, Y., Zhang, H., He, R., Sun, Z.: Learning a high fidelity pose invariant model for high-resolution face frontalization. arXiv preprint arXiv:1806.08472 (2018)

18. Zhou, H., Liu, J., Liu, Z., Liu, Y., Wang, X.: Rotate-and-render: unsupervised photorealistic face rotation from single-view images. In: Proceedings of the IEEE/CVF Conference on Computer Vision and Pattern Recognition, pp. 5911–5920 (2020)

19. Blanz, V., Vetter, T.: A morphable model for the synthesis of 3D faces. In: Proceedings of the 26th Annual Conference on Computer Graphics and Interactive Techniques, pp. 187–194 (1999)

20. Li, S., Li, H., Cui, J., Zha, H.: Pose-aware face alignment based on CNN and 3DMM. In: BMVC, p. 106 (2019)

21. Ferrari, C., Lisanti, G., Berretti, S., Del Bimbo, A.: Effective 3D based frontalization for unconstrained face recognition. In: 2016 23rd International Conference on Pattern Recognition (ICPR), pp. 1047–1052. IEEE (2016)

22. Zhu, X., Lei, Z., Yan, J., Yi, D., Li, S.Z.: High-fidelity pose and expression normalization for face recognition in the wild. In: Proceedings of the IEEE Conference on Computer Vision and Pattern Recognition, pp. 787–796 (2015)

23. Isola, P., Zhu, J.Y., Zhou, T., Efros, A.A.: Image-to-image translation with conditional adversarial networks. In: Proceedings of the IEEE Conference on Computer Vision and Pattern Recognition, pp. 1125–1134 (2017)

24. Zhu, J.Y., et al.: Multimodal image-to-image translation by enforcing bi-cycle consistency. In: Advances in Neural Information Processing Systems, pp. 465–476 (2017)

25. Choi, Y., Choi, M., Kim, M., Ha, J.W., Kim, S., Choo, J.: StarGAN: unified generative adversarial networks for multi-domain image-to-image translation. In: Proceedings of the IEEE Conference on Computer Vision and Pattern Recognition, pp. 8789–8797 (2018)

26. Zhu, J.Y., Park, T., Isola, P., Efros, A.A.: Unpaired image-to-image translation using cycle-consistent adversarial networks. In: Proceedings of the IEEE International Conference on Computer Vision, pp. 2223–2232 (2017)

27. Guo, J., Zhu, X., Yang, Y., Yang, F., Lei, Z., Li, S.Z.: Towards fast, accurate and stable 3D dense face alignment. In: Vedaldi, A., Bischof, H., Brox, T., Frahm, J.-M. (eds.) ECCV 2020. LNCS, vol. 12364, pp. 152–168. Springer, Cham (2020). https://doi.org/10.1007/978-3-030-58529-7_10

28. Pathak, D., Krahenbuhl, P., Donahue, J., Darrell, T., Efros, A.A.: Context encoders: feature learning by inpainting. In: Proceedings of the IEEE Conference on Computer Vision and Pattern Recognition, pp. 2536–2544 (2016)

29. Liu, Z., Luo, P., Wang, X., Tang, X.: Deep learning face attributes in the wild. In: Proceedings of the IEEE International Conference on Computer Vision, pp. 3730–3738 (2015)

30. Wu, X., He, R., Sun, Z., Tan, T.: A light CNN for deep face representation with noisy labels. IEEE Trans. Inf. Forensics Secur. **13**(11), 2884–2896 (2018)

31. Yim, J., Jung, H., Yoo, B., Choi, C., Park, D., Kim, J.: Rotating your face using multi-task deep neural network. In: Proceedings of the IEEE Conference on Computer Vision and Pattern Recognition, pp. 676–684 (2015)
32. Ghodrati, A., Jia, X., Pedersoli, M., Tuytelaars, T.: Towards automatic image editing: learning to see another you. arXiv preprint arXiv:1511.08446 (2015)
33. Luan, X., Geng, H., Liu, L., Li, W., Zhao, Y., Ren, M.: Geometry structure preserving based GAN for multi-pose face frontalization and recognition. IEEE Access **8**, 104676–104687 (2020)

Infrared and Visible Image Fusion Based on Multi-scale Gaussian Rolling Guidance Filter Decomposition

Jiajia Zhang, Pei Xiang, Xiang Teng, Xin Zhang, and Huixin Zhou[✉]

Xidian University, Xi'an 710071, China
1131303762@qq.com

Abstract. With the development of multi-source detectors, the fusion of infrared and visible light images has received close attention from researchers. Infrared images have the advantages of all-day time, and can clearly image temperature-sensitive targets under low or no light conditions. Visible light images have strong imaging capabilities for target details under good lighting conditions. After the two are fused, the advantages of the two imaging methods can be integrated. In this paper, to obtain more valuable scene information in the fused image, an infrared and visible image fusion method based on multi-scale Gaussian rolling guidance filter (MLRGF) decomposition is proposed. First of all, the MLRGF is utilized to decompose infrared images and visible light images into three different scale layers, which are called detail preservation layer, edge preservation layer and energy base layer, respectively. Then, the three different scale layers are respectively fused based on the properties of different scale layers through spatial frequency-based, gradient-based and energy-based fusion strategies. Finally, the final fusion result is obtained by adding the fusion results of the three different scale layers. Experimental results show that the proposed method has achieved excellent results in both subjective evaluation and objective evaluation.

Keywords: Image fusion · MLRGF · Rolling guidance filter · MSMG-PCNN · SF-LV

1 Introduction

Image fusion is an enhancement technology whose purpose is to combine images obtained by different types of sensors to generate an image with rich information for subsequent processing. Infrared image and visible light image fusion is an important branch of image fusion. The fusion of infrared and visible light images is very meaningful, and the research has broad application prospects in the fields of video surveillance, military, aerospace, and low-quality surveys.

Traditional image fusion methods can be roughly divided into the following three types: transform domain based fusion methods, spatial domain based fusion methods and deep learning based fusion methods.

© The Author(s), under exclusive license to Springer Nature Switzerland AG 2022
S. Berretti and G.-M. Su (Eds.): ICSM 2022, LNCS 13497, pp. 75–91, 2022.
https://doi.org/10.1007/978-3-031-22061-6_6

The methods of transform domain are mostly based on the idea of multi-scale decomposition, including Laplacian pyramid (LP) transform [1], dual-tree complex wavelet transform (DTCWT) [2], non-subsampled contourlet transform (NSCT) [3, 4] and other methods. The advantage of these methods is that the detailed information of different scales can be merged, and the information is relatively rich. However, the fusion process is more complicated, the amount of calculation is too large, and the fusion result may have a reduced contrast.

The spatial based domain fusion method directly operates on the pixels without cross-domain transformation. The necessary information of the multi-modal image is directly extracted in the spatial domain. The spatial domain based fusion method has strong binding force on the fusion. The detailed information of the obtained image is also more comprehensive. However, there is a risk of undesirable phenomena such as blocking effect. Traditional spatial domain image fusion methods include segmentation methods based on regional blocks, saliency-based methods, and other spatial domain methods [5, 6].

The fusion method of deep learning is a new way that has emerged in recent years. Deep learning is a research sub-field of machine learning. Artificial Neural Network (ANN) is used as the main architecture of the model. The model fits the end-to-end data mapping well. The deep learning fusion method can not only mine the deep features of the image, but also has good model learning capabilities. But its storage occupancy rate and computational complexity are expensive, which needs high configured hardware equipment. Existing deep learning fusion methods can be roughly divided into three categories: methods based on convolutional sparse representation (CSR), methods based on pre-trained deep networks, and methods based on end-to-end learning deep networks. The method based on convolutional sparse representation replaces the matrix product with convolutional operation in the objective function based on traditional sparse representation. The consistency between different image patches is considered in this method. And the sparse coding coefficients can be obtained in an efficient way. Liu et al. apply Convolutional Sparse Representation (CSR) [7] to image fusion, and perform image decomposition through sparse coding coefficients obtained by CSR. Compared with other traditional methods based on sparse representation, this method retains more detailed information and has better robustness when multi-source image registration is not good. The method based on pre-trained deep network uses convolutional neural network as an intermediate tool to apply part of the steps of image fusion. [8, 9] has done many work on this area. The method based on the end-to-end learning deep network takes multi-modal images as input and fused images as output to generate end-to-end images. Ma et al. proposed the first image fusion method based on generative adversarial network, FusionGAN [11], which transforms the fusion task into an adversarial learning process of infrared and visible image information retention.

There are many kinds of methods mentioned above, and each has advantages and disadvantages. In view of the fact that existing image fusion methods, especially non-deep learning methods, have a weak ability to retain details. In this paper, an infrared and visible light image fusion method based on multi-scale Gaussian rolling guidance filter (MLRGF) decomposition is proposed to get better fusion results. Just as shown in Fig. 1.

The main contributions of this article are as follows:

(1) An effective image decomposition framework called MLRGF is proposed for multi-scale decomposition of infrared and visible images.
(2) Different fusion strategies are adopted for the feature layers of different scales according to the characteristics of the feature layers.
(3) Through the comparison of subjective and objective evaluations of several sets of fusion results, it is proved that the proposed fusion framework has more superior performance in most evaluations.

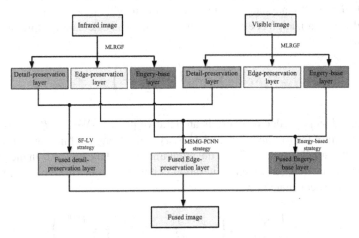

Fig. 1. Framework of the proposed method.

2 Proposed Method

Motivated by the good edge preserving properties in [12], an infrared and visible image fusion method based on MLRGF decomposition is proposed. It uses filters of different performance to get the details with different attributes. Therefore, we can design the fusion strategies according to attributes, which means that the details with different attributes in two images can be better fused.

2.1 Image Decomposition

Infrared and visible light images are respectively decomposed into three different scale layers by MLRGF. The structure of the decomposition model is shown in Fig. 3. In MLRGF, Gaussian filtering and rolling guidance filter are used for multi-scale decomposition. Gaussian filtering is very classic, it won't be described here.

Rolling Guidance Filter. Rolling guidance filter [12] is an effective edge-preserving filtering method. It can ensure the accuracy of object boundaries in large areas when removing and smoothing small and complex areas in the image. Rolling guidance filter can be divided into two steps: the removal of microstructures and the restoration of edges. in the first step, Gaussian filtering is used to filter out small structures. The Expression of the filter is shown as

$$G(i) = \frac{1}{K_i} \sum_{j \in N(i)} \exp\left(-\frac{\|i - j\|^2}{2\sigma_s^2}\right) I(j) \tag{1}$$

where I represents the input image, G represents the output image, σ_s represents the standard deviation of Gaussian filtering, $N(i)$ is the neighborhood centered on pixel i, and K_i is the normalization coefficient of the weight. The calculation expression of K_i is as follows

$$K_i = \sum_{j \in N(i)} \exp\left(-\frac{\|i - j\|^2}{2\sigma_s^2}\right) \tag{2}$$

The second step of edge recovery is an iterative process. We use J^{t+1} to represent the output result of the t-th iteration, and J^1 is set as the Gaussian filtering result G after filtering out the small structure. Given the input image I and the image J^t in the previous iteration, the value of J^{t+1} in the t-th iteration is obtained in the form of joint bilateral filtering. It can be calculated by

$$J^{t+1}(i) = \frac{1}{k_i} \sum_{j \in N(i)} \exp\left(-\frac{\|i - j\|^2}{2\sigma_s^2} - \frac{\|J^t(i) - J^t(j)\|^2}{2\sigma_r^2}\right) I(j) \tag{3}$$

where k_i is the normalization coefficient. The calculation formula of k_i is as follows

$$k_i = \sum_{j \in N(i)} \exp\left(-\frac{\|i - j\|^2}{2\sigma_s^2} - \frac{\|J^t(i) - J^t(j)\|^2}{2\sigma_r^2}\right) \tag{4}$$

where σ_s and σ_r control the spatial weight and the range weight respectively.

Through Eqs. (1) and (3), we use a sliding window to calculate the result from left to right on the entire image. In this paper, the rolling guidance filter process image I iterated n times is expressed as RGF(I, n).

Image Decomposition Based on MLRGF. The ability of a single multi-scale filtering method to extract features is limited, and the extraction of detailed features is not sufficient. Like the classic Gaussian filter, it can well retain the average energy information of the image and filter out edges and tiny structural information. However, due to the different pixel distribution characteristics of the edge and the tiny structure in the image, the same fusion method is used for fusion, and the effect often leads to the neglect of the tiny structure. This will lead to insufficient fusion information. Based on this, in this article, our method is to use filters with different performance to construct an information variance. It can separate the structural features with different attributes to achieve the full fusion of information. An image decomposition method based on MLRGF is proposed, as shown in Fig. 3. In MLRGF, we use two kinds of filters with different performance: Gaussian filter and rolling guidance filter. Gaussian filter can well retain the low-frequency information of the image. Edges and tiny structural information can be filtered out. Rolling guidance filter can filter out small structures while retaining large-scale edge information in the image. By combining the two structures, we can separate the small structure from the large-scale edges through the differential operations. Afterwards, the image fusion is adopted to ensure that the information is fully utilized.

Here we just using Gaussian filter to illustrate the detail preservation. It is shown in Fig. 2

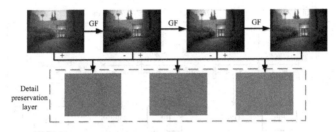

Fig. 2. Structure of MLRGF

In Fig. 3, I is the input image, I_{gn} ($n = 1,2,3$) is the result image of the n-th Gaussian filtering, I_{dn} ($n = 1,2,3$) is the result of the n-th rolling guidance filter, DP_n ($n = 1,2,3$) is the n-th level of detail preservation layer, EP_n ($n = 1,2,3$) is the n-th level of edge preservation layer, and B is the energy base layer.

The calculation process of MLRGF is as follows

$$I_{gn} = \begin{cases} I, & n = 0 \\ GF\left(I_{g(n-1)}, \mu_g, \sigma_g\right), & n = 1, 2, 3 \end{cases} \tag{5}$$

$$I_{dn} = \begin{cases} I, & n = 0 \\ RGF\left(I_{d(n-1)}, m\right), & n = 1, 2, 3 \end{cases} \tag{6}$$

In Eqs. (5) and (6), GF(·) is Gaussian filtering, μ_g and σ_g are the mean and variance of Gaussian filtering, RGF(·) is rolling guidance filter, and m is the number of RGF iterations. In this paper, through a large number of experiments, it is determined that $\mu_g = 0$, $\sigma_g = 20$, and $m = 5$.

After obtaining multiple filtered images, three feature maps of different scales can be calculated as

$$DP_n = I_{g(n-1)} - I_{dn} \tag{7}$$

$$EP_n = I_{dn} - I_{gn} \tag{8}$$

$$B = I_{gn} \tag{9}$$

After the above calculation steps, the multi-scale decomposition process in Fig. 3 is completed. The relationship between the different scale images obtained by this decomposition and the original image is

$$I = \sum_{i=1}^{n} (DP_i + EP_i) + B \qquad (10)$$

Through the relationship of Eq. (10), the calculation method of reconstruction can be obtained. Experiments have proved that when $n = 3$, the processing time and the fusion effect can reach a better result.

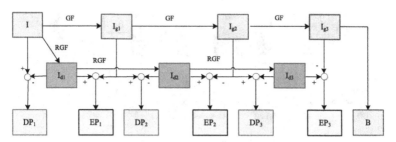

Fig. 3. Structure of MLRGF

2.2 Image Fusion

After the image is decomposed, we adopt different fusion strategies for different scale layers according to the characteristics of the decomposed image. The detail-preserving layer has the tiny structure and detail characteristics of the image, so we adopt the fusion strategy of spatial frequency-local variance (SF-LV) [13] integration. The edge-preserving layer has large-scale edge information and a small amount of small structure information of the image, so it adopts the fusion strategy of multi-scale morphological gradient domain pulse coupled neural network (MSMG-PCNN) [14]. The energy base layer contains a large amount of low-frequency information, so a weighted fusion strategy based on energy weights is adopted.

Fusion Strategy of Detail Preservation Layer
The detail preservation layer after image decomposition contains more small structures and edges. We believe that the more small structures and edges contained in a local area, the more information should be preserved. Therefore, we adopt the fusion strategy of SF-LV. The spatial frequency can be calculated by [15]. Here just using SF(x, y) represents the spatial frequency result at point (x, y).

The calculation formula of local variance is as follows

$$LV(x, y) = \sum_{r=-P/2}^{P/2} \sum_{s=-Q/2}^{Q/2} \frac{(I(x+r, y+s) - \mu)^2}{P \times Q} \tag{11}$$

where P and Q respectively represent the size of the region centered on (x, y). μ is the average value of pixels in the $P \times Q$ area.

The calculation formula of the SF-LV strategy is as follows

$$FuD_i = \begin{cases} DP_i^{IR}, & if \ SF_i^{FIR} > SF_i^{FVIS} \ and \ LV_i^{FIR} > LV_i^{FVIS} \\ DP_i^{VIS}, & otherwise \end{cases} \tag{12}$$

where FuD_i is the fusion result of the detail preservation layer, DP_i^{IR} and DP_i^{VIS} are the detail preservation layers decomposed from visible light image and infrared image respectively, SF_i^{FIR} and SF_i^{FVIS} are the results of the spatial frequency of DP_i^{IR} and DP_i^{VIS}, respectively, LV_i^{FIR} and LV_i^{FVIS} is the calculation result of the local variance of DP_i^{IR} and DP_i^{VIS}. In this paper, the value of i above is 1, 2, 3. Through Eqs. (11) to (12), the fusion of the detail preservation layer is completed.

Fusion Strategy of Edge Preservation Layer. The edge preservation layer has more large-scale edges and a small amount of micro-structure edges, which requires effective extraction of edge structures. Multi-scale Morphological Gradient Domain Pulse Coupled Neural Network (MSMG-PCNN) is an effective edge extraction and fusion strategy. It can comprehensively consider the edge strength and distribution of the corresponding positions in different images. This method has been successfully applied to image fusion [14]. In this article, the fusion strategy is applied.

MSMG

Multi-scale morphological gradient is a very effective edge extractor. The specific calculation steps are as follows. First define the multi-scale morphological gradient operator

$$SE_t = \underbrace{SE_1 \oplus SE_1 \oplus \cdots \oplus SE_1}_{t}, \quad t \in \{1, 2, \ldots, N\} \tag{13}$$

where SE_1 represents the basic structural unit, and t represents the number of scales. In this article, the value of t is 3.

After that, the structural elements of Eq. (13) can be used to calculate the special features.

$$G_t(x, y) = f(x, y) \oplus SE_t - f(x, y) \ominus SE_t \tag{14}$$

where $\oplus$ and $\ominus$ are morphological expansion and morphological corrosion respectively. $f(x, y)$ represents the pixel at position (x, y) in original image.

Finally, the calculation formula of MSMG is

$$M(x, y) = \sum_{t=1}^{N} w_t \cdot G_t(x, y) \tag{15}$$

where w_t represents the gradient weight of the t-th scale, which is calculated by

$$w_t = \frac{1}{2t + 1} \tag{16}$$

MSMG-PCNN

Pulse Coupled Neural Network (PCNN) [15] is a simple neural network. It is created by imitating the working mechanism of the human eye's retina. For the application of PCNN in the field of image fusion, its structure can be simplified to a two-channel PCNN model, and the calculation formulas are as follows.

$$F_{ij}^1(k) = S_{ij}^1(k) \tag{17}$$

$$F_{ij}^2(k) = S_{ij}^2(k) \tag{18}$$

$$L_{ij}(k) = \begin{cases} 1, & \text{if } \sum_{r,t \in S} Y_{rt}(k-1) > 0 \\ 0, & \text{otherwise} \end{cases} \tag{19}$$

$$U_{ij}(k) = \max\left\{ F_{ij}^1(k)\left(1 + \beta_{ij}^1 L_{ij}(k)\right), F_{ij}^2(k)\left(1 + \beta_{ij}^2 L_{ij}(k)\right) \right\} \tag{20}$$

$$Y_{ij}(k) = \begin{cases} 1, & \text{if } U_{ij}(k) \geq \theta_{ij}(k-1) \\ 0, & \text{otherwise} \end{cases} \tag{21}$$

$$\theta_{ij}(k) = \theta_{ij}(k-1) - \Delta + V_\theta Y_{ij}(k) \tag{22}$$

$$T_{ij} = \begin{cases} k, & \text{if } U_{ij}(k) \geq \theta_{ij}(k-1) \\ T_{ij}(k-1), & \text{otherwise} \end{cases} \tag{23}$$

where S_{ij}^1 and S_{ij}^2 represent the pixel values of the two input images in the neural network at point i; L_{ij} represents the link parameter; β_{ij}^1 and β_{ij}^2 represent the link strength; F_{ij}^1 and F_{ij}^2 represent the input feedback. U_{ij} is a dual-channel output. θ_{ij} is the threshold of the step function, D_e is the degree of the threshold decline, V_θ determines the threshold of active neurons, T_{ij} is a parameter that determines the number of iterations, and $Y_{ij}(k)$ is the output of the k-th PCNN.

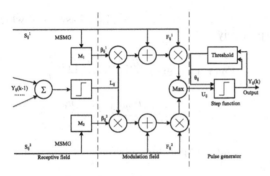

Fig. 4. Model of MSMG-PCNN

MSMG-PCNN replaces the link weights β_{ij}^1 and β_{ij}^2 of the dual-channel PCNN with the output result of MSMG. Which is

$$\beta_{ij}^1 = M_1 \tag{24}$$

$$\beta_{ij}^2 = M_2 \tag{25}$$

where M_1 and M_2 respectively represent the MSMG calculation results of the two images to be fused. They can be calculated by Eq. (15). Based on this, the fusion model is given by the following formula.

$$FuE_i = MSMGPCNN\left(EP_i^{IR}, EP_i^{VIS}, t\right) \tag{26}$$

where MSMG-PCNN(·) represents the MSMG-PCNN operation function. EP_i^{IR} and EP_i^{VIS} represent the i-th edge preservation layer of infrared and visible light images, respectively. t is the multi-scale decomposition parameter of MSMG. In this article, $t = 3$.

The MSMG-PCNN fusion model is shown in Fig. 4.

Fusion Strategy of the Energy Base Layer
The information in the energy base layer is mainly low-frequency information, and the information intensity is relatively strong. Therefore, the fusion of the base layer plays a decisive role in the overall quality of the fused image. Since the base layer contains most of the information of the source image, the energy attribute fusion strategy based on the natural index is used. The fusion strategy is mainly divided into three steps.

Calculate the energy characteristic attributes of the base layer.

$$F_{IR} = \mu_{IR} + M_{IR} \tag{27}$$

$$F_{VIS} = \mu_{VIS} + M_{VIS} \tag{28}$$

where μ_{IR} and M_{IR} represent the mean and median of infrared base layer image. μ_{VIS} and M_{VIS} represent the mean and median of visible light base layer image, respectively.

Calculate the energy characteristic function of the base layer, which is represented by E_{IR} and E_{VIS} respectively.

$$E_{IR}(x, y) = \exp\left(\frac{|B_{IR}(x, y) - F_{IR}|}{\tau}\right) \qquad (29)$$

$$E_{VIS}(x, y) = \exp\left(\frac{|B_{VIS}(x, y) - F_{VIS}|}{\tau}\right) \qquad (30)$$

where $\exp(\cdot)$ represents natural exponential operation. τ represents the gain factor. In this article, $\tau = 0.25$.

Calculate the fused base layer through the weighted fusion method.

$$FuB(x, y) = \frac{E_{IR}(x, y) \times B_{IR}(x, y) + E_{VIS}(x, y) \times B_{VIS}(x, y)}{E_{IR}(x, y) + E_{VIS}(x, y)} \qquad (31)$$

Reconstruction of Fusion Image

The fusion image can be reconstructed by the fusion results of three different scales. According to Eq. (10), the reconstruction process can be calculated by the following equation.

$$Ful = \sum_{i=1}^{n} (FuD_i + FuE_i) + FuB \qquad (32)$$

3   Experimental Results and Analysis

The method proposed will be compared with the comparison algorithms from both subjective and objective aspects. In this paper, GFF [6], ASR [16], and LatLRR [17] are used as comparison algorithms. Ten groups of comparative experiments will be demonstrated to prove the effectiveness of the proposed method.

3.1   Subjective Evaluation

For the fusion of infrared and visible light images, the more detailed information and clearer targets the fused image contains, the better the fusion effect. Based on this criterion, ten comparative experiments in Figs. 5, 6 and 7 were completed.

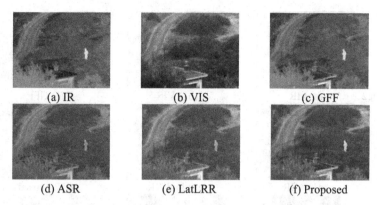

(a) IR   (b) VIS   (c) GFF

(d) ASR   (e) LatLRR   (f) Proposed

Fig. 5. "Camp" images and fused results obtained by different algorithms

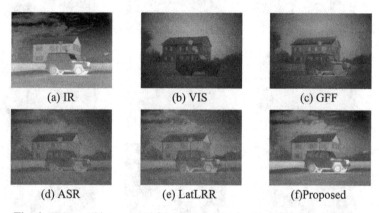

(a) IR   (b) VIS   (c) GFF

(d) ASR   (e) LatLRR   (f)Proposed

Fig. 6. "Marne1" images and fused results obtained by different algorithms

In Figs. 5, 6 and 7, we can find that the proposed method has obvious advantages in the detail preservation of visible and infrared images and the overall contrast of the fused image.

At the subjective evaluation point of view, we believe our proposed method has better performance than the comparison algorithms.

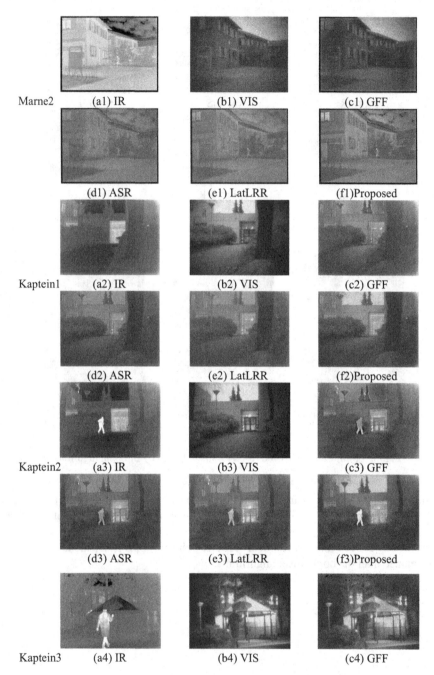

Fig. 7. Other images and fused results obtained by different algorithms

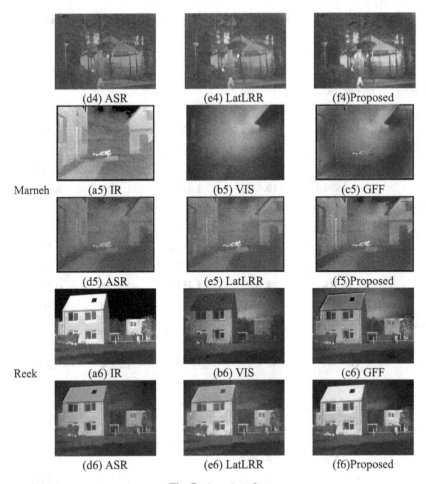

Fig. 7. (*continued*)

3.2 Objective Evaluation

In this paper, four metrics in the field of image fusion are used, including Spatial Frequency (SF) [15], Structural Similarity Index Measure (SSIM) [18], Mutual Information (MI) [19] and Total fusion metrics ($Q^{AB/F}$) [20].

Table 1 shows the objective evaluation of the ten groups of experiments. The best results have been bolded. It can be seen that the proposed method is superior to other state-of-the-art algorithms in most metrics and scene. For the other cases, even if it is not the best fusion performance, the proposed algorithm achieves the second-best performance.

Table 1. Objective evaluation of different fusion methods in different scenes

Scenes	Methods	Metrics			
		SF	SSIM	MI	$Q^{AB/F}$
Camp	Ori_IR	10.1484	/	/	/
	Ori_Vis	10.3898	/	/	/
	GFF	9.5811	1.3271	1.1609	0.3563
	ASR	8.4986	1.3797	1.0173	0.3132
	LatLRR	9.5155	1.4076	1.1871	0.3413
	Proposed	**10.4361**	**1.4106**	**1.1981**	**0.4079**
Marne1	Ori_IR	6.4167	/	/	/
	Ori_Vis	5.4432	/	/	/
	GFF	36.6389	1.3384	1.3716	0.2335
	ASR	22.5328	1.3783	1.1233	0.2119
	LatLRR	24.5023	**1.4520**	1.1756	0.3347
	Proposed	**37.9303**	1.4440	**1.4885**	**0.3694**
Marne2	Ori_IR	15.8361	/	/	/
	Ori_Vis	15.7928	/	/	/
	GFF	16.9725	1.0564	**1.5530**	**0.3552**
	ASR	17.8556	1.0536	0.9365	0.3403
	LatLRR	12.4235	**1.1687**	1.0576	0.2926
	Proposed	**19.3725**	1.1052	1.1057	0.3413
Kaptein1	Ori_IR	9.8562	/	/	/
	Ori_Vis	6.3057	/	/	/
	GFF	**9.2459**	1.4593	1.3578	**0.3479**
	ASR	6.6551	1.5050	1.3472	0.2548
	LatLRR	6.3482	1.5153	1.2274	0.2671
	Proposed	8.4286	**1.5277**	**1.5528**	0.3348
Kaptein2	Ori_IR	8.0487	/	/	/
	Ori_Vis	6.8097	/	/	/
	GFF	7.5548	1.4264	1.4978	**0.3533**
	ASR	6.3323	1.4789	1.4034	0.2522
	LatLRR	6.0002	**1.4842**	1.3313	0.2632
	Proposed	**8.0420**	1.4839	**1.5593**	0.3383
Kaptein3	Ori_IR	7.1915	/	/	/

(continued)

Table 1. (*continued*)

Scenes	Methods	Metrics			
		SF	SSIM	MI	$Q^{AB/F}$
	Ori_Vis	10.5763	/	/	/
	GFF	10.0352	1.3266	**2.0080**	0.3698
	ASR	9.2686	1.3542	1.2070	0.3306
	LatLRR	7.1797	**1.4076**	1.2711	0.2818
	Proposed	**10.2342**	1.3683	1.6813	**0.3812**
Marneh	Ori_IR	14.7345	/	/	/
	Ori_Vis	8.5192	/	/	/
	GFF	11.6621	1.3125	**1.7585**	0.2714
	ASR	12.1138	1.2841	1.0300	0.2482
	LatLRR	8.9203	1.3571	1.0744	0.2610
	Proposed	**12.5406**	**1.4440**	1.3836	**0.3093**
Marne3	Ori_IR	5.8108	/	/	/
	Ori_Vis	5.9779	/	/	/
	GFF	5.2356	1.2504	**1.8952**	0.1949
	ASR	4.5984	1.2515	1.6231	0.2444
	LatLRR	4.2041	**1.2847**	1.3225	0.2349
	Proposed	**6.0173**	1.2822	1.8541	**0.3068**
Reek	Ori_IR	13.4373	/	/	/
	Ori_Vis	8.1043	/	/	/
	GFF	11.9493	1.3206	2.0073	0.3288
	ASR	11.4748	**1.3263**	2.1922	0.3086
	LatLRR	9.3914	1.3038	2.0393	0.2721
	Proposed	**13.7951**	1.2509	**2.6201**	**0.3311**
Tank	Ori_IR	26.5526	/	/	/
	Ori_Vis	30.4033	/	/	/
	GFF	31.0580	**0.8433**	1.5700	**0.5780**
	ASR	27.8283	0.7667	1.4810	0.5144
	LatLRR	20.8130	0.7582	1.4700	0.3545
	Proposed	**32.3896**	0.7756	**1.7924**	0.4827

4   Conclusion

In this paper, an infrared and visible image fusion method based on MLRGF decomposition is proposed. In this method, the input source images are first decomposed into three different scale layers: the detail preservation layer, the edge preservation layer and the energy base layer. According to the respective characteristics of the three scale layers, three different fusion strategies are used for fusion. Finally, the fusion image is obtained by adding the fusion results of three different scales. Experimental results show that the method proposed performs better than other fusion methods in both subjective and objective evaluation. However, the complexity of the method is relatively high, which leads to poor real-time. So in the future work, we will try to simplify the method framework and improve the method performance.

References

1. Burt P.J., Adelson E.H.: The Laplacian pyramid as a compact image code. Readings in Computer Vision. Morgan Kaufmann, pp. 671–679 (1987)
2. Lewis, J.J., O'Callaghan, R.J., Nikolov, S.G., et al.: Pixel-and region-based image fusion with complex wavelets. Inf. Fusion **8**(2), 119–130 (2007)
3. Da Cunha, A.L., Zhou, J., Do, M.N.: The nonsubsampled contourlet transform: Theory, design, and applications. IEEE Trans. Image Process. **15**(10), 3089–3101 (2006)
4. Zhao, C., Guo, Y., Wang, Y.: A fast fusion scheme for infrared and visible light images in NSCT domain. Infrared Phys. Technol. **72**, 266–275 (2015)
5. He, K., Sun, J., Tang, X.: Guided image filtering. IEEE Trans. Pattern Anal. Mach. Intell. **35**(6), 1397–1409 (2012)
6. Li, S., Kang, X., Hu, J.: Image fusion with guided filtering. IEEE Trans. Image Process. **22**(7), 2864–2875 (2013)
7. Liu, Y., Chen, X., Ward, R.K., et al.: Image fusion with convolutional sparse representation. IEEE Signal Process. Lett. **23**(12), 1882–1886 (2016)
8. Liu, Y., Chen, X., Cheng, J., et al.: Infrared and visible image fusion with convolutional neural networks. Int. J. Wavelets Multiresolut. Inf. Process. **16**(03), 1850018 (2018)
9. Li, H., Wu, X.J., Durrani, T.S.: Infrared and visible image fusion with ResNet and zero-phase component analysis. Infrared Phys. Technol. **102**, 103039 (2019)
10. He, K., et al.: Deep residual learning for image recognition. In: Proceedings of the IEEE conference on computer vision and pattern recognition, pp. 770–778 (2016)
11. Ma, J., Yu, W., Liang, P., et al.: FusionGAN: A generative adversarial network for infrared and visible image fusion. Inf. Fusion **48**, 11–26 (2019)
12. Zhang, Q., Shen, X., Xu, L., et al.: Rolling Guidance Filter. In: European conference on computer vision. Springer, Cham, pp. 815–830 (2014)
13. Tan, W., Zhou, H., Song, J., et al.: Infrared and visible image perceptive fusion through multi-level Gaussian curvature filtering image decomposition. Appl. Opt. **58**(12), 3064–3073 (2019)
14. Tan, W., Xiang, P., Zhang, J., et al.: Remote sensing image fusion via boundary measured dual-channel PCNN in multi-scale morphological gradient domain. IEEE Access **8**, 42540–42549 (2020)
15. Xiao-Bo, Q., Jing-Wen, Y., Hong-Zhi, X., et al.: Image fusion algorithm based on spatial frequency-motivated pulse coupled neural networks in nonsubsampled contourlet transform domain. Acta Automatica Sinica **34**(12), 1508–1514 (2008)
16. Liu, Y., Wang, Z.: Simultaneous image fusion and denoising with adaptive sparse representation. IET Image Proc. **9**(5), 347–357 (2015)
17. Li, H., Wu, X.J.: Infrared and visible image fusion using latent low-rank representation. arXiv preprint arXiv:1804.08992 (2018)

18. Wang, Z., Bovik, A.C., Sheikh, H.R., et al.: Image quality assessment: from error visibility to structural similarity. IEEE Trans. Image Process. **13**(4), 600–612 (2004)
19. Qu, G., Zhang, D., Yan, P.: Information measure for performance of image fusion. Electron. Lett. **38**(7), 313–315 (2002)
20. Petrovic, V., Xydeas, C.: Objective image fusion performance characterization. In: Tenth IEEE International Conference on Computer Vision (ICCV'05), vol. 1., IEEE 2, pp. 1866–1871 (2005)

Multi-directional Edge Detection Algorithm Based on Fuzzy Logic Judgment

Xiang Teng, Jiajia Zhang, Zhe Zhang, Sijian Hou, Jun Tang, and Huixin Zhou$^{(\boxtimes)}$

School of Physics and Optoelectronic Engineering, Xidian University, Xi'an 710071, China
850993632@qq.com

Abstract. Aiming at the problems of blurred edges and difficult edge detection of noisy images, a multi-angle morphological edge detection algorithm combining fuzzy logic judgement strategy is proposed, which improves the defects of previous edge detection algorithms relying only on a single gray gradient difference and using fixed directional weights. First of all, the noisy images are morphologically filtered by a morphological binary structure team, then a set of improved multi-directional morphological operators are used for grey-level edge detection, combined with a fuzzy logic inference strategy for boundary enhancement and non-boundary point weakening of the image after morphological edge detection, and finally a two-dimensional K-mean method is used to segment the region containing the edges to generate image edge curves. The experimental results show that the method proposed in this paper has good noise suppression and edge detection effects compared with traditional edge detection algorithms such as Canny operator and fuzzy logic detection.

Keywords: Morphology detection · Fuzzy logic · Two-dimensional K-means

1 Introduction

The edge of an image is one of the basic features, and a clear and accurate edge extraction result is helpful to improve the accuracy of feature extraction and pattern recognition [1]. However, in the real environment, noise often affects the result of image edge detection, making the boundary inconspicuous, so many scholars in this field have conducted corresponding research before.

In the field of image processing, noise and edges often belong to the same high frequency signal, so in order to eliminate noise interference, it is first necessary to filter out the noise signal at the gray gradient level while maintaining the edge information of the target. However, if the filtering is not correct, the edges of the image may also be filtered out at the same time, resulting in the discontinuous edge of the extracted image [2]. Secondly, due to the illumination and the object itself, the edges often exist on different scales. And caused by different grayscale differences, the difference range between the gray level of the edge pixel and the background gray level is often unknown, so the edge of the image also has the characteristics of the edge grayscale gradient

S. Berretti and G.-M. Su (Eds.): ICSM 2022, LNCS 13497, pp. 92–104, 2022.
https://doi.org/10.1007/978-3-031-22061-6_7

difference. Therefore, it is often difficult to detect the complete image edge by the traditional method based on the grayscale difference operator alone [3].

In recent years, with the development of deep convolutional neural networks, methods to detect image edges using neural networks have been proposed continuously. Dense extreme inception network for edge detection (DexiNed) [4], richer convolutional features (RCF) [5], learning to predict crisp boundaries (LPCB) [6] and other models have already achieved good results and these networks provide a new idea for image edge detection. However, due to the imbalance between the height of the background area and the object boundary in the image and the characteristics of the convolutional neural network, the extracted image edges are relatively thick, and the ability to extract the edges of small objects is weak [7]. Traditional multi-directional morphological edge detectors can effectively extract the edges in the image, but for different use scenarios, it takes a lot of time and energy to select the synthesis coefficient, which seriously affects the effect of multi-directional morphological edge detection and the efficiency of the detector [8].

Therefore, aiming at the above problems, according to the characteristics of image noise signal and illumination transformation, a multi-directional morphological edge detection algorithm based on fuzzy logic judgment strategy is proposed in this paper. First, the noisy image is morphologically filtered by the morphological binary structure group, then a set of improved multi-directional morphological operators are used for gray-level edge detection, next the image after morphological edge detection is combined with fuzzy logic inference strategy for boundary enhancement and non-boundary points and interference noise weakening, and finally the region containing edges is segmented by the two-dimensional K-means method to obtain the final image of the target edge profile.

2 Multi-directional Edge Detection Algorithm Based on Fuzzy Logic Inference

2.1 Multi-directional Adaptive Weight Edge Detection

2.1.1 Small-Scale Morphological Filtering

In the operation of morphological filtering, this paper first uses two structural elements A and B to construct mathematical relations such as morphological open and closed operations for noise reduction of the original image respectively, so as to reduce the effect of noise on the edge detection effect. This morphological filtering operation can effectively remove the noise while better preserving the edge information of the image.

$$A = \begin{bmatrix} 1 & 0 & 1 \\ 0 & 1 & 0 \\ 1 & 0 & 1 \end{bmatrix}, B = \begin{bmatrix} 0 & 1 & 0 \\ 1 & 1 & 1 \\ 0 & 1 & 0 \end{bmatrix} \tag{1}$$

Morphological operations use pre-defined structural elements to match images to extract signals and suppress noise. Since in grayscale images, the morphological open operation suppresses the peak (positive pulse) noise in the signal, while the morphological closed operation suppresses the bottom valley (negative pulse) noise in the signal.

Therefore, in the re-Eq. (1) of this paper, the signal is averaged after two opposite operations of open operation and closed operation, which is equivalent to smoothing the signal twice, which has the effect of removing the orphan point of image noise. The morphological filtering operation is defined as

$$F_0 = \frac{1}{2}(f \circ A \cdot B + f \cdot A \circ B) \tag{2}$$

f is the original image and F_0 is the denoised image after morphological filtering. "$\circ$" is on behalf of the open operation and "$\cdot$" is on behalf of the close operation. The filtering effect is shown in Fig. 1.

(a)   (b)

Fig. 1. Image noise filtering effect comparison (a) Original image with salt and pepper noise (b) processed image

2.1.2 Multi-directional Edge Extraction

In the method based on gray gradient detection, in order to effectively detect the edges in different directions in the image and avoid the phenomenon of missed detection, this paper adopts the method of detecting and extracting the edge information of the image in different directions. Four 3×3 linear structural elements C1, C2, C3, C4 with directions of 0°, 45°, 90°, 135° are constructed respectively. The structural elements are shown as (3).

$$C_1 = \begin{bmatrix} 0 & 0 & 0 \\ 1 & 1 & 1 \\ 0 & 0 & 0 \end{bmatrix}, C_2 = \begin{bmatrix} 0 & 0 & 1 \\ 0 & 1 & 0 \\ 1 & 0 & 0 \end{bmatrix}$$
$$C_3 = \begin{bmatrix} 0 & 1 & 0 \\ 0 & 1 & 0 \\ 0 & 1 & 0 \end{bmatrix}, C_4 = \begin{bmatrix} 1 & 0 & 0 \\ 0 & 1 & 0 \\ 0 & 0 & 1 \end{bmatrix} \tag{3}$$

Through performing the edge detection on the image F0 after morphological filtering by using structural elements in 4 directions, and using morphology to perform operations such as corrosion, expansion, opening and closing operations on the image, the edge detection results of the image in four different directions can be obtained respectively.

Firstly, by performing two morphological operations on the grayscale image after noise reduction, this paper obtains the grayscale image y_{di} that preserves the dark detail edges of the original grayscale image and the grayscale image y_{ei} that preserves the bright detail edge of the original grayscale image, as shown in Eqs. (4) and (5). In order to obtain the edge details, the maximum values $E_{\max, i}$ and minimum values $E_{\min, i}$ after edge extraction of each pixel are calculated separately, as shown in Eqs. (7) and (8), and the difference between the two is supplemented by the result of edge extraction as the detail information of the edge, as shown in Eq. (6).

$$y_{d_i} = F_0 \oplus C_i \circ C_i - F_0 \cdot C_i \tag{4}$$

$$y_{e_i} = F_0 \cdot C_i - F_0 \ominus C_i \cdot C_i \tag{5}$$

$$y_{de_i} = y_{d_i} + y_{e_i} + \lambda(E_{\max,i} - E_{\min,i}) \tag{6}$$

$$E_{\max,i} = \max\{y_{d_i}, y_{e_i}\} \tag{7}$$

$$E_{\min,i} = \min\{y_{d_i}, y_{e_i}\} \tag{8}$$

In the formula, the edge information y_{di} and y_{ei} are obtained by constructing two mathematical relations between morphological operations. In order to enhance the details of the edge, calculate the maximum $E_{\max, i}$ and minimum $E_{\min,i}$ of each pixel after edge extraction. The difference between the two parameters is used as a supplement to the details of the edge of the image, and y_{dei} is the detection result obtained after the processing of the structural element C_i. In the formula, the parameter $\lambda = 1$.

The four kinds of structural elements are used to detect the edges of the image respectively, and the edge detection sub-images in four directions can be obtained, thereby completing the multi-directional extraction of the image edges. The multi-directional extraction is shown in Fig. 2.

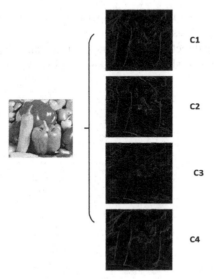

Fig. 2. Multi-directional edge detection sub-image

2.2 Fuzzy Logic Reasoning Strategy

2.2.1 Noise Reduction Image Gradientization

Since the multi-directional edge detection operator detects edges only from a morphological point of view by comparing neighboring pixel intensities, a small intensity difference between two neighboring pixels does not always indicate an edge because uniform regions are often not explicitly defined. Therefore, the multi-directional morphology detection operator sometimes detects some non-boundary pixels, which affects the subsequent image processing process.

To solve the above problems, this paper refers to the inference strategy of fuzzy logic to assist the decision making by establishing the edge affiliation function of pixels in the image, so as to achieve the purpose of boundary enhancement as well as non-boundary pixel suppression for the results after multi-directional edge extraction.

The process of fuzzy logic inference strategy is shown as follows. First, in this paper, the original denoised image is regarded as a two-dimensional discrete function, and a pair of gradient filters G_x and G_y are set to derive this two-dimensional discrete function, which are shown as follows.

$$G_x = [-1, 1], G_y = \begin{bmatrix} +1 \\ -1 \end{bmatrix} \tag{9}$$

The derivative of the two-dimensional discrete function of the image is shown in Eq. (10).

$$\begin{cases} G(x, y) = dx(i, j) + dy(i, j) \\ dx(i, j) = I(i+1, j) - I(i, j) \\ dy(i, j) = I(i, j+1) - I(i, j) \end{cases} \tag{10}$$

For the original noise-reduced image, this paper uses the above gradient filters G_x and G_y to convolute the image I respectively, so that the gradient matrices I_x and I_y along the horizontal and vertical directions of the image can be obtained, and the value domain interval is $[-1,1]$. The flow is shown in the following Fig. 3.

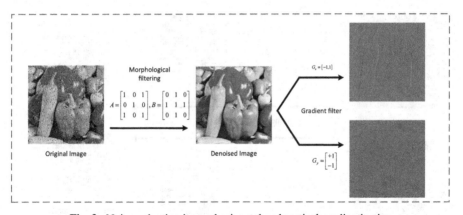

Fig. 3. Noise reduction image horizontal and vertical gradientization

2.2.2 Input Gradient Image Variable Fuzzification

I_X and I_y of two gradient image pixel points are used as inputs to the fuzzy inference system edgefis. For each input, this article uses a 0-mean Gaussian affiliation function to map the gradient value of that point to between 0 and 1, meaning the degree to which the gradient belongs to ZERO. If the pixel has a gradient value of 0, then the degree of affiliation that it belongs to ZERO is 1. The Gaussian affiliation function is shown in Eq. (11).

$$\begin{cases} f(x; \sigma, c) = e^{\frac{-(x-c)^2}{2\sigma^2}} \\ f(y; \sigma, c) = e^{\frac{-(y-c)^2}{2\sigma^2}} \end{cases} \tag{11}$$

The degree to which a pixel point belongs to zero (non-edge pixel point) is defined using the 0 mean Gaussian affiliation function, where σ is the standard deviation and is set to 0.1 and c is the mean value of 0.

Since the two-dimensional matrix calculated by the Gaussian membership function is the likelihood estimation of non-edge pixels, it is further calculated to obtain the edge gradient likelihood estimation of the input image pixels, denoted as G_x and G_y, respectively, as shown in Eq. (12) Show. The input image blur flow chart is shown in Fig. 4.

$$\begin{cases} G_x = 1 - f(x; \sigma, c) \\ G_y = 1 - f(y; \sigma, c) \end{cases} \tag{12}$$

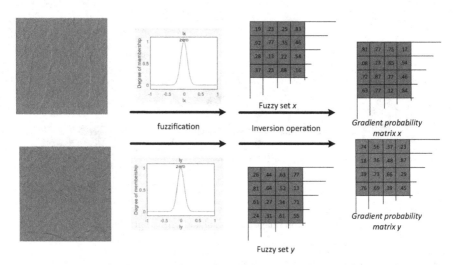

Fig. 4. Flow diagram of input image Ix and Iy blurring

2.3 Fuzzy Logic Enhancement Strategy

In order to enhance the edge strength of the multi-directional gradient map while suppressing the pixel interference of non-edge points, this paper adopts a fuzzy logic [9] inference strategy to assist in the decision making of the edge images, and for the edge detection images in four directions, four different weighting strategies are used to process the multi-directional edge extraction results.

For the edge detection results in the C_1 and C_3 directions, this paper uses the P_x and P_y likelihood probability matrices to weight them to obtain the edge likelihood probabilities $\underline{P}_1$ and P_3, while the joint likelihood probability matrix of P_x and P_y is used to compare C_2 and C_4 The input image of the direction is weighted to obtain P_2 and P_4. The specific formula is shown in (13).

$$\begin{cases} P_1 = P_y \\ P_3 = P_x \\ P_2 = P_4 = \left(P_x + P_y\right)/2 \end{cases} \tag{13}$$

After obtaining the likelihood probabilities of the edges in the four directions, this paper calculates the likelihood probabilities through a third-order function to obtain the gain results of the corresponding pixels on the images in each direction. The third-order gain function is shown in Eqs. (14) and (15).

$$T_i = [k \times (x - th)]^3 + 1 \tag{14}$$

$$G_i = \begin{cases} T_i T_i > 0 \\ 0 T_i < 0 \end{cases} \tag{15}$$

In Eq. (13), T_i represents the gain of each pixel of the image corresponding to the direction i, and k represents the gain adjustment coefficient, which can adjust the strength of the edge and non-edge pixel gain or attenuation, and the value of k is 2.5 in this paper. th represents the critical threshold for judging edges and non-edge pixels. In this paper, the value of th is 0.6. The likelihood probability higher than th will be judged as an edge point, and the pixel value of the corresponding point will be increased by gain. And in Eq. (14). Otherwise, it will be judged as a non-edge point. The value of the corresponding pixel will be attenuated, so as to achieve the purpose of widening the grayscale difference between edge pixels and non-edge pixels. The process of multi-directional edge extraction image enhanced by fuzzy logic is shown in Fig. 5.

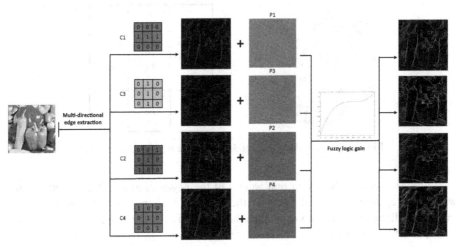

Fig. 5. Multi-directional edge extraction image is enhanced by fuzzy logic

2.4 Direction Adaptive Weight Fusion

After the image is processed by the multi-directional morphological detection algorithm, four directional edge detection results are obtained. When the morphological algorithm is used to extract the edge of the image, the edge of the image can be determined by the difference of the grayscale values obtained after the processing of structural elements [10]. The larger the value is, the greater the probability that there is an image edge in the vertical direction of the structural element is.

Therefore, by calculating the sum of the grayscale value changes of the structural elements of each pixel in different directions in the image, the proportion of edges in different directions can be obtained, and this can be used as the direction weight of the image [11]. As shown in the 3 × 3 image area in Fig. 6, the center pixel of this area is p_5, and d_i is defined as $d_i = |p_i\text{-}p_5|$, where d_i represents the grayscale difference between pixel i and the center point of the image area. The sum of the gray change values of a central pixel in the four directions defined by C_i are D_{c1}, D_{c2}, D_{c3}, D_{c4}, respectively.

The specific definition is shown in (16).

$$
\begin{aligned}
D_{c_1} &= d_1 + d_2 + d_3 + d_7 + d_8 + d_9 \\
D_{c_2} &= d_1 + d_2 + d_4 + d_6 + d_8 + d_9 \\
D_{c_3} &= d_1 + d_4 + d_7 + d_3 + d_6 + d_9 \\
D_{c_4} &= d_2 + d_3 + d_6 + d_4 + d_7 + d_8
\end{aligned}
\tag{16}
$$

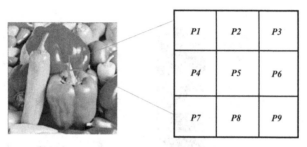

Fig. 6. An example of a set of 3×3 feature regions in an image

In order to obtain the proportion weight of the edges in different directions in the whole image, define D'_{c_1} as the sum of the gray value changes in all directions after traversing all the pixels in the image, then the direction adaptive weight w_i of sub-image fusion is defined in (17).

$$
w_i = \frac{D'_{c_i}}{\sum_{j=1}^{4} D'_{c_j}}, i = 1, 2, 3, 4
\tag{17}
$$

After completing the calculation of multi-directional edge extraction and direction adaptive weight, the multi-directional edge detection result and the direction adaptive weight are weighted and summed to obtain the final edge detection result as shown in (18).

$$
Y = \sum_{i=1}^{4} w_i y_{de_i}
\tag{18}
$$

Through the fusion of sub-images in various directions, the fusion result Y of the proposed adaptive multi-direction anti-noise morphological edge detection algorithm is obtained. The fusion process of multi-direction sub-images is shown in Fig. 7.

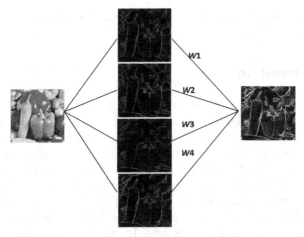

Fig. 7. An example of a set of 3 × 3 feature regions in an image

2.5 Two-Dimensional K-means Adaptive Threshold Segmentation

Next, in order to eliminate the existence of continuous low-frequency noise, the two-dimensional K-means [14] method is adopted to segment the image. The K-means algorithm is an unsupervised classification algorithm based on the proximity rule in pattern recognition, but the classification algorithm based on one-dimensional histogram only considers the gray value information of the image. More sensitive to isolated points, the two-dimensional K-means algorithm increases the influence of pixels and low-frequency noise on the class center value. Let $f(p)$ be the grayscale of pixel $p(x, y)$, and the average grayscale value of the defined area is shown in (19).

$$h(p) = \frac{1}{S(\sigma)} \sum_{p_i \in \sigma} f(p_i) \qquad (19)$$

In the formula, σ is the area centered on the pixel of $p(x, y)$, and $S(\sigma)$ is the area of the area. Image $I_{M \times N}(x, y)$ can generate $M \times N$ data pairs $(f(p), h(p))$. The subsequent processing is similar to the one-dimensional K-means clustering algorithm process, that is, the initial cluster center is selected; the above process is repeated until the position

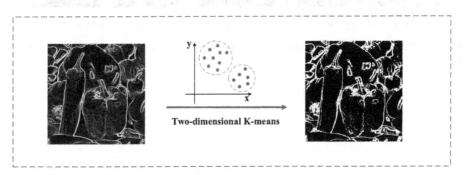

Fig. 8. Adaptive threshold segmentation results based on two-dimensional k-means algorithm

of the cluster center no longer changes. The result of adaptive threshold segmentation is shown in Fig. 8.

3   Simulation and Analysis

In the field of image processing, the evaluation criteria for evaluating image quality are divided into subjective evaluation and objective evaluation to judge the quality of images. Its advantage is that it can truly reflect the visual quality of the image [15], while the objective evaluation index is mainly based on a series of quantitative data to evaluate the image and analyze the data through a formula to derive a judgment on the image quality.In order to intuitively express the edge extraction results of the algorithm used in this paper on low-quality images with noise [16], this paper adds salt and pepper noise and Gaussian noise to the input image to reduce the image quality, and uses classic edge detection algorithms related to this field. The proposed algorithm is processed to achieve the purpose of testing the edge detection results of various algorithms, so as to obtain the subjective visual comparison results of image processing.

In the first set of experiments, the pepper noise is added to the original image, and various edge detection algorithms are adopted to perform image edge extraction detection and compare the extraction results. The image edge extraction results of different algorithms are shown in Fig. 9. And in the second set of experiments, the Gaussian noise is added into the original image, the results of edge detection are shown in Fig. 10.

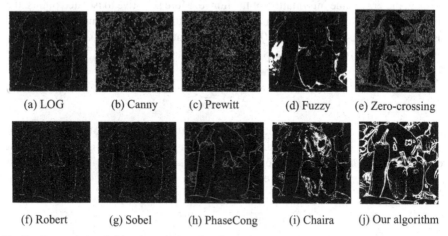

(a) LOG   (b) Canny   (c) Prewitt   (d) Fuzzy   (e) Zero-crossing

(f) Robert   (g) Sobel   (h) PhaseCong   (i) Chaira   (j) Our algorithm

Fig. 9. Comparison of detection results of various algorithms for images with salt and pepper noise

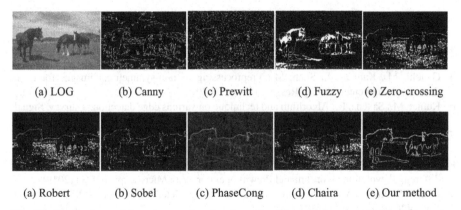

(a) LOG   (b) Canny   (c) Prewitt   (d) Fuzzy   (e) Zero-crossing

(a) Robert   (b) Sobel   (c) PhaseCong   (d) Chaira   (e) Our method

Fig. 10. Comparison of detection results of various algorithms for images with Gaussian noise

Through the results of processing low-quality images with added salt and pepper noise and Gaussian noise through the various algorithms in the above figure, it can be found that traditional classic algorithms such as Canny [7], Prewitt [8], Robert [9] and Sobel [10] will suffer from greater noise when processing low-quality noisy images, which is not conducive to the further processing of subsequent images. The edge detection algorithms such as PhaseCong and Fuzzy have better noise suppression effects, but the detailed information of the image will be lost a lot, and it will also cause the edge of the image to appear discontinuous after detection, Therefore, it will also affect the subsequent image processing. As can be seen from the above figure, the algorithm proposed in this paper has a strong suppression effect on both pepper noise and Gaussian noise, and also has an excellent edge retention effect on both foreground and background, which retains more details. Therefore, the algorithm we proposed has an excellent performance.

4   Conclusion

Based on previous studies, we proposed a multi-angle morphological edge detection algorithm combined with fuzzy logic inference strategy for the problem of edge detection of noisy images, which can achieve edge enhancement as well as noise and non-edge suppression by generating an image two-dimensional fuzzy matrix to assist in judging the images after multi-directional edge detection. As a result, the proposed method can not only effectively reduce the noise interference in the image, but also make the detected edges more clear and continuous, which can achieve better image edge detection results.

References

1. Jiang, H., Zhu, S., Zhao, H., Xu, B., Li, X.: Adaptive regional single-pixel imaging based on the Fourier slice theorem. Opt. Expr. **25**(13), 15118–15130 (2017)
2. Tan, W., Zhou, H., Song, J., Li, H., Yu, Y., Du, J.: Infrared and visible image perceptive fusion through multi-level Gaussian curvature filtering image decomposition. Appl. Opt. **58**(12), 3064–3073 (2019)

3. Wang, X., Fang, Y., Li, C., Gong, S., Yu, L., Fei, S.: Static gesture segmentation technique based on improved Sobel operator. J. Eng. **2019**(22), 8339–8342 (2019)

4. Zhou, J., Liu, S., Qian, H., et al.: Metasurface enabled quantum edge detection. Sci. Adv. **6**(51), eabc4385 (2020)

5. Gandhi, M., Kamdar, J., Shah, M.: Preprocessing of non-symmetrical images for edge detection. Augmented Human Res. **5**(1), 1–10 (2020)

6. Kumar, M., Saxena, R.: Algorithm and technique on various edge detection: a survey. Signal Image Process. **4**(3), 65 (2013)

7. Song, R., Zhang, Z., Liu, H.: Edge connection based Canny edge detection algorithm. Pattern Recognit Image Anal. **27**(4), 740–747 (2017). https://doi.org/10.1134/S1054661817040162

8. Song, Y., Ma, B., Gao, W., Fan, S.: Medical image edge detection based on improved differential evolution algorithm and Prewitt operator. Acta Microscopica **28**(1) (2019)

9. Kang, M., Xu, Q., Wang, B.: A roberts' adaptive edge detection method. J. Xi'an Jiaotong Univ. **42**(10), 1240–1244 (2008)

10. Wang, Y.: Existence and nonexistence of positive solutions for mixed fractional boundary value problem with parameter and-Laplacian operator. J. Funct. Spaces **2018**, 1–16 (2018)

11. Wang, X.: Laplacian operator-based edge detectors. IEEE Trans. Pattern Anal. Mach. Intell. **29**(5), 886–890 (2007)

12. Huang, Y.M., Yan, H.Y., Wen, Y.W., Yang, X.: Rank minimization with applications to image noise removal. Inf. Sci. **429**, 147–163 (2018)

13. Kumar, A., Raheja, S.: Edge detection in digital images using guided L0 smoothen filter and fuzzy logic. Wireless Pers. Commun. **121**, 2989–3007 (2021)

14. Fan, P., Zhou, R.-G., Hu, W.W., Jing, N.: Quantum image edge extraction based on Laplacian operator and zero-cross method. Quantum Inf. Process. **18**(1), 1–23 (2018). https://doi.org/10.1007/s11128-018-2129-x

15. Xiang, P., et al.: Hyperspectral anomaly detection by local joint subspace process and support vector machine. Int. J. Remote Sens. **41**(10), 3798–3819 (2020)

16. Peng, C., Kang, Z., Cai, S., Cheng, Q.: Integrate and conquer: Double-sided two-dimensional k-means via integrating of projection and manifold construction. ACM Trans. Intell. Syst. Technol. **9**(5), 1–25 (2010)

17. Chaira, T., Ray, A.K.: A new measure using intuitionistic fuzzy set theory and its application to edge detection. Appl. Soft Comput. **8**(2), 919–927 (2008)

18. Wang, R., Wu, M., Chen, T., Li, Z., Sun, J.: Edge detection of infrared image based on morphology. J. Beijing Univ. Posts Telecommun. **44**(01), 66–71 (2021)

Multimedia Applications

Security Concerns and Citizens' Privacy Implications in Smart Multimedia Applications

Edgard Musafiri Mimo[(⊠)] and Troy McDaniel

The Polytechnic School, Arizona State University, Mesa, AZ 85212, USA
{emusafir,troy.mcdaniel}@asu.edu

Abstract. Many cities around the world have seen their transformation enabled due to their adoption of many smart multimedia technologies that are available to them and characterize their objectives in many ways. It is important to note that smart multimedia technologies and systems in smart cities play a big role in enabling smart cities initiatives. Thus, the smart multimedia systems and technologies must be thoroughly assessed in view of the various security issues that are discussed in literature to critically address the various privacy concerns among the citizens of smart cities. The privacy issues pertaining to smart multimedia systems and technologies in smart cities cannot be underestimated viewing their complexity because the issue of privacy is first and foremost an individual issue as it pertains to everyone in smart cities, but smart cities must discover means to collectively find a solution that addresses both the privacy issues collectively for smart cities and personally for each citizen of a smart city. This paper discusses privacy concerns pertaining to smart multimedia systems and technologies involved in audio, image, and video processing technology as well as proposes a novel approach to addressing the citizens' privacy concerns in smart multimedia technology data collection through the advancement of machine learning and artificial intelligence by characterizing and personalizing the collected smart multimedia data.

Keywords: Security · Privacy · Multimedia · Smart city · Citizen centered

1 Introduction

The presence of smart multimedia systems and technologies is apparent and ubiquitous in many of the enhancements that enable smart cities, which range from audio, voice, and speech recognition to image and object detection, and all the way to face and emotion recognition technologies. It cannot be overstated that there are benefits that smart multimedia technologies offer to smart cities with respect to audio, voice, image, and face recognition use cases [1, 6, 13] that have quickly become a value additive component to almost all the activities and programs that facilitate the realization of smart cities. Smart multimedia systems and technologies remain one of the main avenues of cyberattacks and threats that pose several security issues to the many multimedia systems within smart cities and the technologies that enable them. There are many security

S. Berretti and G.-M. Su (Eds.): ICSM 2022, LNCS 13497, pp. 107–115, 2022.
https://doi.org/10.1007/978-3-031-22061-6_8

risks associated with smart multimedia technologies that this paper will present but at a high-level due to their extensive coverage in the literature. Thus, for the benefit of highlighting the privacy issues that spring out of the security issues within smart multimedia systems and technologies, this paper focuses mainly on three types of smart multimedia technology systems: audio, sound, and speech detection and recognition [13]; image and object detection and processing [6, 8]; and real-time video analysis [6, 15], detection, and processing that enables many critical surveillance systems [15] that are numerously deployed in smart cities. The main problem with these systems is that they permit avenues that can enable the collection of individual data as one data stream with different individual privacy levels pertaining to the criticality of the data components as one might contain privacy-related information while others may not.

For example, an audio or voice recording IoT device would ideally capture the audio signal from different sound sources that are within the vicinity of its collection point, and as such the captured audio is an overall and cumulative result of the captured audio wavelengths of the different sound sources. The captured audio may contain, e.g., components of a gunshot [7] in the area, citizens' conversations and voices, car noises, and traffic speed sounds all at once if any of these sounds occurred simultaneously. This scenario highlights some of the privacy concerns of citizens in the sense that many citizens might not even consent to their voices and sound signals captured, yet there are many vocal data sets collected from citizens [7, 15] that are being recorded in many places in smart cities without citizens knowing and consenting to their audio being captured.

The same scenario can be experienced with many images and object recognition systems and technologies where a captured image of a potential target captures the target as well as other objects surrounding the target and even individuals that might not necessarily consent to be in the camera's shot or give their image rights altogether. It is evident then that many captured images and video frames that are collected with IoT devices to enable many smart multimedia systems provide a lot of privacy concerns due to the nature of the collected data that can be a constitution of many different elements with different privacy levels. Advancements in machine learning and artificial intelligence provide us with great opportunities in analyzing the components of the captured multimedia data stream and assigning privacy levels to the components with the overall multimedia data stream privacy level to equal the highest privacy level among its components. In this manner, the storage and processing of the components' data can command greater care in the way the captured information is being used.

More examples pertaining to these smart multimedia systems and technologies will be discussed further with the help of literature to better assess the privacy issues surrounding these systems and identify ways to mitigate and lessen the privacy concerns among smart cities' citizens as they use and adopt more smart multimedia-enabled services, technologies, and systems. Accordingly, there is a critical and urgent need among smart cities to address the problem of captured multimedia data that can be a cumulative aggregation of many individual multimedia data such as video streams captured with sound to clearly be determined to address the privacy concerns of citizens fully within smart cities. The privacy issue within smart cities is an individual concern of each citizen based on what information they consider confidential and private. Smart cities must cooperatively find a collective solution for their citizens, and this starts with having a way to associate the collected information data to an individual and not a group of individuals because in a group of individuals there might be different opinions as to what information may be considered private among individuals [1]. Thus, this paper proposes an approach to characterize collected information with the advancement of machine learning and artificial intelligence and associate it with a specific individual and not a group of individuals to facilitate a better assessment of privacy issues and levels pertaining to smart multimedia systems and technologies within smart cities.

2 Related Works

Smart multimedia systems and technologies generate a lot of unstructured data that is part of big data [2, 3], and as such these systems and their data are also the target of many cyber-attacks and threats [3, 5] that must be properly assessed and dealt with. The cyber security threats in this context pertain to both the systems and technologies collecting and generating multimedia big data [4], and the collected multimedia big data itself as any glitch within the overall smart multimedia ecosystem can quickly jeopardize the integrity of the systems and the users' collected data. The security issues pertaining to vision systems provide opportunities that enable smart cities in facilitating advances in facial [9, 15, 16] and emotion [10, 14] detection and recognition systems cannot be underestimated. There are many privacy implications involved in the vision systems space that cannot be overlooked especially as they generate more and more concerns for the citizens.

Likewise, the security issues pertaining to audio processing systems suffer from the same preoccupations regarding the security threats that target these systems and technologies together with their associated unstructured data that is yet to be fully harnessed for the benefit of smart cities. There is a myriad of threats [5, 9, 14, 16] that need to be assessed to ensure the privacy of citizens is preserved within smart cities when it comes to smart multimedia systems and technologies. Security and privacy threats in smart multimedia are available almost everywhere in their collection points [5, 16] through cameras and recording devices, their processing [14, 16, 17] through the point of processing and their storage [2, 5, 17] in data centers. Consequently, there is a necessity to consider the overall privacy implications for the citizens of smart cities especially when the data collected contain information of several individuals within smart cities that might have different privacy considerations.

The incorporation of audio and video systems and technology remains a critical pillar of the many advancements that the smart multimedia surveillance ecosystem [4] relies on to enable different multimedia data streams pertaining to audio, images, video, sensor signals, and textual data that possess critical information about the environment and almost everything in it including people. Machine learning and artificial intelligence work and technologies have been developed over the last decade to enable a clearer interpretation of the environment to drive better decision-making [2, 11]. Many smart multimedia solutions are deployed in smart cities to support and enhance traditional surveillance mechanisms by enabling a distributed architecture to support various kinds of cameras and sensors that facilitate better views that may capture critical confidential information that can easily collect detailed vocal and visual features about citizens like their faces, voices, and mobility [13, 16] which can be used in profiling and identifying an individual.

3 Smart Multimedia Systems

The privacy concerns in this paper focus on image and vision systems; audio, voice, and speech systems; as well as biometric systems [17]. When it comes to the security issues of smart multimedia systems, it cannot be ignored that it is in both the physical and the technological layers. The physical security of smart multimedia IoT devices that collect citizens' data is as important as the technological security that supports them in terms of who can and cannot access or control them. The two layers of security already give rise to many securities-related worries in smart cities that need to be addressed and solutions before citizens begin to fully trust and yield to the smart cities' privacy-aware concept since it will eradicate the risks of security induced privacy-related fears among citizens [1, 17]. The fact that smart multimedia IoT devices are being deployed in smart cities with little or no control whatsoever of the specific component of the data they capture, for example with audio recorders [10, 13] capturing all the audio and sound waves signals from their surroundings regardless of location, poses many privacy concerns as citizens might not consent to be recorded in certain environments.

The same is applicable to camera or video footage that is collected and recorded in smart cities, many times citizens [9, 10] might not even be aware that they are on camera, and therefore, not give consent to allow the collection of their image. So, this is another avenue of potential security-induced privacy concern that smart cities need to address and properly deal with, in the attempt of creating privacy-aware smart cities. Many technologies and systems are being empowered today in smart cities with the availability of smart multimedia IoT devices that present many privacy concerns in facial recognition systems [8], sound detection systems [5, 13], motion-sensing systems [5, 16], thermal sensing systems [10, 12] and many other smart multimedia systems [9]. Privacy concerns arise given these systems' general points of data collection that have the potential to collect data from various individuals at once as one data stream. As such, they require individual component analysis to be able to identify the overall privacy level of the data stream as needed.

Thus, the technological security layer poses an even greater privacy risk should hackers access and control the IoT device at the data collection point; in turn, many citizens' information would be at risk. Many solutions have been proposed to help alleviate the security and privacy burdens caused by IoT devices [17] in processing information near the point of collection, but still, these solutions do not address the core problem: qualifying the technology or IoT's enabled service deployment in smart cities. The potential use of the smart multimedia IoT devices in smart cities remains one of the greatest assets and enablers of possibilities in smart cities, and future smart cities' solutions are heavily relying on the development and deployment of several smarter multimedia IoT devices that can facilitate characterization of different input signal sources.

4 Relevance

Smart multimedia technologies available today provide a lot of benefits that can be harnessed to help solve or lessen privacy concerns among citizens. There are advances in audio, voice, and speech systems that facilitate clear detection and recognition of individual voices with all the distinctive features enough to be able to distinguish between two individuals. This same technological advancement can be used to associate privacy levels to different components of the recorded audio signals, and even attenuate those harmonics and frequencies that contain critically deemed privacy information with digital processing filters [13] as shown in Fig. 1.

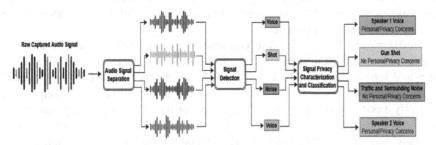

Fig. 1. The process of audio signal privacy characterization and classification.

This approach can also be used to assess the image and video detection and recognition systems as one considers the overall image component as a collection of different pixels of smaller images, and as such the privacy levels of each pixel or a group of pixels would not necessarily be the same as some pixels carry more critical information than others. Thus, a captured image can be considered as an image with different privacy sensitivity sections based on the pixels revealing the confidential identity of several citizens within the image that must be critically assessed as shown in Fig. 2 below.

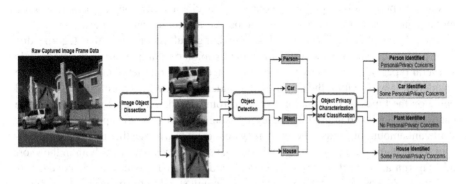

Fig. 2. The process of image frame data privacy characterization and classification.

This approach will command and encourage a new field of research studies in the privacy field to leverage the available advanced technologies and techniques in machine learning and artificial intelligence to categorize different components of multimedia data based on the critical privacy levels as proposed by the 3D value-driven privacy framework [1]. The use of this approach in facial recognition systems is practically critical as the different face feeds can be separated into several micro feeds for each face detected in a video feed, and this way the non-target individuals in the feed can be deleted and not stored for privacy reasons. The overall drive of characterizing and personalizing multimedia data feeds in this manner is to be able to associate the privacy levels of each feed according to the information they possess and reveal. It is critical to avoid the collection of large datasets that might not be needed for anything and that are not used in a way to enable systems or technologies in smart cities. Thus, it is essential to find a better way of dissecting and processing smart multimedia data to provide only the storage of the needed characterized data and avoid the inference of other data in profiling individuals in smart cities.

The 3D privacy framework [1] recommends regulating each individual smart multimedia technology, application, system, and process in view of the citizen's privacy lenses to determine how the privacy of personal data that is collected can be preserved in the long term and not be linked in whatsoever way to other data that can help profile or identify a certain individual in any way, shape, or form [1]. Thus, it is paramount to understand how smart multimedia services are performed and enabled in smart cities especially those that are involved with the collection of personally identifiable data by determining and assessing smart multimedia IoT devices or systems involved, the multimedia data being generated and collected, and the overall information communication and technology data transformation's routes toward decision making as the data may transcend between several systems and through several different connectivity protocols.

5   Implementation

When it comes to implementing such mechanisms at scale and within smart cities, there are several challenges that need to be considered to enable the integration of these mechanisms with the already deployed legacy systems. Some considerations of relevance to practically deploy the AI driven data characterization and personalization among many others include cost, compute power, feasibility, and user experience.

The process of characterizing and personalizing multimedia feeds to categorize different privacy implications will require a lot of capital cost at first due to the magnitude of several types of data namely text, audio, image, video, biometric etc. that need to be considered. It will require the city to better understand how to treat each data type and decide on the types of information to retain from each data type [1]. Additionally, the cost of identifying and upgrading the legacy systems to support AI capabilities require a significant amount of funds. The support system team to oversee that proper functionality of the systems remain an additional cost that smart cities will incur both in the short and long run as more data types are created and more IoT devices are deployed. It is evident that the cost of implementing this system will stand as the main stumbling block for tech companies and smart cities to proliferate the AI enabled data characterization and personalization system in dealing with citizens' personal information.

Compute power of the IoT devices at the point of data collection is critical in facilitating the realization of the proposed AI driven data characterization and personalization with privacy ratings to better handle citizens' personal data. The IoT devices collecting the data need to be upgraded with enough compute power to seamlessly perform the data characterization at the point of collection before the data is sent to be stored in databases. Not all computer chips in IoT devices can compute, process, and perform the AI data characterization and personalization, so there will be a need to upgrade the processors in some point of collection IoT devices. Computer chips are becoming fast and more performant, but that does not necessarily mean that all the IoT devices would need to be updated with the most performing processors for the task. Thus, it is important with the identification of the data types collected in smart cities to upgrade IoT devices used for data collection with processors that provide the optimal compute needed to facilitate the AI data characterization and personalization.

The feasibility challenge of deploying the AI data characterization and personalization swiftly and reliably remains a big concern because of the number of legacy systems already deployed in the smart cities including databases and systems that are already dependent on them. The biggest concerns of feasibility come in the form of all the dependent legacy systems that would still expect data in the legacy form and not pre-processed data with removed contents or added restrictions. The feasibility aspect of the network infrastructure needed may not allow for the integration of legacy devices in ensuring robust reliability and security with optimized data management workflow. It is crucial to understand what aspect of the legacy systems can be useful in the deployment of the AI data characterization and personalization processes to enable the proper reclassification of both data and data types. It is not easily feasible to integrate new technologies with legacy product due to the numerous amounts of innovations that have taken place both in software and hardware.

User experience is another challenge that must be addressed in deploying the AI data characterization and personalization at the point of data collection due to the overhead and latency that occur in processing the data and completing the users' requests. It cannot be underestimated considering that citizens may not approved of additional delays in receiving their responses or processing their requests. User experience would be affected as data being collected needs to be processed before it is sent to the database and trigger the response that users might be waiting for. Thus, adding an intermediary step in the data processing workflow would result in adding processing times to users' requests. This may not be plausible for some users and may result in users giving up on some applications especially if the data collection, characterization, and personalization are complex and require significant amount of processing times. The user experience is a key factor as it plays a major role facilitating citizens to navigate through smart cities innovation and ultimately improve their overall wellbeing and quality of life.

Therefore, all the briefly discussed factors above must be critically evaluated to facilitate a better deployment of the AI driven data characterization and personalization in smart cities to reap the benefit of attributing privacy levels to data and establishing the needed data handling methods.

6 Conclusion

It is paramount to consider how smart multimedia systems-related security risks transcend between different technologies and leave loopholes that must be addressed to avoid creating more privacy concerns for citizens using smart multimedia systems and technologies within smart cities. The proposed approach of characterizing and personalizing multimedia feeds to categorize different privacy implications will help secure and protect the privacy of citizens long before they get the profit of smart multimedia service provided benefits rather than enabling multimedia services without properly addressing privacy concerns that may later arise. The 3D privacy framework can leverage the proposed approach and offer a novel way of addressing security induced privacy concerns for citizens as the smart multimedia application, system, process, and technology that are deployed across smart cities are thoroughly assessed and classified based on the smart multimedia IoT devices that are employed, the Big Data that is generated, and the ICT systems and protocols that are used to fulfill and perform smart multimedia services [1, 15]. As the smart multimedia security and privacy issues are closely addressed and regulated in all the deployed multimedia technologies, applications, and systems in smart cities as recommended by the 3D privacy framework, there will be more privacy-aware smart cities that will begin to form and be established.

Acknowledgment. The authors thank Arizona State University and the National Science Foundation for their funding support under Grant No. 1828010.

References

1. Mimo, E., McDaniel, T.: 3D privacy framework: The citizen value driven privacy framework, pp. 1–7 (2021). https://doi.org/10.1109/ISC253183.2021.9562841
2. Mohanty, S.: Everything you wanted to know about smart cities. IEEE Cons. Electron. Mag. **5**, 60–70 (2016). https://doi.org/10.1109/MCE.2016.2556879
3. Hahn, D., Munir, A., Behzadan, V.: Security and privacy issues in intelligent transportation systems: Classification and challenges. IEEE Intelligent Transportation Systems Magazine, p. 1 (2019). https://doi.org/10.1109/MITS.2019.2898973
4. Cucchiara, R.: Multimedia surveillance systems, p. 3–10 (2005). https://doi.org/10.1145/1099396.1099399
5. Gupta, B.B., Yamaguchi, S., Agrawal, D.P.: Advances in security and privacy of multimedia Big Data in mobile and cloud computing. Multi. Tools Appl. **77**(7), 9203–9208 (2017). https://doi.org/10.1007/s11042-017-5301-x
6. Alvi, S., Afzal, B., Shah, G., Atzori, L., Mahmood, W.: Internet of multimedia things: Vision and challenges. Ad Hoc Networks, p. 33 (2015). https://doi.org/10.1016/j.adhoc.2015.04.006. https://www.sciencedirect.com/topics/engineering/smart-camera
7. Gold, H.: ShotSpotter: Gunshot detection system raises privacy concerns on campuses. Guardian News (2015). https://www.theguardian.com/law/2015/jul/17/shotspotter-gunshot-detection-schools-campuses-privacy
8. Bowyer, K.: Face recognition technology and the security versus privacy tradeo (2021)
9. Adams, A., Sasse, A.: Privacy in multimedia communications: Protecting users, not just data (2001).https://doi.org/10.1007/978-1-4471-0353-0_4
10. Maple, C.: Security and privacy in the internet of things. J. Cyber Policy **2**(2), 155–184 (2017). https://doi.org/10.1080/23738871.2017.1366536
11. Zoonen, L.: Privacy concerns in smart cities. Gov. Inf. Q. **33**, 472–480 (2016). https://doi.org/10.1016/j.giq.2016.06.004

12. Guhr, N., Werth, O., Blacha, P.P.H., Breitner, M.H.: Privacy concerns in the smart home context. SN Appl. Sci. **2**(2), 1–12 (2020). https://doi.org/10.1007/s42452-020-2025-8

13. Noda, K., Hashimoto, N., Nakadai, K., Ogata, T.: Sound source separation for robot audition using deep learning. In: 2015 IEEE-RAS 15th International Conference on Humanoid Robots (Humanoids), pp. 389–394 (2015)

14. Jaiswal, M., Emily, M.P.: Privacy enhanced multimodal neural representations for emotion recognition. Proc. AAAI Conf. Artif. Intell. **34**, 7985–7993 (2020). https://doi.org/10.1609/aaai.v34i05.6307

15. Wang, J., Amos, B., Das, A., et al.: A Scalable and Privacy-Aware IoT Service for Live Video Analytics, pp. 38–49 (2017). https://doi.org/10.1145/3083187.3083192

16. Prabhakar, S., Pankanti, S., Jain, A.: Biometric recognition: Security and privacy concerns. Security Privacy **1**, 33–42 (2003). https://doi.org/10.1109/MSECP.2003.1193209

17. Evans, N., Marcel, S., Ross, A., Teoh, A.: Biometrics security and privacy protection [from the guest editors]. Signal Process. Mag. **32**, 17–18 (2015). https://doi.org/10.1109/MSP.2015.2443271

Metric Learning on Complex Projective Spaces

Yujin Wu[1]([⊠]) and Mohamed Daoudi[2,3]

[1] University Lille, CNRS, Centrale Lille, UMR 9189 CRIStAL,F-59000Lille, France
yujin.wu.etu@univ-lille.fr
[2] IMT Nord Europe, Institut Mines-Télécom, University Lille, Centre for Digital
Systems, F-59000Lille, France
mohamed.daoudi@imt-nord-europe.fr
[3] University Lille, CNRS, Centrale Lille, Institut Mines-Télécom, UMR 9189
CRIStAL, F-59000Lille, France

Abstract. Shape analysis of landmarks is a fundamental problem in computer vision and multimedia. We propose a family of metrics called Fubini-Study distances defined in the complex projective space based on the seminal work of Kendall [11] for metric learning to measure the similarity between shape representations which are modeled directly by the equivalence classes of the 2D landmark configurations. Experiments conducted on the face landmarks for facial expression recognition demonstrate the competitiveness of the proposed method with respect to state-of-the-art approaches. A comparison with the metric defined in the Euclidean space has also been explored, proving that the Fubini-Study metric is more effective and discriminative than the Euclidean metric in identifying facial deformation.

Keywords: Metric learning · Complex projective spaces · Landmarks · Shape

1 Introduction

Recently landmark detection and tracking methods from human faces and bodies became reliable and accurate. Shape Analysis of landmarks is a fundamental problem in many computer vision and multimedia applications such face verification [5], person re-identification [12], facial expression recognition [2,9,20] and action recognition [15]. Due to the need for suitable metrics to measure the similarity between a set of landmarks, a large number of metric learning algorithms have been proposed for various scenarios, However, most of the conventional metric learning methods for these visual tasks are designed based on linear data. On the other hand, the set of landmarks detected in the video frames or static images is a natural choice to model the facial shape or body shape for computer vision applications. In the work of [10], the face and body landmark sequences were parametrized as trajectories on the Riemannian manifold of PSD matrices of fixed-rank, and a metric was defined on the manifold for the similarity-based

S. Berretti and G.-M. Su (Eds.): ICSM 2022, LNCS 13497, pp. 116–127, 2022.
https://doi.org/10.1007/978-3-031-22061-6_9

classification. Daoudi *et al.* [1] used the Gram matrix of 2D landmarks as the representation of the body shape on a PSD manifold with a Riemannian metric to assess the depression severity of a patient. Szczapa *et al.* [16] encoded the facial movement as trajectories on the PSD manifold for pain intensity estimation using a Riemannian metric. However, none of the methods discussed above explored the potential of adapting the used metric in order to improve the accuracy of the regression or classification task. In these solutions, optimization of the learning task entirely relied on training some ML module operating on a predefined metric, whose adoption was motivated by the geometry of the manifold of the data.

Based on the seminal work on shape analysis of Kendall [11], we propose a new family of metrics -*Fubini-Study metrics* in the complex projective space. A metric learning algorithm based on Large Margin Nearest Neighbors (LMNN) is exploited to improve the discriminative power of the proposed metric. Finally, the validity of the proposed metric is verified in facial expression recognition scenario, where the facial shape can be represented directly by the equivalence classes of 2D landmarks for the similarity-based classification. Moreover, a comparative study between the Fubini-Study metrics and Euclidean metrics is also provided. In summary, the main contribution of the paper are:

- A family of metrics defined in the complex projective space is proposed for shape analysis of 2D landmarks;
- A metric learning algorithm to find the optimal metric that parameterized by a positive-definite Hermitian matrix;
- A comparison between the Fubini-Study metric and the Euclidean metric. This comparison shows the effectiveness of the proposed metric for landmarks shape analysis;
- An evaluation of the learned Fubini-Study metric for the facial expression recognition based on landmarks shape analysis, also in comparison with state-of-the-art solutions.

2   Related Work

The purpose of metric learning is to find an adequate metric to better capture the potential similarity relationship hidden in the data. The optimal metric obtained from metric learning can bring samples from the same class as close as possible, while keeping the differently labeled samples far from each other. In the following we focus on reporting on previous work that adopted this learning model for shape analysis tasks.

Euclidean Metric Learning. Most of existing approaches seek to find the optimal metric in the Euclidean space. In the work of Wan *et al.* [20], spontaneous facial expression recognition task was reformulated as a maximum likelihood based metric learning problem with the Mahalanobis distance defined in the

Euclidean space. The spatially close (distant) data points have a higher probability of being in the same class in the learned feature space, thereby facilitating KNN-based classification. Kacem *et al.* [9] exploited a chart of barycentric coordinates to map affine equivalence classes of facial landmarks to the Euclidean space and then applied metric learning to the family of Euclidean metrics for the facial expression recognition task. However, their method relies on the stability of a reference triangle, which is used to obtain an affine-invariant shape representation for metric learning.

Non-Euclidean Metric Learning. Some non-linear methods were also proposed to learn a more discriminant metric, which can encode intrinsic geometry of manifold. Huang *et al.* [6] proposed a Fisher LDA-like framework for video based recognition. After mapping the data from the original Grassmann manifold to a low-dimensional, more discriminative one, the projection Metric can be directly learned from it. However, there exists a parameter p that needs to be set to construct the linear subspace shape representation for the underlying manifold. Daoudi *et al.* [2] propose a metric learning over a family of simple metrics on the space $\mathbb{R}^{2 \times n}_* / SO(2)$. The complex projective space proposed in this paper can be seen as a compactification of the set of all oriented ellipsed enclosing one unit of area, which tells us that we are dealing basically with the same space in [2] if take dilations into account. However, the Fubini-Study metrics are complete, have very simple geodesics (they are all circles), and the gradient of the distance square in terms of the Hermitian metric is simple in the case of diagonal matrices.

3   Complex Projective Space

A 2D landmark configuration $\mathbf{z}$ consists of n ordered points (x_1, y_1), ..., (x_n, y_n) on the plane. Expressing the points as complex numbers (i.e., writing $z_j := x_j + iy_j$ instead of (x_j, y_j)), we identify $\mathbf{z}$ as the vector $(z_1, \ldots, z_n) \in \mathbb{C}^n$. We will consider two landmark configurations $\mathbf{z}$ and $\mathbf{w}$ to be *equivalent* if the points w_j $(1 \leq j \leq n)$ in the second configuration are obtained from the points z_j in the first by means of a common translation, rotation, and dilation. This set of equivalence classes is the *shape space* in which we wish to work. Because of the geometry of complex numbers, the equivalence of two landmark configurations $\mathbf{z}$ and $\mathbf{w}$ translates into the existence of a nonzero complex number a and a complex number v so that $w_j = az_j + v$, $\forall j \in \{1, \ldots, n\}$. If we consider only *centered* configurations where $z_1 + \cdots + z_n = 0$, then two centered configurations $\mathbf{z}$ and $\mathbf{w}$ are equivalent if and only if there exists a nonzero complex number a for which $\mathbf{w} = a\mathbf{z}$. This is precisely the definition of *complex projective space.*

Definition 1. *Complex projective space of (complex) dimension n, $\mathbb{CP}^n$, is the set of equivalence classes of nonzero vectors in $\mathbb{C}^{n+1}$ with the equivalence relation*

$$(z_0, \ldots, z_n) \sim (az_0, \ldots, az_n)$$

for any nonzero complex number a.

A centered landmark $\mathbf{z}$ with coordinates $(z_1, \ldots, z_n)$ gives rise to an equivalence class $[\mathbf{z}]$ which can also be described by its *homogeneous coordinates* $[z_1 : z_2 : \ldots : z_n]$. The equivalence class $[\mathbf{z}]$ is a point in $\mathbb{CP}^{n-1}$, which, moreover, lies in the projective hyperplane given by the homogeneous equation $z_1 + \cdots + z_n = 0$.

Fubini-Study Metrics. A useful feature of complex projective space is that it carries a family of simply-defined and well-studied metrics: let A be a positive-definite Hermitian matrix, the *Fubini-Study distance*, d_A, associated to A between points $\mathbf{z} = [z_1 : \ldots : z_n]$ and $\mathbf{w} = [w_1 : \ldots : w_n]$ in $\mathbb{CP}^{n-1}$ is defined by

$$\frac{|\langle A\mathbf{z}, \mathbf{w}\rangle|^2}{\langle A\mathbf{z}, \mathbf{z}\rangle\langle A\mathbf{w}, \mathbf{w}\rangle} = \cos^2(d_A(\mathbf{z}, \mathbf{w})). \tag{1}$$

We recall that the expression $\langle \mathbf{z}, \mathbf{w}\rangle = z_1\bar{w}_1 + \cdots z_n\bar{w}_n$ is the standard complex-valued inner product on $\mathbb{C}^n$.

Euclidean Metrics. Given a positive-definite Hermitian matrix, its associated Fubini-Study metric allows us to define the distance between two equivalence classes of landmark configurations. However, there exists also a simple alternative. Assume that for all of our landmark configurations, the i-th landmark is different from the origin. For each landmark configuration $\mathbf{z}$, instead of considering the point in $\mathbb{CP}^{n-1}$ given by $\mathbf{z} = [z_1 : \ldots : z_n]$, we can consider the vector

$$\mathbf{z}_i := \left(\frac{z_1}{z_i}, \ldots, \frac{z_{i-1}}{z_i}, \frac{z_{i+1}}{z_i}, \ldots, \frac{z_n}{z_i}\right) \in \mathbb{C}^{n-1}.$$

Note that any two landmark configurations $\mathbf{z}$ and $\mathbf{z}'$ are equivalent if and only if $\mathbf{z}_i = \mathbf{z}'_i$. In this way, our shape space, or rather the piece of it that interests us, becomes the linear space $\mathbb{C}^{n-1} = \mathbb{R}^{2n-2}$. This enables us to use different Euclidean metrics to measure the distance between equivalence classes of landmark configurations.

Recall that each Euclidean metric in $\mathbb{R}^{2n-2}$ is determined by a positive-definite $(2n-2) \times (2n-2)$ real matrix M and that the associated distance is given by the formula

$$D_M(\mathbf{x}_i, \mathbf{y}_i) = (\mathbf{x}_i - \mathbf{y}_i)^T M(\mathbf{x}_i - \mathbf{y}_i), \tag{2}$$

where $\mathbf{x}_i$ and $\mathbf{y}_i$ are coordinates in $\mathbb{C}^{n-1} = \mathbb{R}^{2n-2}$ associated to the equivalence classes $[\mathbf{x}]$ and $[\mathbf{y}]$ in $\mathbb{CP}^{n-1}$.

4   Metric Learning with Fubini-Study Metrics

4.1   Distance Function for Fubini-Study Metrics

Given the formula for the Fubini-Study distance in Eq. (1), it will be more convenient to work with a distance function $f(t) = 1 - \cos^2(t)$ as part of the loss function for metric learning. This function is smooth, satisfies $f(0) = 0$, and it is strictly increasing in the interval $[0, \pi/2]$. As we are particularly interested on the dependency of the positive-definite Hermitian matrix A, the distance function is represented by the term F(A),

$$F(A) = 1 - \frac{|\langle A\mathbf{z}_i, \mathbf{z}_j \rangle|^2}{\langle A\mathbf{z}_i, \mathbf{z}_i \rangle \langle A\mathbf{z}_j, \mathbf{z}_j \rangle} \tag{3}$$

for a number of chosen landmark configurations $\mathbf{z}_1, \ldots, \mathbf{z}_k$ in $\mathbb{CP}^{n-1}$. The gradient of F(A) with respect to the variable A is then given by

$$\nabla_A F(A) = \frac{-1}{(\langle A\mathbf{z}_i, \mathbf{z}_i \rangle \langle A\mathbf{z}_j, \mathbf{z}_j \rangle)^2} \times \left[\langle A\mathbf{z}_i, \mathbf{z}_i \rangle \langle A\mathbf{z}_j, \mathbf{z}_j \rangle (\langle A\mathbf{z}_j, \mathbf{z}_i \rangle \mathbf{z}_i \mathbf{z}_j^* + \langle A\mathbf{z}_i, \mathbf{z}_j \rangle \mathbf{z}_j \mathbf{z}_i^*) \right.$$
$$\left. - |\langle A\mathbf{z}_i, \mathbf{z}_j \rangle|^2 \left(\langle A\mathbf{z}_j, \mathbf{z}_j \rangle \mathbf{z}_i \mathbf{z}_i^* + \langle A\mathbf{z}_i, \mathbf{z}_i \rangle \mathbf{z}_j \mathbf{z}_j^* \right) \right]$$

which will be used for the optmization of the Fubini-Study metrics.

4.2   Metric Learning Algorithm

Large Margin Nearest Neighbors (LMNN) [22] was selected as the metric learning algorithm. The purpose of this algorithm is to reduce the distance between each sample and its *target neighbors*, which are the k pre-selected nearest neighbor samples of the same class, while trying to keep it away from its *imposters*, which are differently labeled samples that invade the margin established by those target neighbor. Assuming that the target neighbor set has been selected (the nearest neighbors of each sample are calculated by the Fubini-Study distance), the loss function of LMNN consists of two terms. The first term is the target neighbors pulling term, given by

$$\varepsilon_{pull}(A) = \sum_{j \rightsquigarrow i} F_{ij}(A) = 1 - cos^2(d_A(\mathbf{z}_i, \mathbf{z}_j)) \tag{4}$$

where $F_{ij}(A)$ is the distance function in Eq. (3) between complex-valued samples $\mathbf{z}_i$ and $\mathbf{z}_j$ corresponding to Hermitian matrix A. $j \rightsquigarrow i$ iff jth sample is a target neighbor of ith sample. The second term is the impostors pushing term, given by

$$\varepsilon_{push}(A) = \sum_i \sum_{j \rightsquigarrow i} \sum_l (1 - y_{il})[1 + F_{ij}(A) - F_{il}(A)]_+ \tag{5}$$

where $y_{il} = 0$ if ith sample and lth sample are differently labeled, and 1 otherwise, the term $[\cdot]_+$ is defined as $[z]_+ = max\{z, 0\}$. Finally, the loss function is given by,

$$\varepsilon(A) = (1 - \mu)\varepsilon_{pull}(A) + \mu\varepsilon_{push}(A), \mu \in [0, 1] \tag{6}$$

where μ is the weighting parameter for balancing push and pull effects. The gradient of the loss function can be formulated as

$$\nabla_A(\varepsilon(A)) = (1 - \mu)\nabla_A(\varepsilon_{pull}(A)) + \mu\nabla_A(\varepsilon_{push}(A)), \mu \in [0, 1] \qquad (7)$$

where,

$$\nabla_A(\varepsilon_{pull}(A)) = \sum_{j \rightsquigarrow i} \nabla_A F_{ij}(A) \qquad (8)$$

and,

$$\nabla_A(\varepsilon_{push}(A)) = \sum_i \sum_{j \rightsquigarrow i} \sum_l (1 - y_{il})[\nabla_A F_{ij}(A) - \nabla_A F_{il}(A)] \qquad (9)$$

The Riemannian Steepest Descent algorithm in the toolbox Pymanopt [19] was implemented using the cost function and gradient in Eqs. (6) and (7), respectively, to seek the optimal solution for positive-definite Hermitian matrix A. The main procedure for our metric learning on complex projective space is given in Algorithm 1.

Algorithm 1: Metric learning in Complex Projective Space $\mathbb{CP}^{n-1}$

Data: N training samples $\mathcal{Z} = \{(Z^i, y^i)\}_1^N$ with their associated labels, k is the number of *target neighbors*.

Result: The optimal positive-definite Hermitian matrix A^*

1   $A \leftarrow I(n)$, identity complex matrix of dimension n. ;

2   **for** $i = 1 \ldots N$ **do**

3   Define k *target neighbors* $\mathcal{N}_s^A(i)$ and their corresponding *impostors* $\mathcal{N}_o^A(i)$ for each sample Z^i using the Fubini-Study metric in Eq. (1) ;

   end

4   $Cost \leftarrow \varepsilon(A, \mathcal{N}_s^A, \mathcal{N}_o^A)$, $\varepsilon(.)$ is given by Eq. (6);

5   $Grad \leftarrow \nabla_A(\varepsilon(A))$, $\nabla_A(.)$ is given by Eq. (7);

6   $A^* \leftarrow SteepestDescent(A, Cost, Grad)$, $SteepestDescent(.)$ is the Riemannian Steepest Descent optimization algorithm in $\mathbb{CP}^{n-1}$;

---

5   Experiments

To investigate the effectiveness of the proposed metric acting in the complex domain, we conducted several experiments for the facial expression recognition task. First, we introduced the CK+ dataset used in the experiments. All details of the experimental setup are then reported in Sect. 5.2. In the end, classification results using the proposed metric and comparative results with the Euclidean metric and other state-of-the-art approaches in facial expression recognition are presented in Sect. 5.3.

5.1 Datasets

The **Cohn-Kanade Extended (CK+) dataset** [13] consists of 327 annotated frontal video sequences performed by 118 subjects, wherein each subject is required to exhibit seven facial expressions during the experiment, namely – *anger, contempt, disgust, fear, happy, sad* and *surprise*. One sequence contains images from neutral expression (first frame) to peak expression (last frame). For the classification experiments, we exploit only the last frame, where the intensity of facial expression attains its peak.

The **Oulu-CASIA dataset** [23] consists of frontal facial videos from 80 subjects captured under three illumination conditions:dark, normal and weak normal, during which the subjects were required to imitate six classic facial expressions:anger, disgust, fear, happiness, sadness and surprise. Similar to the CK+ dataset, the last frame of each seauence has the highest expression intensity. Therefore, we use the last frame of the 480 video sequences under normal illumination for the classification task.

5.2 Experimental Setting

For each landmark configuration, we excluded the 17 points of face contour. The 2D centered landmarks are then written in complex form to represent the facial shape. Inspired by the work of [9,10], we adopted the *pairwise proximity function SVM* (ppfSVM) [3,4] for classification. The strategy of ppfSVM is to perform classification on proximity data, where each sample is represented by its similarity with all samples in the data set. A conventional SVM is then applied to classify the transformed data. In our case, the proposed Fubini-Study metric is used to construct the input for classification. For a fair comparison, we performed the same experimental protocol in the literature [2,8,10,18], namely 10-fold subject-independent cross-validation on CK+ dataset and Oulu-CASIA dataset. All subjects were divided into 10 groups in ascending order of ID, 9 groups were used for training, and the remaining group was used for testing. The number of target neighbors parameter k was selected during the cross validation. Figure 1 shows the effect of different k values on the accuracy of facial expression classification using the proposed metric. A multi-class SVM with Gaussian kernel was selected for classification and its hyper parameters were determined by a grid search method. In the end, the average classification performance is reported in the following sections.

5.3 Results and Discussion

Table 1(a) presents the average classification accuracy of the Fubini-Study metric for facial expressions recognition before and after distance metric learning when using the last peak frame of each video sequence, respectively. It can be seen from the table that the learned Fubini-Study metric obtained a maximum gain of about 8% in classification accuracy compared to the original metric, which proves that the LMNN metric learning algorithm performs well on the CK+ dataset and Oulu-CASIA dataset. To further investigate the validity of the proposed metric

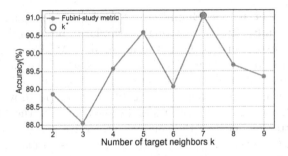

Fig. 1. Average classification accuracy on the CK+ dataset using the Fubini-Study metric when varying the target neighbour parameter k.

for identifying each emotion, we provide also the confusion matrix obtained using the learned Fubini-Study metric for CK+ dataset in Table 2(a). The *Disgust*, *Happy*, and *Surprise* expressions can be well recognized with an accuracy rate of over 95%, while the main confusions happened in the *Contempt* expression. This can be explained by its slight facial changes and its smaller sample size compared to other expressions. Figure 2(a), (b) show the 2D visualization of the original CK+ data before and after metric learning, respectively, using the t-SNE [14] with the proposed metric. We can observe that the original Fubini-Study metric has been able to well distinguish the two types of expressions *Happy* and *Surprise*. After metric learning, the more difficult expressions like *Angry*, *Contempt*, *Fear* and *Sad* are also differentiated by the optimal metric.

Table 1. Overall accuracy (Acc %) of our static optimal solution compared to the identity matrix on CK+ dataset and Oulu-CASIA. I: identity matrix; $[\cdot]_{opt}$: optimal matrix.

Dataset	(a) Fubini-Study Metric		(b) Euclidean Metric	
	$A = I$	$A = A_{opt}$	$M = I$	$M = M_{opt}$
CK+	87.53	**91.05**	79.97	89.15
Oulu-CASIA	55.62	**63.75**	50.49	61.84

Table 2. Confusion matrix using the learned metrics for the CK+ dataset. *Angry* (An); *Contempt* (Co); *Disgust* (Di); *Fear* (Fe); *Happy* (Ha); *Sad* (Sa); *Surprise* (Su)

(a) Fubini-Study metric

	An	Co	Di	Fe	Ha	Sa	Su
An	**84.4**	2.2	4.4	0.0	0.0	8.9	0.0
Co	5.5	**61.1**	5.6	11.1	5.6	11.1	0.0
Di	0.0	0.0	**98.3**	0.0	0.0	1.7	0.0
Fe	0.0	4.0	0.0	**80.0**	4.0	4.0	8.0
Ha	0.0	0.0	0.0	1.4	**98.6**	0.0	0.0
Sa	14.3	3.6	0.0	0.0	0.0	**82.1**	0.0
Su	0.0	1.2	0.0	0.0	0.0	0.0	**98.8**

(b) Euclidean metric

	An	Co	Di	Fe	Ha	Sa	Su
An	**84.4**	2.2	4.4	0.0	0.0	8.9	0.0
Co	11.1	**50.0**	0.0	5.6	0.0	27.8	5.6
Di	0.0	0.0	**100.0**	0.0	0.0	0.0	0.0
Fe	0.0	0.0	0.0	**72.0**	16.0	8.0	4.0
Ha	0.0	0.0	0.0	0.0	**100.0**	0.0	0.0
Sa	14.3	0.0	3.6	0.0	0.0	**78.6**	3.6
Su	1.2	1.2	0.0	1.2	0.0	0.0	**96.4**

Fubini-Study Metric vs Euclidean Metric. A comparison with the Euclidean metric in Eq. (2) in measuring the similarity between the shape representations is also executed on CK+ datset and Oulu-CASIA dataset to show the effectiveness of the proposed metric. From the results of the Euclidean metric in Table 1(b), we can observe that the accuracy of shape representation in complex projective space is improved by about 2% − 8% compared with performing analysis in Euclidean space. These results demonstrate that the facial deformation can be better modeled in the complex projective space to capture facial expressions more accurately. Table 2(b) gives the confusion matrix of Euclidean metric for the CK+ dataset. The comparison shows that emotions with gentle facial movements such as *Sad, Contempt* and *Fear* can be better captured by the Fubini-Study metric. The 2D visualization obtained by using the original Euclidean metric in Fig. 2(c) shows an approximately linear relationship, and the points of different emotions are mixed together and cannot be distinguished. As shown in Fig. 2(d), the learned Euclidean metric cannot separate the complicated emotions–*Sad, Contempt* and *Fear* as well as the Fubini metric. These results are consistent with the confusion matrices in Table 2.

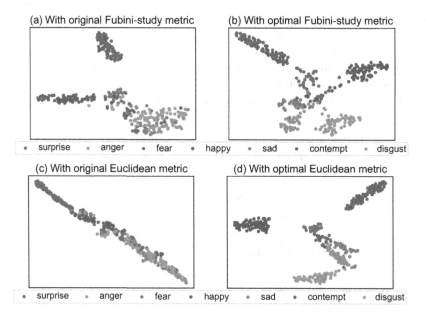

Fig. 2. 2D visualization of original CK+ data using t-SNE method with the proposed metrics. (a)(b) show the visualization results before and after metric learning with Fubini-Study metric. (c)(d) show the visualization results before and after metric learning with Euclidean metric.

Comparison with the State-of-the-Art. We also compared the proposed method with state-of-the-art approaches for CK+ dataset, with the details reported in Table 3. For a fair comparison, only the geometry-based methods that use facial landmarks were considered in the comparative evaluation. We divided these methods into two categories, depending on whether the input is a static image or a dynamic video sequence. The only comparable static image based approach [2] used the Gram matrix computed from landmarks as the shape representation and a metric defined on the the manifold of oriented ellipses centered at the origin in an Euclidean n-space was selected for metric learning. The results show that our method performs better, both in terms of the original metric and the optimal metric. For dynamic approaches, the ranked-first approach is the work of [10], where the facial landmark sequences were modeled as parametrized trajectories on the Riemannian manifold of positive semi-definite matrices of fixed-rank. However, their metric contains a parameter k, which needs to be carefully selected for each data set. According to the performance ranking, the following method [18] encoded time-varying facial shapes and used deep neural networks for classification. Compared with these work, our method using solely the last frame also achieves competitive results on facial expression recognition.

Table 3. Comparison with state-of-the-art geometric methods on the CK+ dataset.

Input	Methods	Accuracy
Video sequence	Shape velocity on $\mathcal{G}$ [17]	82.80
Video sequence	LDCRFs [7]	85.84
Video sequence	ITBN [21]	86.30
Video sequence	Intrinsic SCDL - SVM [18]	89.43
Video sequence	Extrinsic SCDL - Bi-LSTM [18]†	95.73
Video sequence	**Shape trajectories on $\mathcal{S}^+(d,n)$ [10]**	**96.87**
Static image	original metric on $\mathcal{S}_c^+(2,n)$ [2]	85.83
Static image	**original metric on $\mathbb{CP}^{n-1}$ (ours)**	**87.53**
Static image	optimal metric on $\mathcal{S}_c^+(2,n)$ [2]	90.53
Static image	**optimal metric on $\mathbb{CP}^{n-1}$ (ours)**	**91.05**

6 Conclusions

In this paper, a non-Euclidean metric learning method is proposed and applied to facial expression recognition, where the facial shapes are directly encoded by the equivalent class of the landmarks in the complex projective space, and the similarity between them can be measured by Fubini-Study metrics. The experimental results show that, compared to the metric defined in the Euclidean space, the Fubini-Study metrics are more effective and discriminative for changes in facial expressions. The latter one also achieves comparable performance to the state-of-the-art approaches. In future work, we plan to extend the proposed method to dynamic scenarios, where the sequences of landmarks will be explored to model the temporal relationship.

Acknowledgements. The proposed work was supported by the French State, managed by the National Agency for Research (ANR) under the Investments for the future program with reference ANR-16-IDEX-0004 ULNE. And we thank Prof. J-C. Alvarez Paiva from University of Lille for fruitful discussions on the formulation of the distance between shapes in complex projective spaces.

References

1. Daoudi, M., Hammal, Z., Kacem, A., Cohn, J.F.: Gram matrices formulation of body shape motion: an application for depression severity assessment. In: 2019 8th International Conference on Affective Computing and Intelligent Interaction Workshops and Demos (ACIIW), pp. 258–263 (2019). https://doi.org/10.1109/ACIIW.2019.8925009
2. Daoudi, M., Otberdout, N., Paiva, J.-C.Á.: Metric learning on the manifold of oriented ellipses: application to facial expression recognition. In: Del Bimbo, A., (eds.) ICPR 2021. LNCS, vol. 12666, pp. 196–206. Springer, Cham (2021). https://doi.org/10.1007/978-3-030-68780-9_18
3. Graepel, T., Herbrich, R., Bollmann-Sdorra, P., Obermayer, K.: Classification on pairwise proximity data. In: Kearns, M., Solla, S., Cohn, D. (eds.) Advances in Neural Information Processing Systems, vol. 11. MIT Press (1999)
4. Gudmundsson, S., Runarsson, T.P., Sigurdsson, S.: Support vector machines and dynamic time warping for time series. In: 2008 IEEE International Joint Conference on Neural Networks (IEEE World Congress on Computational Intelligence), pp. 2772–2776 (2008). https://doi.org/10.1109/IJCNN.2008.4634188
5. Hu, J., Lu, J., Tan, Y.P.: Discriminative deep metric learning for face verification in the wild. In: Proceedings of the IEEE Conference on Computer Vision and Pattern Recognition (CVPR), pp. 1875–1882 (2014)
6. Huang, Z., Wang, R., Shan, S., Chen, X.: Projection metric learning on grassmann manifold with application to video based face recognition. In: Proceedings of the IEEE Conference on Computer Vision and Pattern Recognition (CVPR), pp. 140–149 (2015)
7. Jain, S., Changbo Hu, Aggarwal, J.K.: Facial expression recognition with temporal modeling of shapes. In: 2011 IEEE International Conference on Computer Vision Workshops (ICCV Workshops), pp. 1642–1649. IEEE, Barcelona, Spain (2011). https://doi.org/10.1109/ICCVW.2011.6130446. https://ieeexplore.ieee.org/document/6130446/
8. Jung, H., Lee, S., Yim, J., Park, S., Kim, J.: Joint fine-tuning in deep neural networks for facial expression recognition. In: 2015 IEEE International Conference on Computer Vision (ICCV), pp. 2983–2991 (2015). https://doi.org/10.1109/ICCV.2015.341
9. Kacem, A., Daoudi, M., Alvarez-Paiva, J.C.: Barycentric representation and metric learning for facial expression recognition. In: 2018 13th IEEE International Conference on Automatic Face Gesture Recognition (FG 2018), pp. 443–447 (2018). https://doi.org/10.1109/FG.2018.00071

10. Kacem, A., Daoudi, M., Amor, B.B., Berretti, S., Alvarez-Paiva, J.C.: A novel geometric framework on gram matrix trajectories for human behavior understanding. arXiv:1807.00676 (2018)
11. Kendall, D.G.: Shape manifolds, procrustean metrics, and complex projective spaces. Bull. Lond. Math. Soc. **16**(2), 81–121 (1984). https://doi.org/10.1112/blms/16.2.81
12. Liao, S., Hu, Y., Zhu, X., Li, S.Z.: Person re-identification by local maximal occurrence representation and metric learning. In: Proceedings of the IEEE Conference on Computer Vision and Pattern Recognition (CVPR), pp. 2197–2206 (2015)
13. Lucey, P., Cohn, J.F., Kanade, T., Saragih, J., Ambadar, Z., Matthews, I.: The extended Cohn-Kanade Dataset (CK+): a complete dataset for action unit and emotion-specified expression. In: 2010 IEEE Computer Society Conference on Computer Vision and Pattern Recognition - Workshops, pp. 94–101 (2010). https://doi.org/10.1109/CVPRW.2010.5543262
14. van der Maaten, L., Hinton, G.: Visualizing data using t-SNE. J. Mach. Learn. Res. **9**(86), 2579–2605 (2008)
15. Shao, L., Liu, L., Yu, M.: Kernelized multiview projection for robust action recognition. Int. J. Comput. Vision **118**(2), 115–129 (2016). https://doi.org/10.1007/s11263-015-0861-6
16. Szczapa, B., Daoudi, M., Berretti, S., Pala, P., Bimbo, A.D., Hammal, Z.: Automatic estimation of self-reported pain by interpretable representations of motion dynamics. In: 2020 25th International Conference on Pattern Recognition (ICPR), pp. 2544–2550 (2021). https://doi.org/10.1109/ICPR48806.2021.9412292
17. Taheri, S., Turaga, P., Chellappa, R.: Towards view-invariant expression analysis using analytic shape manifolds. In: Face and Gesture 2011, pp. 306–313. IEEE, Santa Barbara, CA, USA (Mar 2011). https://doi.org/10.1109/FG.2011.5771415
18. Tanfous, A.B., Drira, H., Amor, B.B.: Sparse coding of shape trajectories for facial expression and action recognition. arXiv:1908.03231 (2019)
19. Townsend, J., Koep, N., Weichwald, S.: Pymanopt: a python toolbox for optimization on manifolds using automatic differentiation. J. Mach. Learn. Res. **17**(137), 1–5 (2016)
20. Wan, S., Aggarwal, J.: Spontaneous facial expression recognition: a robust metric learning approach. Pattern Recogn. **47**(5), 1859–1868 (2014). https://doi.org/10.1016/j.patcog.2013.11.025
21. Wang, Z., Wang, S., Ji, Q.: Capturing complex spatio-temporal relations among facial muscles for facial expression recognition. In: 2013 IEEE Conference on Computer Vision and Pattern Recognition, pp. 3422–3429. IEEE, Portland, OR, USA (2013). https://doi.org/10.1109/CVPR.2013.439
22. Weinberger, K.Q., Saul, L.K.: Distance metric learning for large margin nearest neighbor classification. J. Mach. Learn. Res. **10**, 207–244 (2009)
23. Zhao, G., Huang, X., Taini, M., Li, S.Z., Pietikäinen, M.: Facial expression recognition from near-infrared videos. Image Vis. Comput. **29**(9), 607–619 (2011). https://doi.org/10.1016/j.imavis.2011.07.002

Gamified Smart Grid Implementation Through Pico, Nano, and Microgrids in a Sustainable Campus

Citlaly Pérez[1]([✉]) [iD], Juana Isabel Méndez[1] [iD], Antonio Rivera[1] [iD], Pedro Ponce[1] [iD], Sergio Castellanos[2] [iD], Therese Peffer[3] [iD], Alan Meier[4] [iD], and Arturo Molina[1] [iD]

[1] Institute of Advanced Materials for Sustainable Manufacturing, Tecnologico de Monterrey, Monterrey, MX 64849, USA
{A01336766,isabelmendez,A01337294,pedro.ponce,armolina}@tec.mx
[2] Department of Civil, Architectural and Environmental Engineering, The University of Texas at Austin, Austin, TX, USA
sergioc@utexas.edu
[3] Institute for Energy and Environment, University of California, Berkeley, CA 94720, USA
tpeffer@berkeley.edu
[4] Energy and Efficiency Institute, University of California, Davis, CA 95616, USA
akmeier@ucdavis.edu

Abstract. Deploying a smart city is becoming a trend toward achieving sustainability and enabling a better lifestyle for its inhabitants. Energy plays a leading role in smart cities, as most of our everyday activities and environment are related to energy sources. Integrating renewable energy into the electric power system is also challenging due to the intermittency and security of supply; a solution to this challenge is a decentralized system, which integrates renewable energies, reduces fossil fuel usage, and increases eco-efficiency. Furthermore, research suggests including social interaction between the inhabitants and the smart grid through interfaces with game elements that motivate and educate individuals about the smart grid. Thus, smart grids require including new distributed entities that have not existed previously, such as picogrids, nanogrids, and microgrids. In addition, the internet of things (IoT) applications provide advanced monitoring and control in the smart grid in case of an outage or disturbances. Therefore, this paper presents a microgrid, nanogrid, and picogrid integration using a building facility at Tecnologico de Monterrey, Mexico City Campus. Besides, a solar photovoltaic array was analyzed to understand the energy consumption and its interaction with the Campus microgrid. The building considers three groups: specialized laboratories, computer rooms, and classrooms. Additionally, a gamified platform was deployed to teach the Campus community the differences between these grids and how each picogrid had different consumption.

Keywords: Sustainable Campus · IoT Campus · Microgrid · Picogrid · Nanogrid · Renewable sources · Gamified platform

S. Berretti and G.-M. Su (Eds.): ICSM 2022, LNCS 13497, pp. 128–143, 2022.
https://doi.org/10.1007/978-3-031-22061-6_10

1   Introduction

Non-renewable sources such as fossil fuels, including coal, oil, and natural gas, have been powering economies for over a century and a half by supplying about 80% of the world's energy [1]. Nonetheless, they require millions of years to regenerate; thus, they negatively impact the environment [2]. Moreover, the atmosphere's carbon dioxide and other greenhouse gases come from burning fossil fuels to generate energy [1]. Consequently, those gases are the primary cause of climate change and reduce the quality of life (QoL).

On the other hand, renewable sources, including solar, wind, hydro, biofuels, and others, help transition into a less carbon-intensive and more sustainable energy ecosystem [2]. Table 1 depicts the characteristics of some renewable sources and their technology type categorization. There are two categories of renewable energy technologies [3]:

- Dispatchable.
- Non-dispatchable or Variable Renewable Energy requires meeting the following characteristics for their integration into the current power system:

  - Variability is due to the temporal availability of resources.
  - Uncertainty is due to unexpected changes in resource availability.
  - Local specific properties due to the geographical availability of resources
  - Low marginal costs since the resources are freely available.

Furthermore, due to the Paris Agreement, over 700 cities worldwide have shifted to renewable energy [4]. These shiftings conceptualize and target their transition through decarbonized projects, net-zero energy buildings, or carbon-neutral outlines to move towards more renewable and sustainable cities [5].

By 2050, New York expects an 80% reduction in CO_2 emissions [6]. Copenhagen released an official plan to target CO_2 neutrality by 2025 [7]. In 2013, Vejle, Denmark, became part of the 100 Resilient Cities and proposed four pillars [8]:

- A Co-creating City: Multidisciplinary collaborations between public and private sectors to address the city's challenges.
- A Climate Resilient City: Their future development uses sustainable resources, renewable energy, and green transport by taking water and climate change as their drivers for their development.
- A Socially Resilient City: Future generations are the key to social and economic cohesion. Urban spaces and social housing aim to strengthen social resilience and community adherence.
- A Smart City: An efficient society is a product of smart technologies that promote social interactions and social inclusivity. The multiple collaborations, public accessibility, and digital technologies support youth education to create a digital society.

Therefore, a smart city improves citizens' QoL by providing them with services, technologies, and products that react more quickly and efficiently to their needs [5].

Furthermore, these services require a structure that provides solutions to the following challenges: healthcare, public safety, traffic and mobility, education, transportation, security, and energy [9–13].

Table 1. Non-dispatchable and dispatchable renewable energy sources.

Renewable energy	Characteristics	Category
Solar [14]	Photovoltaic (PV) panels convert solar energy into direct current electricity using semiconducting materials. Solar PV combines two advantages: module manufacturing in large plants, which allows economies of scale, and modular technology that deploys fewer PV panels	Non-dispatchable
Wind [15]	Wind turbines extract power from an airflow to produce mechanical or electrical power. Eolic energy is a mature technology with various system sizes, producing cheap energy at the utility scale. However, such technology is expensive on a small scale. In addition, due to wind's high unpredictability, turbines are commonly accompanied by other energy sources or storage systems when used in small applications	Non-dispatchable
Biomass [16]	Biomass is mainly a stored source of solar energy. Plants produce biomass through photosynthesis. Biomass can be burned directly for heat or converted to renewable liquid and gaseous fuels through various processes	Dispatchable
Hydropower [17]	Hydropower relies on the typically fast-moving water in a large river or rapidly descending water from a high point. This source converts the force of that water into electricity by spinning a generator's turbine blades	Dispatchable
Geothermal [18]	Geothermal energy derives from the thermal energy flux from the earth's center and is used only for thermal production or cogeneration. Geothermal electricity is very cheap when the proper ground conditions are met, although not many places have the suitable soil characteristics	Dispatchable

In this regard, smart renewable energy systems positively affect urban environments and citizens by influencing their well-being and increasing their QoL [5]. Thus, these systems are critical in generating clean energy and producing zero emissions as they require the integration of the power grid to ensure stability, protection, and operational restrictions. However, the power grid interaction with renewable sources must ensure stable operation after fault, load changes, and other network disturbances within a smart city [19].

As a result, decentralized power systems help solve relevant tasks such as optimizing and stabilizing power systems due to their flexibility in integrating renewable energy sources and intelligent control centers in power production and distribution [19]. Furthermore, a decentralized energy system seeks to put power sources closer to the end-user; thus, renewable sources, combined heat, and power can interact with the grid and help reduce fossil fuel use to increase eco-efficiency [19].

1.1 Smart Grid

Although there is a variety of smart grid definitions, the Electric Power Research Institute defines it as [20]:

- *The Smart Grid incorporates information and communications technology into every aspect of electricity generation, delivery, and consumption to minimize environmental impact, enhance markets, improve reliability and service, and reduce costs and improve efficiency.*

Thus, Smart Grids requires to fulfill the following [19, 21]:

- Enhance the operation of the legacy high voltage grid, for instance, by using synchrophasors.
- Improve grid-customer interaction by providing smart metering or real-time pricing.
- Include the new distributed entities that have not existed previously. Consider the picogrids, nanogrids, microgrids, and active distribution networks.

Furthermore, the common goals of power management in smart grids include minimizing electricity costs, reducing the peak to optimal ratio, maximizing user comfort, minimizing consolidated energy consumption, and integrating renewable energy [19]. Besides, the smart grid needs to provide a dynamically interactive real-item infrastructure encompassing the many visions of diverse power system stakeholders [21].

Consequently, citizens play a relevant role in the smart grid interaction, as they need to be aware of the importance of changing the ancient electric delivery methods for a sustainable one and the benefits it has to their economy [19, 21]. Furthermore, the feedback between the power plant and the customer is possible thanks to the two-way communication implemented around the smart grid.

Decentralized Energy System. Decentralized energy system conceptualization led to new grid proposals of arrangements based on size and configuration. Therefore, three sizes of grids based on small to medium off-grid or grid-connected systems are picogrid, nanogrid, and microgrid [19]. These grids use renewable energy sources and can operate independently of grid-supplied power. The pico, nano, and microgrids are easily installed, flexible, and can, in time, be connected to main power grids if and when such networks expand [22].

1.2   Picogrid, Nanogrid, and Microgrid

Picogrid's size is around 1 kW, and its key components are generation plus single-phase distribution [19]. Therefore, picogrids imply an appliance-level power distribution network with an ultra-low-power demand [22]. For instance, a powered USB hub becomes its nanogrid, independent (for power) from the upstream computer. Laptops, smartphones, tablets, or sensor networks fall into this category of picogrid [22].

Nanogrids can be interconnected with each other and with microgrids or macrogrids and include a controller, loads, gateways, and storage possibility; they are limited to a single building structure considered a primary load. It is also limited to a network of off-grid loads below 5 kW. This category includes: distributed generators, batteries, electric vehicles, and smart loads capable of islanding with some level of intelligent distributed energy resource management controls [19].

Microgrids have well-defined electrical boundaries and interconnected loads and distributed energy resources [23]. Their purpose is to harness distributed and renewable energy sources in medium and low voltage environments [23]; hence, customers have access to use them directly [19]. Furthermore, microgrids can connect and disconnect from the grid to allow both grid-connected or island modes. Microgrids have a generation capacity of 1–50 kW that serves consumers and uses distribution lines [19].

Microgrids comprise power converters, battery energy storage systems, control and protection systems [23]. In addition, these configurations involve small-scale electricity generation, which serves a limited number of consumers via a distribution grid that can operate in isolation from the national transmission network [23]. Microgrids are appearing on campuses, such as universities, hospitals, military establishments, and business parks in urban environments [19].

Notwithstanding, deploying a smart grid is a complex process requiring all its components to work together as a whole [21]. To achieve this, communication between all the elements is crucial and must be resilient [8]; thus, available Internet of Things (IoT) technologies can gather data by sensing and monitoring citizens and their environment to understand and propose solutions to their needs. Moreover, these grids interact with the power grids and IoT services to accomplish high-functional grid operation and effective electricity usage [24].

Picogrid, nanogrid, and microgrid involved in a Sustainable Campus. An example of a microgrid on Campus is the University of California, San Diego [25], which serves a campus community of more than 45,000 people, 13 million ft2 in 450 buildings, and 1200 acres. The Campus can be seen as a microgrid, each building represents a nanogrid, and each floor's building represents a picogrid. It generates 80% of the electricity used on Campus annually. It has gas turbines that generate 213.5 MW, a 3 MW steam turbine, PV panels that generate 1.2 MW, and a Power Purchase Agreement for fuel cell power that uses methane from a wastewater treatment plant. The microgrid can connect to the larger electric grid and work independently.

1.3   Internet of Things

A smart grid involves transportation and delivery of electricity; every system must be thoroughly efficient and reliable in all its steps [21]. Internet of things (IoT) applications are used to reach this goal because those technologies allow advanced monitoring and control in case of an outage or disturbances, among other issues [24].

The IoT has encouraged the development of devices connected to the internet that allow monitoring and control services. IoT devices are in charge of measuring specific parameters according to their application; those application areas include healthcare, smart building, smart home, and smart grids [21].

In every area where IoT is present, there are essential requirements that the devices must accomplish: low power consumption, small size, low cost, and durability [24]. Alongside those requirements, network features must be considered: low latency, enough bandwidth, resilience, and scalability [24]. Also, crucial security issues must be prevented, such as DoS, DDoS, malware, and phishing.

Three main parts are involved in a general IoT implementation: IoT embedded devices, gateways, and clouds [24]. The first one is related to all the sensors and actuators included in the application, which must be connected by short-distance communication technologies such as Bluetooth Low Energy (BLE), ZigBee, or WiFi. Secondly, the gateways or routers are the linkages between the devices and the cloud service. These routers have different technologies depending on the connection conditions or the distance between devices. Finally, all the data gathered from the IoT devices is sent to the cloud using long-distance communication technologies (WAN) such as NB-IoT, LoraWAN, and Sigfox. In the cloud infrastructure, the data is stored, processed, and made logical decisions [24].

1.4 Gamification in Smart Grids

Gamification uses game elements to enhance user experience and user engagement in real contexts and applications [26]. Gamification motivates and increases user activity and retention by adding gaming elements, enticing users, and encouraging specific types of behavior, creating a significant driving force to induce desirable user behavior. Gamification has been widely used in different fields, such as productivity, finance, health, and sustainability [26].

Marques and Nixon [27] suggested that Smart Grids need to ease the interaction between the users and the grid to communicate load adjustment and understand human motivational psychology. They outlined that the gamified applications should promote and motivate continual change in individuals' behavior regarding the smart grid through three cores: energy education, social interaction, and energy conservation.

Konstantakopoulos et al. [28] proposed an HMI for a social game that encourages energy-efficient behavior among smart building occupants. The social game was employed for residential housing single-room apartments at the Nanyang Technological University campus. This interface deployed IoT sensors to monitor real-time room lighting systems and HVAC usage. The interface included feedback, points, rewards, random rewards through coins and daily energy usage. Besides, they suggested incorporating game applications in smart grid management.

Common gamification design principles include goals and challenges, personalization, rapid feedback, visible feedback, freedom of choice, freedom to fail, and social engagement. Gamification elements include points, scoring, leaderboards, progress bars, ranks, rewards, or incentives [26].

The goal of an interface is to make users feel in control of their experience. Graphical interfaces for gamification purposes allow for specifying goals, rules, settings, context, and types of interactions. The basic mechanics of gamification interfaces are closely related to game design: addressing the human desire for socializing, learning, mastery, competition, achievement, status, self-expression, altruism, or closure [26].

Smart grids appeared to respond to human energy needs and improve electric energy conditions [21]. However, as human energy needs are involved, it requires understanding how humans behave to understand their pattern's consumption. It is complex to know in a smart grid about the type of users; therefore, gamification is a way to teach the community [13]. Thus, this paper proposes to employ gamification techniques to provide an interactive interface about the different types of grids involved in a Campus facility. The microgrid, nanogrid, and picogrid are better explained in a Campus, as it is similar to understanding a community or how a city behaves [12, 13].

2   Proposal

Following UC San Diego's proposal, a great example of a microgrid is a university campus and its application into the Mexico City context. In 2018, Tecnologico de Monterrey, Mexico City Campus (Tec CCM) began its reconstruction due to the 2017 earthquake. The new Sustainable Campus considers renewable sources as photovoltaic arrays, including urban parks and plazas, the mitigation of annual flooding with bioclimatic considerations such as naturally ventilated buildings, pergola for heat gain reduction, and reduction of carbon footprint [12, 13]. Currently, there are six constructed buildings.

This paper focuses on nanogrid and picogrid using CEDETEC building as a case study. Besides, a solar PV array is analyzed to understand better the energy consumption and its interaction with the Campus microgrid. Thus, Table 2 sections CEDETEC by room type, floor, and grid type. Each floor column presents the grid type as picogrid, whereas the nanogrid is presented as a merged row of the four floors. Due to the privacy of the building consumption, estimated energy consumption was proposed using templates that accepted energy simulators have, such as EnergyPlus software [29]. Thus, the equipment density in Watts per meter considered for each room type was as follows: Specialized laboratory (Open lab): 43.1 W/m^2; Computer Room: 20 W/m^2; and Classroom: 10 W/m^2. The estimated occupancy hours considered 12 daily hours during 43 weeks at 80% of occupancy (2064 h) and nine weeks with 20% occupancy (108 h), giving a total annual equipment usage of 2172 h.

Table 2. CEDETEC room type by floor.

Room type	First floor	Second floor	Third floor	Fourth floor
Specialized laboratory	11	1	0	2
Computer room	3	6	3	3

(*continued*)

Table 2. (*continued*)

Room type	First floor	Second floor	Third floor	Fourth floor
Classroom	4	6	12	9
Grid type	Picogrid	Picogrid	Picogrid	Picogrid
	Nanogrid			

Figure 1 shows the microgrid, picogrid, and nanogrid applied into the Sustainable Campus Tec CCM. These are the grid distribution:

- Microgrid: Interconnected loads from each building.
- Nanogrid: A single building capable of being self-sufficient through PV arrays and smart loads.
- Picogrid: Each building level is composed of lighting, appliances, security, and HVAC systems, among others.

In addition to the PV array for the nanogrid, this paper also proposes to integrate solar energy into other applications within the university campus, such as:

- E-bikes with a motor powered by solar energy for the mobility of students and personnel within the Campus.
- Solar lamps.
- Solar chargers for low electricity consumption devices.
- Greenhouses powered by solar energy produce food for consumption within the Campus.

Fig. 1. Sustainable Campus as an example of Smart City using microgrid, nanogrid, and picogrid (Tec CCM).

Besides, Fig. 2 describes the gamification elements required in each grid type. For the microgrid, exploratory tasks and unlockable content are displayed because the community has access to the general level and visualizes the levels available for each building (or nanogrid). Then, the second level is the nanogrid, where the students learn the consumption on each floor and how the four levels of the CEDETEC building represent the nanogrid. Finally, on the third level, each floor represents the picogrid; the community has access to random rewards, social competition, feedback, and challenges. In [12, 13], they proposed for the picogrid level and specifically in each classroom how to provide gamified elements that engage students and teachers in a dynamic for becoming energy aware.

The main specifications of the modules and the inverters used for the PVsyst program simulation are described in Table 2 [30]. The location of the building is (19.28°N, -99.14°W). The roof considered was the east and west sides with a roof area of (800 m^2), and the inclination and azimuth for the PV modules (19°/0°) were considered for the PV modeling proposal of the nanogrid building (CEDETEC, Table 3).

3   Results

Table 4 estimates the annual energy consumption of each floor based on the type of room and equipment density described in the previous section. The annual consumption of this nanogrid building is estimated to be 2220 MWh. With an estimated usage by floor of 65.5%, 10.2%, 12.4%, and 11.9%.

Fig. 2. Pyramid of types of grids involved in the Campus. Microgrid, nanogrid, and picogrid have different gamification elements.

Table 3. PV models and inverters suggested for the CEDETEC's nanogrid.

PV modules		Inverters	
Manufacturer	Generic PVsyst	Manufacturer	PVsyst
Model	Poly 30 Wp 36 cells	Model	3 kWac inverter
Module size	0.360 × 0.650 m^2	**Input characteristics (PV array side)**	

(continued)

Table 3. (*continued*)

PV modules		Inverters	
TRef	25 °C	Vmin	125 V
GRef	1000 W/m^2	Vmax	440 V
Voc	21.70 V	Vmax array	550 V
Isc	2.40 A	Vmin@Pnom	188 V
Vmpp	17.30 V	Pnom DC	4.0 kW
Impp	1.74 A	**Output characteristics (AC grid side)**	
Pmpp	30.10 W	Grid voltage Monophased	230 V
ISC temperature coefficient	1.40 mA/°C	Grid frequency	50/60 Hz
Pnom	30 Wp	Pnom AC	3.0 kWac
		Pmax AC	3.0 kWac
		Inom AC	13.0 A
		Imax AC	16.0 A
		Max. Efficiency	97.0%

Table 4. CEDETEC annual energy consumption for the equipment.

Room type	First floor	Second floor	Third floor	Fourth floor
Specialized laboratory	643.4 kW	2 kW	0 kW	7.9 kW
Computer room	4.1 kW	60.1 kW	14.8 kW	14.4 kW
Classroom	22 kW	42.4 kW	112 kW	99.2 kW
Total picogrid	669.5 kW	104.5 kW	126.8 kW	121.5 kW
Total nanogrid	1022.3 kW × 2172 h = **2220.4 MWh/year**			

Table 5 shows the results from the PVSyst simulation. The produced energy by the proposed PV array will be 204.4 MWh/year with a specific production of 2003 kWh/kWp/year and a performance ratio of 88.21%.

Table 5. PV Array Characteristics suggested for the CEDETEC's nanogrid [30].

PV modules		Inverters	
Number of PV modules	3402 units	Number of inverters	27 units
Nominal (STC)	102 kWp	Total power	81.0 kWac

(*continued*)

Table 5. (*continued*)

PV modules		Inverters	
Module	162 strings × 21 in series	Operating voltage	125–440 V
At operating conditions (50 °C)		Pnom ratio (DC:AC)	1.26
Pmpp	99.8 kWp		
Vmpp	293 V		
Impp	340 A		
Total PV power		**Total inverter power**	
Nominal (STC)	102 kWp	Total power	81 kWac
Total	3402 modules	Number of inverters	27 units
Module area	796 m^2	Pnom ratio	1.26

Considering CEDETEC's energy consumption (2220.4 MWh/year), the proposed PV array will be able to feed 9.2% of this demand. The proposal is that this nanogrid feeds picogrids located in the floors that do not have a large energy consumption, such as the second, third or fourth floor; another proposal would be feeding picogrids located in computer rooms and classrooms with low energy demand. For instance, this PV array will provide power for the computer rooms located on the third and fourth floors and the classrooms on the first and second floors entirely, which represent a demand of (203.3 MWh/year).

Figure 3 shows the IoT implementation applied to Sustainable Campus. First, the sensors and actuators collect and sense the data from the buildings, renewable sources, laboratories, utilities, transportation, and services. Then, the gateway links these devices with the cloud service. Thus, the collected data is sent to the cloud using WAN technologies. In addition, the end-users become active participatory sensors that send information to the cloud, sensors, and actuators. In other words, the first group marked represents all the sensors installed in the classrooms, laboratories, and computer rooms. All those small spaces are the local area networks (LAN), so technologies like Bluetooth, Zigbee, or even WiFi fit correctly.

The sensors should contribute to the correct energy management, including measuring lighting, HVAC, and other loads connected to the energy supply. Also, people's presence, sunlight intensity, and temperature must be detected, among others, to control the systems properly. The data collected on each LAN can be shown to the end-users with an indicator display or a mobile app (fourth group). However, sending the information to the cloud service using WAN protocols is mandatory to use the routers installed along with the Campus (second group). The third group represents the cloud services that will process, store and make decisions depending on the data received, so two-way communication must be established between the end-user and the systems controlling all the actuators.

Furthermore, Fig. 4 depicts a gamified platform deployed in the Genial.ly platform [31]. This gamified platform aims to teach the Campus community the differences

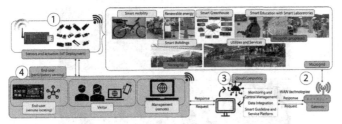

Fig. 3. Sustainable Campus and IoT integration for monitoring and sensing the microgrid, nanogrid, and picogrid at CEDETEC (Tec CCM).

between grids. The game starts with the microgrid level; it depicts a mission and provides exploratory tasks and unlockable content. It allows the community to learn how the four buildings have specific picogrid floors. The players can interact directly with CEDETEC and explore each floor. Besides, the players interact with this building. At the nanogrid level, they can access two new activities to explore the building through a virtual tour that opens a new window.

The other activity belongs to each picogrid, and the community can put the mouse above each floor and learn about the type of rooms available in each picogrid. Thus, information regarding the number of specialized laboratories, computer rooms, and classrooms is depicted. Then, the players select the next step, and a floor building picogrid image is depicted. The objective is to select the connected thermostat and visualize the differences between the professors' and the students' interface.

These interfaces depict how the teacher can send a message to the students by suggesting, for instance, increasing by 1 °C the thermostat setpoint to save energy and how they can save it by increasing it. Literature shows that by increasing 1 °C the setpoint, at least 6% of energy can be saved [13]. Finally, random rewards are deployed so players can select one of the three rewards.

4   Discussion

This paper followed the four pillars that Vejle, Denmark, proposed for Resilient Cities [8]. These pillars applied to the Sustainable Campus include:

- A co-creating City: The proposal provides multidisciplinary collaboration and activities through boot camps with the government or activities that connect the end-users with the Campus [32].
- A Climate Resilient Campus: the services integrate solar energy services in services like mobility, transportation, recreation or leisure, or even in the manufacturing process or agro production as greenhouses powered by solar energy. Empower the local consumption through these greenhouses.

- A Socially Resilient Campus: New educational models that strengthen social resilience and community adherence have been launched since 2019 with the novel Tec-21 model [10, 12, 13].

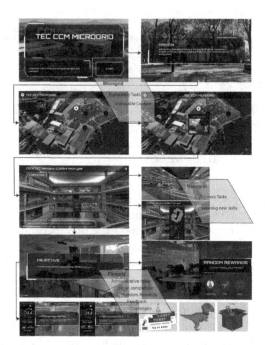

Fig. 4. Gamified platform of Tec CCM microgrid available at [31].

- A Smart Campus: Dividing the Camus into microgrid, nanogrid, and picogrid provides a decentralized energy system that is easily installed, flexible, and can, in time, be connected to main power grids if and when such networks expand [33]. Besides, the inclusion of IoT allows the community to become an active sensor that continuously senses the daily activities that help predict possible failures or know the students' energy usage patterns through the people's presence, sunlight intensity, temperature setpoints, illumination, or equipment usage [24].

Besides, a gamified platform was deployed following Konstantakopoulos et al. [28] and Marques and Nixon [27] suggestions. Thus, a gamified platform that provides inter-action between the users and the grid was deployed to understand the differences between grids. Education through gamification has been used to teach students specific topics. Innovative laboratories in education teach students specific topics like electricity usage [13]. Besides, gamification emerges as a solution to teach students in a ludic manner how to learn particular topics, for example, mathematics [13]. Therefore, there is a potential to apply gamification techniques to teach students the different types of grids and how the consumption reflects in the smart grid. Different game elements are proposed depending on the type of grid because it depends on the level of control the university community has. For example, in the picogrid it is better to use an approach where the student can learn about the use of thermostats and how the setpoint reduces or increases the electrical consumption [12, 13]. The nanogrids are related more to display exploratory tasks to help them become energy aware and identify when they can help reduce electrical consumption. The microgrid enables comparing buildings to teach the community how, depending on the type of building and activity, the type of electrical consumption and learn when they can reduce energy. Teaching the Campus community how to become aware provides them with the knowledge they can apply in their homes to promote these energy reductions.

Future work includes providing real-time interaction in real-time of the microgrid, focusing more on the picogrid, to teach the community how, for instance, managing the thermostat of the setpoint affects the energy and CO_2 consumption directly. Besides, an energy monitor of the Campus could be deployed using the community Villach monitor as a guideline [34].

5 Conclusion

This paper proposes a solar PV array for a building with three types of rooms on each floor: fourteen specialized laboratories, fifteen computer rooms, and thirty-one classrooms. Due to private access to the information, the annual energy consumption was estimated based on templates that energy models' simulators offer, such as EnergyPlus. Hence, the annual energy consumption was estimated at 2220 MWh. The PV array was simulated using the PVSyst software and estimated based on the roof and location at 204.4 MWh/year with a specific production of 2003 kWh/kWp/year and a performance ratio of 88.21%. With this information, the proposed PV array for the nanogrid will be able to feed 9.2% of the building's energy consumption, which can be distributed in the following two options: first, supply power for the second, third, or fourth floor, or second, feed picogrids that will provide energy for the computer rooms of the third and fourth floors and the classrooms of the first and second floors entirely.

To increase the QoL in the sustainable Campus, this paper also proposes the integration of renewable energy sources into services and products in mobility, transportation, recreation, or leisure. These services include E-bikes powered by solar energy, solar lamps and chargers for low electricity consumption devices such as mobile phones, and solar energy greenhouses for personalized food production.

Thus, the communication incorporates information in every grid about electricity generation, delivery, and consumption to improve efficiency and social interaction between each grid and the community members. Consequently, the inclusion of microgrid, picogrid, and nanogrid in a Campus facilitates the monitoring changes after proposing activities or goals such as energy reduction or thermal comfort analysis. As a result, the smart campus community plays a primary role in the smart grid interaction, as they need to be aware of the importance of modifying energy consumption behavior.

Therefore, a gamified platform is deployed to sensitize the community members on the differences between microgrid, nanogrid, and picogrid. This game shows the CEDETEC building as an example of a nanogrid and how the picogrid affects each floor. Likewise, an example of a thermostat display from the student's and the teacher's perspective is deployed to show the community how the teacher can send a message to the student to increase or decrease the setpoint and how its action affects energy consumption.

Acknowledgments. Research Project supported by Tecnologico de Monterrey and CITRIS under the collaboration ITESM-CITRIS Smart thermostat, deep learning, and gamification project (https://citris-uc.org/2019-itesm-seed-funding/).

References

1. EESI: Fossil Fuels | EESI. https://www.eesi.org/topics/fossil-fuels/description. Accessed 24 May 2022
2. IEA: Renewables – Fuels & Technologies. https://www.iea.org/fuels-and-technologies/ren ewables. Accessed 24 May 2022
3. IEA-ETSAP, IRENA: Renewable Energy Integration in Power Grids (2015)
4. Barbière, C.: 700 cities promise renewable energy transition by 2050. https://www.euractiv. com/section/climate-environment/news/700-cities-promise-renewable-energy-transition-by-2050/. Accessed 24 May 2022
5. Thellufsen, J.Z., et al.: Smart energy cities in a 100% renewable energy context. Renew. Sustain. Energy Rev. **129**, 109922 (2020). https://doi.org/10.1016/j.rser.2020.109922
6. Shorris, A.: The Plan for a Strong and Just City, p. 354 (2015)
7. The City of Copenhagen: The CPH 2025 Climate Plan | Urban Development. https://urband evelopmentcph.kk.dk/node/5. Accessed 13 Sept 2021
8. Vejle, K.: Vejle's Resilience Strategy. (2016)
9. Visvizi, A., Lytras, M. (eds.): Smart Cities: Issues and Challenges. 1st ed. Elsevier (2018)
10. Ponce, P., Mendez, J.I., Medina, A., Mata, O., Meier, A., Peffer, T., Molina, A.: Smart cities using social cyber-physical systems driven by education. In: 2021 IEEE European Technology and Engineering Management Summit (E-TEMS), pp. 155–160, IEEE, Dortmund, Germany (2021). https://doi.org/10.1109/E-TEMS51171.2021.9524889
11. Méndez, J.I., et al.: Human–machine interfaces for socially connected devices: From smart households to smart cities. In: McDaniel, T., Liu, X. (eds.) Multimedia for Accessible Human Computer Interfaces, pp. 253–289. Springer, Cham (2021). https://doi.org/10.1007/978-3-030-70716-3_9
12. Mendez, J.I., Ponce, P., Medina, A., Peffer, T., Meier, A., Molina, A.: A smooth and accepted transition to the future of cities based on the standard ISO 37120, Artificial Intelligence, and gamification constructors. In: 2021 IEEE European Technology and Engineering Management Summit (E-TEMS). pp. 65–71. IEEE, Dortmund, Germany (2021). https://doi.org/10.1109/E-TEMS51171.2021.9524900
13. Méndez, J.I., Ponce, P., Peffer, T., Meier, A., Molina, A.: A gamified HMI as a response for implementing a smart-sustainable university campus. In: Camarinha-Matos, L.M., Boucher, X., Afsarmanesh, H. (eds.) PRO-VE 2021. IAICT, vol. 629, pp. 683–691. Springer, Cham (2021). https://doi.org/10.1007/978-3-030-85969-5_64
14. Calvillo, C.F., Sánchez-Miralles, A., Villar, J.: Energy management and planning in smart cities. Renew. Sustain. Energy Rev. **55**, 273–287 (2016). https://doi.org/10.1016/j.rser.2015.10.133
15. Brenden, R.K., Hallaj, W., Subramanian, G., Katoch, S.: Wind energy roadmap. In: PICMET '09 – 2009 Portland International Conference on Management of Engineering Technology, pp. 2548–2562 (2009). https://doi.org/10.1109/PICMET.2009.5261810
16. Payne, J.E.: On biomass energy consumption and real output in the US. Energy Sources Part B **6**, 47–52 (2011). https://doi.org/10.1080/15567240903160906
17. Shinn, L.: Renewable Energy: The Clean Facts, https://www.nrdc.org/stories/renewable-ene rgy-clean-facts. Accessed 26 Aug 2021

18. Hammons, T.J.: Geothermal power generation worldwide. In: 2003 IEEE Bologna Power Tech Conference Proceedings, vol. 1, p. 8 (2003). https://doi.org/10.1109/PTC.2003.1304115

19. Shah, Y.T.: Hybrid Power: Generation, Storage, and Grids. CRC Press, Boca Raton (2021). https://doi.org/10.1201/9781003133094

20. EPRI: Smart Grid Demonstration Project Media Brief. http://mydocs.epri.com/docs/Corpor ateDocuments/MEDIAKITS/SmartGridmediabrief9-23-08.pdf. Accessed 24 May 2022

21. Ponce, P., Molina, A., Mata, O., Ibarra, L., MacCleery, B.: Power System Fundamentals. CRC Press, Boca Raton (2017). https://doi.org/10.1201/9781315148991

22. IRENA: Off-grid renewable energy systems: Status and methodological issues, 36 (2015)

23. Rezkallah, M., Chandra, A., Singh, B., Singh, S.: Microgrid: Configurations, control and applications. IEEE Trans. Smart Grid 10, 1290–1302 (2019). https://doi.org/10.1109/TSG. 2017.2762349

24. Siozios, K., Anagnostos, D., Soudris, D., Kosmatopoulos, E.: IoT for Smart Grids. Springer (2019)

25. UC San Diego Microgrid. https://the-atlas.com/projects/uc-san-diego-microgrid. Accessed 29 Sept 2021

26. Chou, Y.: Actionable Gamification Beyond Points, Badges, and Leaderboards. CreateSpace Independent Publishing Platform (2015)

27. Marques, B., Nixon, K.: The gamified grid: Possibilities for utilising game-based motivational psychology to empower the Smart Social Grid. In: 2013 Africon, pp. 1–5, IEEE, Pointe-Aux-Piments, Mauritius (2013). https://doi.org/10.1109/AFRCON.2013.6757748

28. Konstantakopoulos, I.C., Barkan, A.R., He, S., Veeravalli, T., Liu, H., Spanos, C.: A deep learning and gamification approach to improving human-building interaction and energy efficiency in smart infrastructure. Appl. Energy 237, 810–821 (2019). https://doi.org/10.1016/ j.apenergy.2018.12.065

29. EnergyPlus | EnergyPlus. https://energyplus.net/. Accessed 3 June 2021

30. PVsyst – Logiciel Photovoltaïque. https://www.pvsyst.com/. Accessed 29 Sept 2021

31. Méndez, J.I.: Microgrid – Tec CCM. https://view.genial.ly/62848533c151340012d988e8/int eractive-content-microgrid-tec-ccm. Accessed 24 May 2022

32. SECTEI, Tecnológico de Monterrey: "Bootcamp De Emprendimiento Científico Y Tecnológi-co": Modalidad En Línea. https://www.ingenieria.unam.mx/planeacion/eg/documentos/SEC TEI_Bootcamp.pdf (2021)

33. Sadiku, M.N.O., Adebo, P.O., Musa, S.M., Ajayi-Majebi, A.: Nanogrid: An introduction. Int. J. Eng. Res. Technol. 10 (2021)

34. Smart City Villach. https://www.smartcities.at/city-projects/smart-cities-en-us/vision-step-i-en-us/. Accessed 28 Nov 2020

Product Re-identification System in Fully Automated Defect Detection

Chenggui Sun[1], Li Bin Song[1], and Lihang Ying[2]

[1] Department of Computing Science, University of Alberta,
Edmonton, Alberta T6G 2E8, Canada
{chenggui,libin3}@ualberta.ca
[2] ZeroBox Inc., Edmonton, Alberta, Canada
leo@zerobox.ai

Abstract. Product re-identification (Re-ID) is one of the essential functions of the fully automated product defect detection system and has its unique challenges than other re-identification problems (such as person Re-ID). It needs to identify a nearly identical object and decide its product category. Our proposed product Re-ID system includes two key parts: neural networks for image feature extraction, and feature search and retrieval engine. We used Vearch as the feature search and retrieval engine, and extended AlexNet to design a novel AlphaAlexNet for feature extraction, which improved the accuracy of the product Re-ID. We created and used a dataset that consists of 400 images of 18 types of water bottles. Although this dataset is small, our proposed solution shows inspiring results and demonstrates its capability to solve the product Re-ID problem.

Keywords: Product re-identification · Feature extraction · Image search

1 Introduction

An image search/retrieval system is a software system to find similar images to a query image from a set of image databases. Such systems have been widely used and studied in E-commerce, fashion industries, and person Re-ID [1–4]. Image search is a query-based image similarity matching technique. This technique is evaluated by three major metrics: accuracy, performance on responding time, and scalability. An image search system has two major sub-processes: query processes and indexing processes. In the query process, an image is given to the system and the system returns one or multiple images that are similar to the query image. The indexing process is usually a batch process. It loads the features of multiple pre-existing images to the database. By combining the query processes and indexing processes, an image search system can return similar images as a response to a user's request by extracting images' features using images' index numbers. Figure 1 demonstrated a generalized image search system based on queries [5]. As it shows, images are preprocessed for further analysis at first. Then, feature vectors are extracted, and when it is necessary, dimension reductions are performed. The obtained indexes of images are applied to perform

S. Berretti and G.-M. Su (Eds.): ICSM 2022, LNCS 13497, pp. 144–156, 2022.
https://doi.org/10.1007/978-3-031-22061-6_11

feature matching by measuring the similarity distances between the query image and the store images in the dataset. In the end, the image(s) that matches the query image are displayed [6].

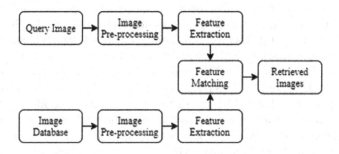

Fig. 1. Scheme of image search process

Image features are generally the visual descriptor of images to humans and include color, shape, text, and face features [7]. Color is one of the most used visual features of images. Color images are typically represented by RGB (Red, Green, Blue) space. The distributions of colors in RGB space in images can be represented by histograms, in which one histogram is assigned to each pixel value. The quantities of histograms' bins determine the color quantization. The spatial relations of pixels should also be used in conjunction with color features. Shape feature includes edges, contours, which can be achieved by various methods. Image texture refers to the visual patterns, which have properties of homogeneity or arrangement of colors and their intensity. Structural and statistical representation methods have been used for texture representation in computer vision.

In this work, a product identification system was developed based on Vearch, which is a visual search system developed by JD E-commerce Platform [8]. The main contributions of our work are:

– Introduced a feature extraction and feature-distance-based image search method to perform product identification. The features were extracted using pre-trained VGG16 and AlexNet, then fed to image search engine - Vearch to perform image indexing and search;
– Developed new AlphaAlexNet for feature extraction. AlphaAlexNet is an improved model with more channels that was developed on top of AlexNet. The deployment of AlphaAlexNet could improve the product identification accuracy.

2   Related Work

For a fully automated production Re-ID system, the main tasks are to build an image search/retrieval engine and extract features. In the domain of object identification, much research on person identification and fashion recommendation systems has been conducted. Most of these methods are built upon Convolution Neural Networks (CNNs). During the past few years, the application of Vision Transformer (ViT) in image classification and identification has been attracting more and more attentions. Wang et al. proposed a new Pyramid Vision Transformer (PVT) method that overcomes the ViT's problem in producing low-resolution outputs, high computation and RAM costs [9]. In their experiment, the PVT method shows promising results on object detection and semantic and instance segmentation tasks with better performance. Sharma et al. highlight their work on top of the ViT and PVT research [10]. They proposed a Locally Aware Transformer (LA-Transformer) method and applied it on person Re-ID. The person Re-ID problem needs to detect a person and identify the same person from images from various sources (e.g., independent security cameras). This research produced excellent results on person Re-ID. Jeff Johnson et al. proposed a design of better-utilizing GPUs to perform similarity searches [11]. In their paper, they proposed not only a k-selection algorithm that can operate in fast register memory and be fusible with other kernels as a result of its flexibility but also a near-optimal algorithmic layout for exact and approximate k-nearest neighbor search. They applied their design for brute-force, approximate and compressed-domain search based on product quantization in different similarity search scenarios. Their algorithm and algorithmic layout enabled better utilization of GPU computation, which outperformed previous art by a large margin. F. Li et al. introduced a solution for e-commerce platforms to classify and search productions based on images [4]. They trained neural networks to conduct image classification by using transfer learning with pre-trained models such as VGG19 and using autoencoders and cosine similarity to search similar images.

The object identification can also be conducted with image search engine. The image search workflow involves two subprocesses, indexing and matching. Facebook AI team created a library - Facebook AI Similarity Search (Faiss) [12]. They proposed to combine graph traversal and compact representations to accomplish the similarity search tasks. This approach requires less memory and lower computation costs. It can handle billions of image indexing and searching with a single machine. Based on Faiss, JD E-commerce Platform developed Vearch, which can provide similar production recommendations on top of over 100 billion product images. Vearch has a high-efficiency, scalable distributed system for efficient similarity search architecture [8].

The open source implementation of Vearch integrated pre-trained VGG and AlexNet as feature extraction methods. There are many approaches to extract features, including low-level local features (such as edges and corners), and global features [13]. Among these approaches, deep learning based-methods such as VGG and AlexNet, have been very popular and powerful. A deep learning method is essentially a neural network. The output of its previous layer is the input of the next layer, and these layers act as feature extractors. A deep learning model is an end-to-end learning model, its fully trained layers can perform much better feature extractions [14]. Besides AlexNet and VGGNet, ResNet and U-net are the other two most popular neural networks for feature extraction [15–17].

AlexNet consists of three types of layers: five convolutional layers, three fully-connected layers with a final 1000-way softmax, max-pooling layers that follow some of the convolutional layers. Its 60 million parameters and 650,000 neurons make overfitting become a concern. In order to reduce overfitting in the fully-connected layers, AlexNet employed "dropout" as regularization [16]. Lv et al. proposed DMS-Robust AlexNet, a neural network designed based on AlexNet architecture, to recognize maize leaf disease [18]. Their network combined dilated convolution and multi-scale convolution to improve its capability to extract features, used batch normalization to prevent over-fitting, and PRelu activation function and Adabound optimizer to improve both convergence and accuracy. Wang and Han proposed a modified AlexNet convolutional neural network for face feature point detection [19]. This modified network has 4 convolution layers and 3 fully connected layers. It removes one layer of convolutional layers of traditional networks and adds the Batch-Normalization layers. The model's input is three sub-images of a face image, which is divided by the model, overlap with each other and have a color channel of their own. The model output the coordinates of the face feature points. Yuan and Zhang used AlexNet and Caffer framework to extract features for image retrieval [20]. The Inria Holidays and Oxford Buildings datasets were used for their experiments, which revealed that the fusion feature could contribute to the improvement of image retrieval.

VGGNet addresses the very important aspect of ConvNet architecture design - depth. It pushes the depth of the network to 16–19 weight layers by adding more (3×3) convolution filters in all layers, resulting in significantly more accurate ConvNet architectures [17]. Khaireddin1 and Chen adopted the VGGNet architecture to perform facial emotion recognition on the FER2013 dataset [21]. Their network has 4 convolutional stages that consist of two convolutional blocks and a max-pooling layer individually, and 3 fully connected layers. The convolutional stages extract features for training the fully connected layers to classify the inputs. Their model achieves state-of-the-art single-network accuracy of 73.28% on FER2013 by tuning the model and its hyperparameters. Zhou et ai. proposed to combine different granularity features from the block1, block2, block3, block4, and block5 in VGG to construct a new network architecture [22]. A local fully connected layer was added after each block to reduce the dimensionality of the features. The combined five different granularity features are fed to the first of three global fully connected layers as input. By doing this, the information flows from a lower layer directly to a fully connected layer and the feature reuse can be increased. Their network architecture also reduces the number of parameters by removing some neurons in two global fully connected layers. This architecture was examined by using CIFAR-10 and MNIST datasets and achieved better performance than traditional VGGs.

ResNet introduces a deep residual learning framework to address the degradation problem and shortcut connections to simply perform identity mapping, which adds neither extra parameter nor computational complexity. It's easy to optimize and gain the increases of accuracy from greatly increased depth [15]. Corbishley et al. utilized the ResNet-152 CNN with super-fine attributes to recognize human attributes from surveillance video for person Re-ID [23]. After re-annotating gender, age and ethnicity of images from an amalgamation of 10 Re-ID datasets - PETA, their work performed significantly better to retrieve images than conventional binary labels did: a 11.2 and 14.8% mAP improvement for gender and age, further surpassed by ethnicity. Kwok, S. presented a framework that made use of Inception-Resnet-v2 to classify breast cancer whole slide images into regions of: normal tissue, benign lesion, in-situ carcinoma and invasive carcinoma [24]. His work won first place in ICIAR 2018 Grand Challenge on Breast Cancer Histology Images. Chen et al. proposed a modularized Dual Path Network (DPN), which consists of Residual Network (ResNet) and Densely Convolutional Network (DenseNet), for image classification [25]. This DPN architecture can reuse and explore new features with ResNet and DenseNet. It performed better than DenseNet and ResNet alone on the ImagNet-1k dataset, the PASCAL VOC detection dataset, and the PASCAL VOC. Huang et al. used this ResNet-based DPN architecture to extract features and classify breast cancer images and achieved great results [26].

3   Proposed Method

In this project, the combination of Vearch and VGG16, AlexNet and AlphaAlexNet is employed to perform the image similarity search.

3.1   Dataset

Eighteen classes of water bottle images from production surveillance videos of a factory with the help of Zerobox Inc. In order to reduce the complexity of experiments, 17 classes of bottles didn't have logos on them, except one class. The images in Fig. 2 demonstrate the 18 types of water bottles used in our research. These images only illustrate the colors and shape of the bottles, not the ratio of their actual size for visual purposes. The production backgrounds of these bottles have been removed in our pre-process stage.

3.2   Method

Our feature-distance-based image similarity search works as follows and is illustrated in Fig. 3:

- Use neural networks to extract features;
- Store extracted features in Vearch;
- Send query/test images from dataset or cameras to feature extraction models;
- Send extracted features to Vearch to perform an image search.

The tested neural networks for feature extraction are listed as follows:

- VGG16: the pre-trained VGG16 was used to extract features from input images;

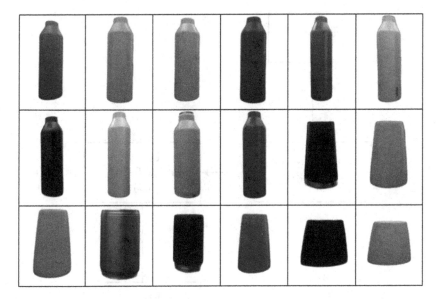

Fig. 2. Bottle Images Provided by Zerobox Inc. The colors of the bottles are (from right to left, from top to bottom): babyblue01, white02, beige01, red01, lavender01, yellow02, blue, white03, white, red02, black tumbler, yellow03, white01, silver, black bottles, babyblue02, black cup, white cup. (Color figure online)

- AlexNet: the pre-trained AlexNet was used to extract features from input images;
- AlphaAlexNet: six extra alpha channels were added to basic AlexNet to help the model to put more weight on shapes and colors. The definition of alpha follows the convention of the image alpha channels.

Vearch, a visual search system developed by JD E-commerce Platform, is used as the image search similarity engine. Vearch consists of two subsystems to perform indexing and search [8]. Indexing is the core subsystem of Vearch and stores extracted image features as image databases. In the original Vearch system, the indexes for all the images are periodically built. The product update events, which include the addition, deletion, and modification of a product image, can trigger Vearch to update the indexes immediately. In our design, if a new image is detected, it will be sent to the indexing process and added to the image database. Another subsystem of Vearch is the search subsystem, which has three key components: Blender, Broker, and Searcher. When a query from a user is forwarded to one of the blenders. The blender will send the query to each broker to contact a subset of searchers to search similar images from a partition of the entire image set in parallel. The top k most similar images will be sent back to the requesting broker, which combines the results from its searchers then sends them to the blender. The combined results will be ranked by the blender and be returned to the user. Vearch's three-level architecture makes it scalable to different levels of tasks of image indexing and searches.

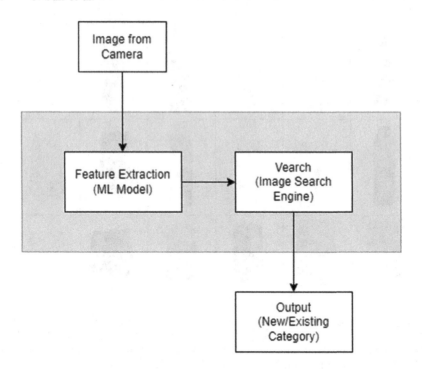

Fig. 3. Layout of bottle Re-ID (query) system

4   Results and Discussion

Our feature-distance-based image search is completed by using pre-trained VGG16, pre-trained AlexNet and our new model - AlphaAlexNet to extract image features, storing the indexed features in the Vearch and using Vearch to perform image similarity search.

The performance of pre-trained VGG16, pre-trained AlexNet, and AlphaAlexNet on 18 bottle identifications are evaluated by confusion matrices. The detailed results about confusion matrices are shown in Table 1 to 3. In these tables, the column on the left lists the actual color of each bottle class and the top row lists the predicted colors. The descriptions of the classes of bottles are listed under each table caption. The number after each class name indicates the minor differences in the colors of bottles. For example, white01, white02, and white03 mean that the colors of three classes of bottles are generally white, however, there are minor differences among those three white colors. The correctly identified numbers of bottles of each class is on the diagonal of each matrix and marked by yellow color, the false identification results are marked by red color.

From Table 1, it can be found that there 3 classes were mislabeled as other classes by the combination of pre-trained VGG16 and Vearch among 25 test bottles:

– one white01 bottle was mislabeled as a babyblue02 bottle,

– one white02 was mislabeled as a babyblue01 bottle,
– one white03 bottle was mislabeled as a beige01 bottle.

Table 1. Confusion Matrix of Image Identification: Pre-trained VGG16 + Vearch. *Classes C01 - C18: 1 Babyblue01, 2 Babyblue02, 3 Beige01, 4 Black bottle, 5 Black cup, 6 Black tumbler, 7 blue, 8 lavender01, 9 red01, 10 red02, 11 silver, 12 white, 13 white01, 14 white02, 15 white03, 16 white cup, 17 yellow02, 18 yellow03*

No.	C1	C2	C3	C4	C5	C6	C7	C8	C9	C10	C11	C12	C13	C14	C15	C16	C17	C18
C1	1	0	0	0	0	0	0	0	0	0	0	0	0	0	0	0	0	0
C2	0	1	0	0	0	0	0	0	0	0	0	0	0	0	0	0	0	0
C3	0	0	2	0	0	0	0	0	0	0	0	0	0	0	0	0	0	0
C4	0	0	0	4	0	0	0	0	0	0	0	0	0	0	0	0	0	0
C5	0	0	0	0	1	0	0	0	0	0	0	0	0	0	0	0	0	0
C6	0	0	0	0	0	1	0	0	0	0	0	0	0	0	0	0	0	0
C7	0	0	0	0	0	0	1	0	0	0	0	0	0	0	0	0	0	0
C8	0	0	0	0	0	0	0	1	0	0	0	0	0	0	0	0	0	0
C9	0	0	0	0	0	0	0	0	2	0	0	0	0	0	0	0	0	0
C10	0	0	0	0	0	0	0	0	0	2	0	0	0	0	0	0	0	0
C11	0	0	0	0	0	0	0	0	0	0	1	0	0	0	0	0	0	0
C12	0	0	0	0	0	0	0	0	0	0	0	2	0	0	0	0	0	0
C13	0	1	0	0	0	0	0	0	0	0	0	0	0	0	0	0	0	0
C14	1	0	0	0	0	0	0	0	0	0	0	0	0	0	0	0	0	0
C15	0	0	1	0	0	0	0	0	0	0	0	0	0	0	0	0	0	0
C16	0	0	0	0	0	0	0	0	0	0	0	0	0	0	1	0	0
C17	0	0	0	0	0	0	0	0	0	0	0	0	0	0	0	1	0
C18	0	0	0	0	0	0	0	0	0	0	0	0	0	0	0	0	1

Table 2 shows that there were 4 classes that were mislabeled as other classes by the combination of pre-trained AlexNet and Vearch among 25 test bottles:

– one beige01 bottle was mislabeled as a white02 bottle,
– one white01 bottle was mislabeled as a babyblue02 bottle,
– one white02 bottle was mislabeled as a babyblue01 bottle,
– one white03 bottle was mislabeled as a white02 bottle.

Table 3 shows that there were 3 classes that were mislabeled as other classes by the combination of AlphaAlexNet and Vearch among 25 test bottles:

– one black bottle was labeled as a silver bottle,
– one white01 bottle was labeled as a babyblue02 bottle,
– one white03 bottle was labeled as a white02 bottle.

Table 2. Confusion Matrix of Image Identification: Pre-trained AlexNet + Vearch. *Classes C01 - C18: as listed in Table 1*

No.	C1	C2	C3	C4	C5	C6	C7	C8	C9	C10	C11	C12	C13	C14	C15	C16	C17	C18
C1	1	0	0	0	0	0	0	0	0	0	0	0	0	0	0	0	0	0
C2	0	1	0	0	0	0	0	0	0	0	0	0	0	0	0	0	0	0
C3	0	0	1	0	0	0	0	0	0	0	0	0	0	1	0	0	0	0
C4	0	0	0	4	0	0	0	0	0	0	0	0	0	0	0	0	0	0
C5	0	0	0	0	1	0	0	0	0	0	0	0	0	0	0	0	0	0
C6	0	0	0	0	0	1	0	0	0	0	0	0	0	0	0	0	0	0
C7	0	0	0	0	0	0	1	0	0	0	0	0	0	0	0	0	0	0
C8	0	0	0	0	0	0	0	1	0	0	0	0	0	0	0	0	0	0
C9	0	0	0	0	0	0	0	0	2	0	0	0	0	0	0	0	0	0
C10	0	0	0	0	0	0	0	0	0	2	0	0	0	0	0	0	0	0
C11	0	0	0	0	0	0	0	0	0	0	1	0	0	0	0	0	0	0
C12	0	0	0	0	0	0	0	0	0	0	0	2	0	0	0	0	0	0
C13	0	1	0	0	0	0	0	0	0	0	0	0	0	0	0	0	0	0
C14	1	0	0	0	0	0	0	0	0	0	0	0	0	0	0	0	0	0
C15	0	0	0	0	0	0	0	0	0	0	0	0	0	1	0	0	0	0
C16	0	0	0	0	0	0	0	0	0	0	0	0	0	0	0	1	0	0
C17	0	0	0	0	0	0	0	0	0	0	0	0	0	0	0	0	1	0
C18	0	0	0	0	0	0	0	0	0	0	0	0	0	0	0	0	0	1

Table 3. Confusion Matrix of Image Identification: AlphaAlexNet + Vearch. *Classes C01 - C18: as listed in Table 1*

No.	C1	C2	C3	C4	C5	C6	C7	C8	C9	C10	C11	C12	C13	C14	C15	C16	C17	C18
C1	1	0	0	0	0	0	0	0	0	0	0	0	0	0	0	0	0	0
C2	0	1	0	0	0	0	0	0	0	0	0	0	0	0	0	0	0	0
C3	0	0	2	0	0	0	0	0	0	0	0	0	0	0	0	0	0	0
C4	0	0	0	4	0	0	0	0	0	0	0	0	0	0	0	0	0	0
C5	0	0	0	0	1	0	0	0	0	0	0	0	0	0	0	0	0	0
C6	0	0	0	0	0	0	0	0	0	0	1	0	0	0	0	0	0	0
C7	0	0	0	0	0	0	1	0	0	0	0	0	0	0	0	0	0	0
C8	0	0	0	0	0	0	0	1	0	0	0	0	0	0	0	0	0	0
C9	0	0	0	0	0	0	0	0	2	0	0	0	0	0	0	0	0	0
C10	0	0	0	0	0	0	0	0	0	2	0	0	0	0	0	0	0	0
C11	0	0	0	0	0	0	0	0	0	0	1	0	0	0	0	0	0	0
C12	0	0	0	0	0	0	0	0	0	0	0	2	0	0	0	0	0	0
C13	0	1	0	0	0	0	0	0	0	0	0	0	0	0	0	0	0	0
C14	0	0	0	0	0	0	0	0	0	0	0	0	0	1	0	0	0	0
C15	0	0	0	0	0	0	0	0	0	0	0	0	0	1	0	0	0	0
C16	0	0	0	0	0	0	0	0	0	0	0	0	0	0	0	1	0	0
C17	0	0	0	0	0	0	0	0	0	0	0	0	0	0	0	0	1	0
C18	0	0	0	0	0	0	0	0	0	0	0	0	0	0	0	0	0	1

From the mislabeled bottles, it can be observed that it's very difficult for neural networks to distinguish bottles that have close colors and same shapes. For example, the shapes of white02 and white03 were almost the same, their colors were very close; the shapes of white01 and babyblue02 were very close. Because the added six extra alpha channels put more weight on shapes and colors, the combination of AlphaAlexNet and Vearch had better performance to match similar images.

Table 4. Identification accuracy of vearch with different feature extraction methods

Feature Extraction	Pre-trained VGG16	Pre-trained AlexNet	AlphaAlexNet
Identification Accuracy	88%	84%	88%

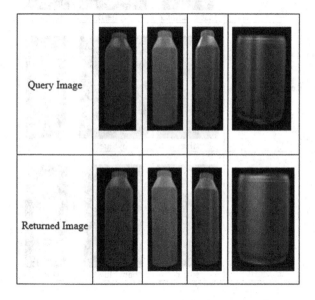

Fig. 4. Real search examples - correct return

The overall identification accuracies of each neural networks are summarized in Table 4. When the image features extracted by the pre-trained VGG16 were fed to Vearch, Vearch performed much better to match images than what it did when the pre-trained AlexNet was used. Its accuracy value is 88% which is higher than the 84% of pre-trained AlexNet. This might result from the fact that VGG has more layers [27]. When the AlphaAlexNet was used to extract image features, Vearch had an accuracy of 88% in image similarity search. In the AlphaAlexNet, 6 extra alpha channels were added to the model. These 6 channels of information came from inverse colors and rotations of images. This helped make the model distinguish close colors and less sensitive to the small variations of shapes. The performance of the AlphaAlexNet with Vearch was also illustrated here with search samples. Figure 4 shows the correctly returned images when a query image was sent to Vearch. Figure 5 shows the false search results when a query image was sent to Vearch. The false returned search results indicate that it's a great challenge to search the correct images if the shape or the colors of the water bottles were close.

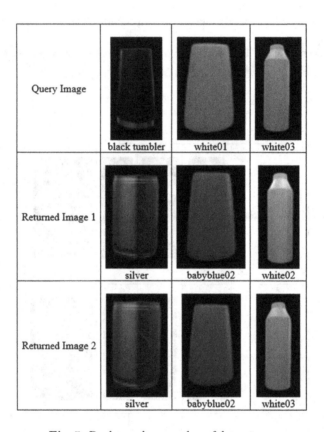

Fig. 5. Real search examples - false return

The results shown in Table 4 and illustrated in Fig. 4 and 5 demonstrate that the combination of Vearch and neural networks didn't perform perfectly or reach our expectations in terms of search accuracy. This should have resulted from the fact that the subtle differences among the color and shape of bottles are not well captured by neural networks. However, the AlphaAlexNet model, which was developed based on existing AlexNet, could improve Vearch's similarity search performance. This implies that the combination of Vearch and neural networks have the potential to have much better accuracy if more efficient neural networks could be developed to extract features or bigger datasets could be obtained to train the neural networks. The potential of Veach's commercialization in the product Re-ID could also be supported by the work of JD e-commerce [8]. The combination of Vearch and neural networks has many advantages over neural network classifications. It can perform image search at a very quick speed, and the speed doesn't deteriorate significantly when dataset size and object classes increase. In future, more work can be done to improve the neural network for better identification accuracy.

5 Conclusion

In this paper, we investigated solutions to tackle a nearly identical object Re-ID problem, which needs to identify an object and determine its product category. We created and used a dataset consisting of 400 images of 18 different types of water bottles. The combination of image feature extraction neural networks, such as AlexNet and VGG16 and image search engine - Vearch, were used to perform Re-ID of objects. Our new model - AlphaAlexNet, an improved model based on AlexNet, could improve the object identification accuracy of Vearch. In future, more efficient feature extraction neural networks can be developed with a focus on the color and shape features of objects that have simple geometric and textural properties.

References

1. Wieczorek, M., Rychalska, B., Dabrowski, J.: On the unreasonable effectiveness of centroids in image retrieval. In: International Conference on Neural Information Processing, pp. 212–223. Springer, Cham (2021). https://doi.org/10.1007/978-3-030-92273-3_18
2. Liu, Z., Yan, S., Luo, P., et al.: Fashion landmark detection in the wild. In: European Conference on Computer Vision, pp. 229–245. Springer, Cham (2016). https://doi.org/10.1007/978-3-319-46475-6_15
3. Park, S., Shin, M., Ham, S., et al.: Study on fashion image retrieval methods for efficient fashion visual search. In: Proceedings of the IEEE/CVF Conference on Computer Vision and Pattern Recognition Workshops (2019)
4. Fengzi, L., Kant, S., Araki, S., et al.: Neural Networks for Fashion Image Classification and Visual Search. Available at SSRN 3602664 (2020)
5. Zhao, J., Sihao, Z., Jing, Z.: Review of the sparse coding and the applications on image retrieval. In: 2016 International Conference on Communication and Electronics Systems (ICCES), pp. 1–5. IEEE (2016)
6. Alaei, F., Alaei, A., Blumenstein, M., et al.: A brief review of document image retrieval methods: recent advances. In: 2016 International Joint Conference on Neural Networks (IJCNN), pp. 3500–3507. IEEE (2016)
7. Oussalah M.: Content based image retrieval: review of state of art and future directions[C]//2008 First Workshops on Image Processing Theory, Tools and Applications. IEEE, 1–10 (2008)
8. Li, J., Liu, H., Gui, C., et al.: The design and implementation of a real time visual search system on JD e-commerce platform. In: Proceedings of the 19th International Middleware Conference Industry, pp. 9–16 (2018)

9. Wang, W., Xie, E., Li, X., et al.: Pyramid vision transformer: a versatile backbone for dense prediction without convolutions. In: Proceedings of the IEEE/CVF International Conference on Computer Vision, pp. 568–578 (2021)

10. Sharma, C., Kapil, S.R., Chapman, D.: Person re-identification with a locally aware transformer. arXiv preprint arXiv:2106.03720 (2021)

11. Johnson, J., Douze, M., Jégou, H.: Billion-scale similarity search with GPUs. IEEE Trans. Big Data **7**(3), 535–547 (2019)

12. Douze, M., Sablayrolles, A., Jégou, H.: Link and code: fast indexing with graphs and compact regression codes. In: Proceedings of the IEEE Conference on Computer Vision and Pattern Recognition, pp. 3646–654 (2018)

13. Nixon, M., Aguado, A.: Feature extraction and image processing for computer vision. Academic Press (2019)

14. Li, X., Li, C., Rahaman, M.M., et al.: A comprehensive review of computer-aided whole-slide image analysis: from datasets to feature extraction, segmentation, classification and detection approaches. Artif. Intell. Rev., 1–70 (2022). https://doi.org/10.1007/s10462-021-10121-0

15. He, K., Zhang, X., Ren, S., et al.: Deep residual learning for image recognition. In: Proceedings of the IEEE Conference on Computer Vision and Pattern Recognition, pp. 770–778 (2016)

16. Krizhevsky, A., Sutskever, I., Hinton, G.E.: ImageNet classification with deep convolutional neural networks. Adv. Neural Inf. Process. Syst. **25** (2012)

17. Simonyan, K., Zisserman, A.: Very deep convolutional networks for large-scale image recognition. arXiv preprint arXiv:1409.1556 (2014)

18. Lv, M., Zhou, G., He, M., et al.: Maize leaf disease identification based on feature enhancement and DMS-robust Alexnet. IEEE Access **8**, 57952–57966 (2020)

19. Huai, W., Zhuo, H.: An improved AlexNet model with multi-channel input images processing for human face feature points detection. In: 2020 12th International Conference on Communication Software and Networks (ICCSN), pp. 246–251. IEEE (2020)

20. Yuan, Z.W., Zhang, J.: Feature extraction and image retrieval based on AlexNet. In: Eighth International Conference on Digital Image Processing (ICDIP 2016). International Society for Optics and Photonics, 10033, 100330E (2016)

21. Khaireddin, Y., Chen, Z.: Facial emotion recognition: State of the art performance on FER2013. arXiv preprint arXiv:2105.03588 (2021)

22. Zhou, Y., Chang, H., Lu, Y., et al.: Improving the performance of VGG through different granularity feature combinations. IEEE Access **9**, 26208–26220 (2020)

23. Martinho-Corbishley, D., Nixon, M.S., Carter, J.N.: Super-fine attributes with crowd prototyping. IEEE Trans. Pattern Anal. Mach. Intell. **41**(6), 1486–1500 (2018)

24. Kwok, S.: Multiclass classification of breast cancer in whole-slide images. In: International Conference Image Analysis and Recognition, pp. 931–940. Springer, Cham (2018). https://doi.org/10.1007/978-3-319-93000-8_106

25. Chen, Y., Li, J., Xiao, H., et al.: Dual path networks. Adv. Neural Inf. Process. Syst. **30** (2017)

26. Huang, C.H., Brodbeck, J., Dimaano, N.M., et al.: Automated breast cancer image classification based on integration of noisy-and model and fully connected network. In: International Conference Image Analysis and Recognition, pp. 923–930. Springer, Cham (2018). https://doi.org/10.1007/978-3-319-93000-8_105

27. Simonyan, K., Zisserman, A.: Very deep convolutional networks for large-scale image recognition. arXiv preprint arXiv:1409.1556 (2014)

Multimedia for Medicine and Health-Care

A Real-Time Fall Classification Model Based on Frame Series Motion Deformation

Nasim Hajari[1]([✉]) and Irene Cheng[2]

[1] Concordia University of Edmonton, Edmonton, Canada
nasim.hajari@concordia.ab.ca
[2] University of Alberta, Edmonton, Canada
locheng@ualberta.ca

Abstract. Fall is common among the elderly and patients with severe health conditions. It can be life threatening, especially if the person does not received the required help to recover from fall on time. Therefore, an automatic real-time fall detection system is very desirable and has been the focus of research in the last couple of years. Traditional computer vision (CV) based fall detection systems require less infrastructure, is cheaper and can be more efficient comparing to systems based on wearable sensors. The robustness and efficiency of CV techniques depend on the extracted feature set from the surveillance video sequences. Generally, the efficiency and accuracy of more recent learning based CV systems rely heavily on the statistical characteristics of training dataset. Acquiring a balanced, comprehensive and representative training data that covers all the necessary aspects of the problem, including viewing direction of the camera and illumination condition of the environment is quite challenging. The problem would be even more serious when the training dataset does not have representative features as the surveillance area. In this paper, we propose a robust, real-time, CV based fall detection technique that can work in different settings. The propose system requires only a single affordable RGB camera. The proposed method works at frame level and only uses two significant feature points for classification, therefore occlusion would not influence the system. We performed experiments on different publicly available datasets such as le2i, UR and multiple camera fall detection datasets. The result shows that the proposed technique can distinguish fall from everyday activities, e.g., sitting down and sleeping and has a higher accuracy, recall and specificity comparing to other methods. The proposed method performs well in indoor environments with different lighting conditions and different viewing directions of the camera.

Keywords: Human fall detection · Action recognition · Computer vision

1 Introduction

The life expectancy of senior people has been increased due to advancements in health care services. As World Health Organization (WHO) [1] and Carone and

Costello [7] suggest, the world population over 60 years old will double by 2050, which will be a total number of 2 billions. It is obvious that all countries will face the impacts of this huge growth and should adjust their health and social systems accordingly. The priority for senior population is living safely and independently either in their homes or in assisted care facilities. A major threat with severe physical and emotional consequence on an elderly is accidental fall [19]. This is because of their deteriorating motor control. The situation can be worse if the person is alone and cannot seek help on time. Therefore designing real-time and reliable fall detection methods are gaining increasing attention during the past decade.

Fall detection approaches can be classified into three main groups based on the underlying platform [31]. The first group of techniques use wearable devices [9,34]. Sensors such as accelometers or gyroscopes are either embedded in clothing or some specific devices such as smart watches. As the sensors are in close contact with the users, the detection is very accurate. However, this approach often becomes ineffective if the user forgets to put on or for other reason, e.g., dead battery, the device stops working. Besides, not all elder people like to wearing specific devices.

The second group of fall detection techniques use ambient devices such as presure sensors on the floor [22,43] or in the living environment. It is quite costly to install or modify these sensors to cover the whole surveillance area. Research shows that this approach can easilly generates false positives [31].

The third group use computer vision-based approaches. These techniques can use a single RGB camera, multiple cameras or depth cameras such as Kinect, and extract a set of features to detect fall [31]. More about vision-based techniques will be presented in Sect. 2. Vision-based approaches have advantages over the previous two groups of techniques. First, the hardware (camera) is easy and inexpensive to set up. One or a set of cameras are used for the whole monitoring region. Second, it is not the responsibility of the user to wear any extra devices. However, most people do not like to be monitored by cameras in their living space due to privacy issues. A solution to this problem is automatic blurring and reduction to a compact form, without visual appearance or disclosing physical identity. Only when emergency situation occurs, necessary data is disclosed to authorized personnel. In emergency situations, an alert will be sent to the appropriate receiver automatically and the person in danger can receive the required help on time.

This paper presents a vision-based fall detection technique using a single RGB camera. The proposed method detects target motions in each video frame and classifies a small segment of frames into fall or non-fall. The technique is suitable for monitoring rooms in long-term care facilities, where seniors spend most of their time as residents carrying out daily activities. The rest of this paper is organized as follow: related work on vision-based fall detection techniques is presented in Sect. 2. Detail of the proposed method is explained in Sect. 3. Section 4 describes the experimental setup and the results of our proposed method. The last section concludes our work.

2   Literature Review

The research on object detection, target tracking, pose estimation and action recognition has been well established in the field of computer vision. State-of-the-art methods include person detection based on HOG features (Histogram of Orientation Gradient) [11] and Pictorial Structure [3]. A common characteristics of existing person detection techniques focus on people in an upright position. Detecting fall requires a different algorithm design and is more challenging than simply detecting a person.

Vision-based fall detection algorithms can be classified into two different types: single view, and multi-view with depth information.

2.1   Single View

Some techniques used a single RGB camera to capture the surveillance area. Most of them generate a bounding box of the moving person, extract the associated features and then classify the action based on the extracted features.

Rougier et al. [37] analyzed human shape deformation through video sequence by studying the person's silhouette sequence through out the motion. They used a shape matching approach to track the silhouette of the person. The idea behind their approach is that during fall the changes in human silhouette will occur rapidly. The authors classified an action into normal or abnormal activity using Gaussian Mixture Model (GMM). The weakness of this approach is that fast silhouette changes can be caused by other actions intentional movements. They mentioned in their paper that this approach could run at 5 frames/sec, which is not adequate for a real-time surveillance application.

The fall detection method proposed by [42] is also based on a single camera. They analyzed the silhouette of the person by subtracting the background using a code book model. This implies the need of a sufficiently large training dataset, to detect foreground accurately. For surveillance applications, specially in smart homes, this is not feasible due to substantial combinations of background objects. The authors described the posture of the person in 2D environment using fitted ellipse and shape structure.

Another single camera approach is proposed by [30]. They incorporated the velocity feature as well as the posture features, to make the detection more robust. They used shadow changes to compute velocity. However, the shadow of a person may not be visible due to lighting direction. Other elements in the environment, such as reflections, can be mistaken as shadows. Thus, the shadow-based velocity data is not reliable.

[38] assumed a clear model with minimal occlusion of the background to perform background subtraction. They used features that describe the position and the velocity. Head detection is an important process. They estimated head based on the position and orientation of the moving bounding box, which can be distorted if the person is carrying an object, such as a chair, where the width and height of the bounding box cannot accurately represent the person.

Unlike previous methods that directly studied the silhouette and bounding box of the moving object, [10] identified three important points for each person. Based on the position and orientation of the point composition, they detected fall. However, to get these points they need to remove noises and extract the foreground accurately. Another disadvantage of this approach is the lack of velocity analysis.

2.2   Multi-view with Depth Information

The other type of techniques is based on depth camera or multiple cameras to provide higher dimentional information of the surveillance area. Researchers often used Kinect [2,5,16,24,29,41] to get the 3D position of human body parts or joints. Several challenges are associated with using Kinect cameras. Kinect is extremely sensitive to lighting condition, capture range and noise.

[5] used the detected joints from a Kinect camera as features to detect fall. They used SVM to classify the actions into fall or non-fall. [41] also used Kinect camera, but they focused on extracting the torso vector. They detected fall based on torso angle. This approach still suffers from not considering velocity of the action. [29] suggested that tracking head only is adequate for fall detection. They detected head and ground, and calculated the distance between head and floor plane in each frame to predict fall. This approach does not work if the ground is bumpy, has stairs or with objects lying around. Some reserachers such as [16] also fuse video and sound data together to validate the detection.

Applying Convolutional Neural Network (CNN) and deep learning to detect fall has become popular in recent years. [15,17,25,26] used different networks such as RNN, faster R-CNN and so on to detect fall. However, a common problem with these methods is the effort required to create a training dataset. If the training set is not representative or insufficiently large, the result is not accurate. If the test conditions are different from the training data, e.g., different illuminations and viewing directions, the detection will fail. Also, adding new conditions or parameters require retraining the model with the new labelled data. Despite the recent advancement in pose estimation techniques [6,14,40] there are still cases where these approaches fail to accurately estimate the pose, specially if the illumination condition of the environment is poor like a room with dim lighting [18]. This becomes more troublesome to monitor home and senior residence environment in the evening or at night when the rooms can be relatively dark.

3   Proposed Method

Most of the vision-based fall detection techniques in the literature focus on action level [15,17,25,26,29,30]. These techniques require breaking the captured video into smaller action segments and detecting fall in short clips. There are a number of disadvantages with this approach. For example, what are the criteria to divide the video into smaller action clips? How to analyze an action spanning beyond the clip border? These issues suggest that detecting fall at frame level, instead

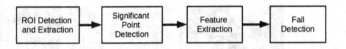

Fig. 1. General pipeline for the proposed fall detection system. The first step is ROI extraction, the second step is head and center of mass detection, the third step is feature extraction and the last step is classification.

of action level, can be a better solution. The frame-level fall detection technique proposed in this paper is real-time. It contains four processing stages as shown in Fig. 1.

The first stage is Region of Interest Detection and Extraction. Different from other single-camera based fall detection techniques [30,37,38,42] and [16], we introduce a new background subtraction algorithm in order to increase the accuracy of person detection and extraction.

The second stage is Significant Points Detection. Only two significant points are needed and they are used at a later stage to compute the necessary set of features for fall detection. We focus on the head, h, and center of mass, c, as demonstrated in our experiments.

The third stage is Feature Extraction based on the two significant points and extracted ROI. There are four computed features in our method including:

- Magnitude of the vector **ch** connecting c and h, denoted by $||\mathbf{ch}||$ in this paper.
- Angle of **ch** with respect to the ground level (horizontal line), represented by $\hat{\mathbf{ch}}$.
- Average angular velocity of **ch**, denoted by $v_{\hat{\mathbf{ch}}}$.
- The width and height ratio of the bounding box, (r_1, r_2), around the region of interest.

By taking in consideration of a number of significant features, including velocity of movement, to analyze the posture of the target, our approach is more robust to false positive as demonstrated in our results.

The last step is Fall Detection. We used unsupervised classification based on thresholding to detect contingent activities, e.g., Falls, which require immediate attention.

3.1 ROI Detection and Extraction

Background subtraction or foreground segmentation is an important step for many computer vision applications, such as surveillance. [21] compares the different techniques proposed in the literature. Based on [21] and our testing, we adopt the Gaussian Mixture Model (GMM) method proposed in [39]. Since a mixture of K Gaussian distribution is used to model the temporal histogram of

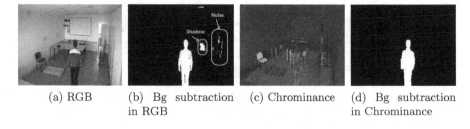

(a) RGB   (b) Bg subtraction   (c) Chrominance   (d) Bg subtraction
                 in RGB   in Chrominance

Fig. 2. Comparison between background subtraction in both RGB space and chrominance space in a frame. The same GMM-based background subtraction is applied on (a) and (c). The chrominance space is more robust to illumination changes.

each pixel throughout the video, this model is very suitable for a dynamic background in fall detection. The GMM method requires a training phase in order to estimate the background, but its execution time is quite fast and is suitable for real-time applications.

The variations in intensity value of each pixel location (i, j) across frames can be due to object movements, background changes or noises present in the scene. For example, changes in illumination due to shadows, highlights or small movements of surrounding objects in the scene caused by unexpected forces, such as wind, which can change the intensity value of a pixel. We use a set of different Gaussian functions to model the intensity value of each pixel. After the training phase, the intensity value of each pixel is compared with the Gaussian mixtures, and the algorithm probabilistically decides whether or not a particular pixel belongs to a moving object.

Although GMM-based background subtraction is suitable for dynamic scenes, it can fail if there are significant undesirable movements. For instance, the shadow of a moving person can be mistaken as a part of foreground. This can be problematic for fall detection, where we are only interested in movement of the person and not the shadow.

To address this problem we seperate the color space into luminance and chrominance components as suggested by [20]. Some color space like HSV or YcbCr gives a better estimation of chromaticity and luminance. Background subtraction in chromaticity will eliminate the detection of movements due to illumination changes. Figure 2(b) shows the result of applying GMM background subtraction on the original RGB frame. The yellow region is the shadow of the person on the wall and the green area shows noises created by changes in illumination caused by the shadow. Figure 2(d) shows that by applying GMM background subtraction on chromaticity component, we can eliminate these unwanted noises.

Another issue is the movements of undesirable objects. The undesirable objects can be smaller or bigger than the target person in the scene. We want to get rid of the undesirable motions and only focus on the person. In other words,

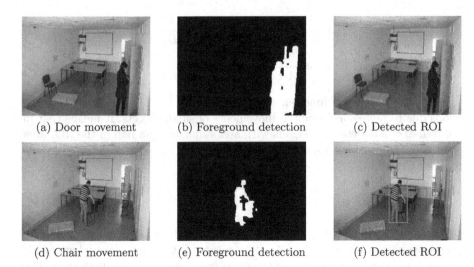

(a) Door movement   (b) Foreground detection   (c) Detected ROI

(d) Chair movement   (e) Foreground detection   (f) Detected ROI

Fig. 3. (c) and (f) show the final results of fusing person detection and background subtraction. As shown by the bounding boxes, the movements of chair and door are accurately excluded from the region of interest in our method.

we are interested in detecting and extracting the Region of Interest (ROI) only, rather than the entire foreground. [38] proposed to remove the unwanted regions by keeping only the region with biggest area. However, this technique fails if some bigger objects are moving in the room such as the door. Therefore we propose a new ROI detection technique using two constraints: area of region and a penalty function. We fuse the background subtraction and person detection results, and remove errors or noises based on a penalty function. Note that person detection and tracking techniques such as those proposed in [11,12], work only on people in upright postures and fail if the person is sitting or lying down. Hence, it is not as robust as our method, which can handle a person in non-upright pose.

Equation 1 summarizes our new ROI detection technique, where PD is the region detected by the person detector, and FD is the region detected by foreground detector. w_1 and w_2 are their corresponding weights, where $0 < w_1 < 1$, $0 < w_2 < 1$ and $w_1 + w_2 = 1$. We call $w_1 PD$, *effective person* and $w_2 FD$ *effective foreground*. ϵ is the penalty function.

$$ROI = (w_1 PD + w_2 FD) - \epsilon \qquad (1)$$

Equations 2 and 3 explain how to calculate the weights. If the algorithm detects no movement in the scene and only detects a person, then w_2 will be set to zero and w_1 will be one. Similarly if it only detects movement and no person, w_1 will be set to zero and w_2 will be one.

$$w_1 = \frac{Area(PD)}{Area(FD) + Area(PD)} \tag{2}$$

$$w_2 = \frac{Area(FD)}{Area(FD) + Area(PD)} \tag{3}$$

In Eq. 1, the penalty function ϵ is computed by Eq. 4. ϵ removes any small regions due to noises or unwanted movements of small objects. It means that in each frame, k, any detected regions with areas smaller than half of the ROI area detected in the previous frame are removed. R is the closed area detected by the proposed technique.

$$\epsilon = (\sum_{i \in R} i < \frac{area_{k-1}}{2}) \tag{4}$$

We used HOG features to extract PD. This basically detects people in upright position as proposed by [11]. It is reasonable to assume that when people are lying down or sitting they don't carry other objects, like walking cane. The formulation is given in Eq. 5, where $\mathbf{X}$ represents the intensity value of pixels $\mathbf{X}$ located in bin c and there are a total of 9 histogram bins.

$$HOG = \sum_{i=1}^{9} (\frac{\partial \mathbf{X}_c}{\partial x}, \frac{\partial \mathbf{X}_c}{\partial y})_i \tag{5}$$

In order to extract FD, we use GMM-based foreground detector proposed by [39]. Each pixel in the image is presented by a mixture of k Gaussian distribution as shown in Eq. 6. The most probable Gaussian distributions belong to the background and the least probable ones represent the foreground.

$$P(\mathbf{X}_t) = \sum_{i=1}^{k} \alpha_i \eta(\mathbf{X}_t \mu_i, \sigma_i) \tag{6}$$

Here α_i is the weight. μ_i and σ_i are the mean and standard deviation of the ith Gaussian distribution.

3.2   Significant Point Detection

In order to detect fall, we need to derive a number of features. Previous studies either used the posture features extracted from the bounding regions ([30, 37, 42]) or analyzed trajectory of some important points or vectors ([10, 29, 36, 38]). In this work, we combine bounding box, trajectory and velocity data of the feature vector. To compute this feature vector, we use head h and center of mass c possitions as two key-points. Note that c is essentiaally the centrre of the bounding box aaround ROI. While having the bounding box information helps in detecting falls, which occur along the viewing direction of the camera, using the velocity information can distinguish falling down from sitting down or lying down. Note that tracking only one key-point, like [36], can create error in 2D space, i.e., video, due to loosing one degree of freedom.

Head Detection. The second key point is head position. Detecting head is a crucial step for surveillance application and as [36] suggested it is very important for fall detection. [28] proposed a head detection approach, which can work on sideview and backview of the head as well. However, their approach is very time consuming and not suitable for a real-time application. Some reserachers suggested that human head can be approximated by circle or eclipse [27,35], and they used different feature extractor to detect imperfect circles or eclipses. Here we use circular Hough Transformation on the detected ROI to find circular curves and potential head position. As [13] proposed, Hough Transformation is based on a voting system and therefore it can detect several imperfect circles with the specified radius. The centres of these circles are the potential positions for the head. Figure 4(a) shows an example of the potential head positions in red crosses. in order to find the actual head position, first we need to filter the set of candidates and remove the positions which are: 1) outside the ROI, or 2) inside the ROI, but more than half of the area of the formed circle is outside the ROI. Now we have a set of final candidate points h'. We use a two-step iterative optimization to find the best head position. The optimization only keeps centres with largest Euclidean distance from c and its reflection point with respect to c is inside the ROI. An example of potential head position and final head position are shown in Fig. 4(a) and Fig. 4(b) respectively. Equations 7 and 8 summarize the optimization process.

(a) Candidate head positions (b) Final head position, side walking (c) Final head positions, perpendicular fall (d) Final head position, back walking

Fig. 4. (a) The candidate locations for had using circular hough transformation. (b,c,d) Final result for head detection with different actions and different people.

$$h = \max_{h'}(\sqrt{(j'_h - j_c)^2 + (i'_h - i_c)^2}) \qquad (7)$$

$$((2i_c - i_h), (2j_c - j_h)) \in ROI \qquad (8)$$

Figure 4 (c,d) shows more examples of head detection applied on different people with different postures. This figure shows that the proposed head detection algorithm is accurate and robust, and is useful for different surveillance applications. It is fast and can work on any position and direction of the head

even in a low resolution video. We have tested our algorithm on two different video resolutions. On the average, it takes 0.017 sec/frame to find the head
position for a lower resolution video of 320 × 240 pixels. However, for a Higher
resolution video of 850 × 480 pixels, it takes 0.048 sec/frame, which is still real-
time. Note that in order to optimize the time performance without affecting the
accuracy, the video should be reduced to an appropriate low resolution without
losing significant features.

3.3   Feature Extraction

The next step is defining distinct features that can capture the characteristics of
falling down. As [32] showed, falling down is a fast action, and its critical phase
is between 300 to 500 milliseconds. Even for cases when a person is gradually
falling down, like from a bending/sitting position or holding onto something
when falling, there still has a fast and critical phase. In this paper we supplement
posture with velocity information to detect fall. We use four different features.
The first feature is the width-height ratio of the ROI bounding box, which is
given in Eq. 9. Almost all the previous vision based fall detectors use bounding
box information, e.g., area or ratio, as a mean of posture feature. It is because
computing bounding box information is simple, fast and reliable for real-time
applications.

$$f_1 = \frac{r_2}{r_1} \quad \text{and} \quad f_1 \in (0, Y] \tag{9}$$

where r_2 is the height and r_1 is the width of the bounding box and Y is the
vertical resolution of the video. The ratio of the bounding box is a real number
between 0 and video vertical resolution. Please note that the upper and lower
bounds are very extreme and will not happen usually. Theoretically the smallest
r_1 is 1 pixel and the biggest r_2 is the vertical resolution of the video. This does
not happen in real cases as the width of the bounding box around the person, r_1,
is always greater than one. Since the camera is usually mounted on the ceiling,
the biggest value for r_2 is not very probable. f_1 greater than 1 mostly represent
an upright posture, whereas f_1 less than 1 can represent a lying posture. An
exception can be falling down in the viewing direction of the camera (z-fall),
where f_1 can be greater than or equal to 1, as can be seen in Fig. 4(e).

To address this issue, we use $\|\mathbf{ch}\|$, which is the vector linking center of
mass and head as explained in Sect. 3.2. The intuition is that due to perspective
projection, the length of the person would not change in standing and perpendicular lying position, assuming the distance of the person to the camera does
not change. However, for all the other lying positions with the same distance
to the camera, the length of the person would be shorter. This feature needs to
be normalized. Equation 10 is used to get the second feature, where $y/2$ is half
of the height of the bounding box generated by the person detector explained
in Sect. 3.1. As discussed before, person detector only detects person in upright
position, when the person if on the lying position thee previous value for $y/2$

will be used. In other words $y/2$ only updates whenever person detector detects an upright person in the scene.

$$f_2 = \frac{||\mathbf{ch}||}{y/2} \quad \text{and} \quad \in [0,1] \tag{10}$$

These two features are not sufficient to capture posture information for different actions such as bending or sitting down. Therefore, we use the angle of significant vector, $\mathbf{ch}$, wrt horizontal line (ground level). Since the direction of the action is not important, we consider this angle to be between $[0, \frac{\pi}{2}]$. This feature is presented in Eq. 11, where (x_c, y_c) and (x_h, y_h) shows the x and y coordinates of the center of mass and head respectively.

$$f_3 = atan(\frac{|y_c - y_h|}{|x_c - x_h|}) \quad \text{and} \quad f_3 \in [0, \frac{\pi}{2}] \tag{11}$$

The last feature captures the velocity of the action. The velocity is crucial to distinguish between intentional lying and falling. We use the average angular velocity of $\mathbf{ch}$ between each five frames. This corresponds to 200 ms or 0.2 s for a 24 frames per second video. This feature is shown in Eq. 12.

$$f_4 = |f_{3_k} - f_{3_{k+5}}| \quad \text{and} \quad f_4 \in [0, \frac{\pi}{2}] \tag{12}$$

3.4   Fall Detection

The last step of the proposed method is detecting fall based on the extracted features. In this paper, we label each frame as *non-fall* or *potential fall* by thresholding on the features. To find the best threshold for each feature we use grid search. If a frame is classified as *potential fall* for five consecutive frames, which is equivalent to 200 ms, then a fall is reported. The reason to choose five frames is to detect faster falls as well as falls that happen slower, due to holding onto something for instance. The summarized conditions for fall is shown in Eq. 13. Angular velocity is an important feature to distinguish fall from other actions. The videos provided in the additional materials show the robustness of our combination of features; even some errors in significant point detection does not affect the result of fall detection.

$$fall = (f_1 < 1.8)and((f_2 < 0.8)or(f_3 < \frac{\pi}{3}))and(f_4 > 15) \tag{13}$$

4   Experiments and Results

Table 1 compares the results of the proposed method with other techniques that applied on the Le2i, UR and Multicam datasets. Please note that we show the average value for accuracy, recall and specificity for Le2i dataset. Also there are not many works that reported specificity and recall.

Table 1. Comparison of the proposed fall detection system with other techniques on Le2i, UR fall detection datasets.

	Le2i	UR
[8] (Ac,Re,Sp)	(99.54%, 99.7%, 97%)	–
[33]	(97%, 99%, –)	(95%, 100%, 92%)
[38]	**(99.61%, –, –)**	–
Proposed method	**(99.36%, 97.93%, 99.67%)**	**(98.8%, 100%, 97.7%)**

Figure 5 shows the result of the proposed method for different scenarios of Le2i dataset and some of the failed cases. As Fig. 5(d,o,s) show, there might be cases where the detected head position is wrong. It can happen for various reasons like head is not completely visible. However, in such cases we can still detect fall or no fall action accurately based on our robust set of features. Also, when the person is in lying position, the enclosed bounding box might be bigger than expected as shown in Fig. 5(a,b). This happens because of the movement of the pad but again our robust features can detect fall accurately. There are situations that our algorithm fails. In fact, all cases of *False Negative* belongs to Z-fall. We can detect Perpendicular-fall and Regular-fall with 100% recall. The proposed method can correctly label not-fall action for almost all activities including bending and scouting.

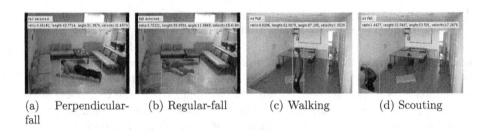

(a) Perpendicular-   (b) Regular-fall   (c) Walking   (d) Scouting
fall

Fig. 5. The result of proposed detection system on Le2i dataset.

5 Conclusion

In this paper, we propose a real-time automatic fall detection system based on motion deformation and velocity of the action. We found that fall detection at frame level is more reliable than at action level due to the challenge in defining the starting point of an action, and breaking a long video into smaller action segments. Our proposed method detects the region of interest (person) through a novel and robust ROI detection technique, associated with significant points extraction (head and center of the mass). We then use the angle and size of the formed vector, as well as its angular velocity, to detect fall based on a set of thresholds obtained by grid search. Experimental results show that our method is accurate and robust to illumination changes and lens distortion. The most frequent false negative happens for Z-fall. In future work, we will use depth information to address the issues for Z-fall.

References

1. World health organization: Global report on falls prevention in older age. www.who.int/ageing/publications/Falls_prevention7March.pdf
2. Alzahrani, M.S., Jarraya, S.K., Salamah, M.A., Ben-Abdallah, H.: FallFree: multiple fall scenario dataset of cane users for monitoring applications using Kinect. In: 2017 13th International Conference on Signal-Image Technology Internet-Based Systems (SITIS), pp. 327–333, December 2017
3. Andriluka, M., Roth, S., Schiele, B.: Pictorial structures revisited: people detection and articulated pose estimation. In: 2009 IEEE Conference on Computer Vision and Pattern Recognition, pp. 1014–1021, June 2009
4. Auvinet, E., Rougier, C., Meunier, J., St-Arnaud, A., Rousseau, J.: Multiple cameras fall dataset. DIRO-Université de Montréal, Technical report 1350 (2010)
5. Bian, Z.P., Hou, J., Chau, L.P., Magnenat-Thalmann, N.: Fall detection based on body part tracking using a depth camera. IEEE J. Biomed. Health Inform. 19(2), 430–439 (2015)
6. Cao, Z., Simon, T., Wei, S.E., Sheikh, Y.: Realtime multi-person 2D pose estimation using part affinity fields. In: CVPR (2017)
7. Carone, G., Costello, D.: Can Europe afford to grow old? 43, September 2006
8. Charfi, I., Miteran, J., Dubois, J., Atri, M., Tourki, R.: Optimized spatio-temporal descriptors for real-time fall detection: comparison of support vector machine and adaboost-based classification. J. Electron. Imaging 22, 22–22-18 (2013)
9. Chen, J., Kwong, K., Chang, D., Luk, J., Bajcsy, R.: Wearable sensors for reliable fall detection. pp. 3551–3554, January 2005
10. Chua, J.L., Chang, Y.C., Lim, W.K.: A simple vision-based fall detection technique for indoor video surveillance. SIViP 9(3), 623–633 (2015)
11. Dalal, N., Triggs, B.: Histograms of oriented gradients for human detection. In: 2005 IEEE Computer Society Conference on Computer Vision and Pattern Recognition (CVPR 2005), vol. 1, pp. 886–893, June 2005
12. Dollar, P., Wojek, C., Schiele, B., Perona, P.: Pedestrian detection: an evaluation of the state of the art. IEEE Trans. Pattern Anal. Mach. Intell. 34(4), 743–761 (2012)
13. Duda, R.O., Hart, P.E.: Use of the hough transformation to detect lines and curves in pictures. Commun. ACM 15(1), 11–15, January 1972
14. Fang, H.S., Xie, S., Tai, Y.W., Lu, C.: RMPE: regional multi-person pose estimation. In: ICCV (2017)
15. Feng, Q., Gao, C., Wang, L., Zhang, M., Du, L., Qin, S.: Fall detection based on motion history image and histogram of oriented gradient feature. In: 2017 International Symposium on Intelligent Signal Processing and Communication Systems (ISPACS), pp. 341–346, November 2017
16. Galvão, Y.M., Albuquerque, V.A., Fernandes, B.J.T., Valença, M.J.S.: Anomaly detection in smart houses: monitoring elderly daily behavior for fall detecting. In: 2017 IEEE Latin American Conference on Computational Intelligence (LA-CCI), pp. 1–6, November 2017

17. Ge, C., Gu, I.Y.H., Yang, J.: Human fall detection using segment-level CNN features and sparse dictionary learning. In: 2017 IEEE 27th International Workshop on Machine Learning for Signal Processing (MLSP), pp. 1–6, September 2017
18. Huang, Z., Liu, Y., Fang, Y., Horn, B.K.P.: Video-based fall detection for seniors with human pose estimation. In: 2018 4th International Conference on Universal Village (UV), October 2018
19. Igual, R., Medrano, C., Plaza, I.: Challenges, issues and trends in fall detection systems. Biomed. Eng. Online 12(1), 66 (2013)
20. KaewTraKulPong, P., Bowden, R.: An Improved Adaptive Background Mixture Model for Real-time Tracking with Shadow Detection, pp. 135–144. Springer, US, Boston, MA (2002). https://doi.org/10.1007/978-1-4615-0913-4_11
21. Sehairi, K., Chouireb, J.M.F.: Comparative study of motion detection methods for video surveillance systems. J. Electron. Imaging 26, 26–26-29 (2017)
22. Klack, L., Möllering, C., Ziefle, M., Schmitz-Rode, T.: Future care floor: a sensitive floor for movement monitoring and fall detection in home environments. In: Lin, J.C., Nikita, K.S. (eds.) Wireless Mobile Communication and Healthcare, pp. 211–218. Springer, Berlin Heidelberg (2011). https://doi.org/10.1007/978-3-642-20865-2_27
23. Kwolek, B., Kepski, M.: Human fall detection on embedded platform using depth maps and wireless accelerometer. Comput. Methods Prog. Biomed. 117(3), 489–501 (2014)
24. Lahiri, D., Dhiman, C., Vishwakarma, D.K.: Abnormal human action recognition using average energy images. In: 2017 Conference on Information and Communication Technology (CICT), pp. 1–5, November 2017
25. Li, X., Pang, T., Liu, W., Wang, T.: Fall detection for elderly person care using convolutional neural networks. In: 2017 10th International Congress on Image and Signal Processing, BioMedical Engineering and Informatics (CISP-BMEI), pp. 1–6, October 2017
26. Lie, W.N., Le, A.T., Lin, G.H.: Human fall-down event detection based on 2D skeletons and deep learning approach. In: 2018 International Workshop on Advanced Image Technology (IWAIT), pp. 1–4, January 2018
27. Liu, H., Qian, Y., Lin, S.: Detecting persons using hough circle transform in surveillance video. In: VISAPP (2), pp. 267–270 (2010)
28. Marin-Jimenez, M.J., Zisserman, A., Eichner, M., Ferrari, V.: Detecting people looking at each other in videos. Int. J. Comput. Vis. 106(3), 282–296 (2014)
29. Merrouche, F., Baha, N.: Fall detection using head tracking and centroid movement based on a depth camera. In: Proceedings of the International Conference on Computing for Engineering and Sciences, pp. 29–34. ICCES 2017 (2017)
30. Mirmahboub, B., Samavi, S., Karimi, N., Shirani, S.: Automatic monocular system for human fall detection based on variations in silhouette area. IEEE Trans. Biomed. Eng. 60(2), 427–436 (2013)
31. Mubashir, M., Shao, L., Seed, L.: A survey on fall detection: Principles and approaches. Neurocomputing 100, 144–152 (2013), special issue: Behaviours in video
32. Noury, N., Rumeau, P., Bourke, A., ÓLaighin, G., Lundy, J.: A proposal for the classification and evaluation of fall detectors. IRBM 29(6), 340–349 (2008)
33. Núñez-Marcos, A., Azkune, G., Arganda-Carreras, I.: Vision-based fall detection with convolutional neural networks. Wireless Commun. Mob. Comput. (2017)
34. Patel, S., Park, H., Bonato, P., Chan, L., Rodgers, M.: A review of wearable sensors and systems with application in rehabilitation. J. Neuroeng. Rehabil. 9(1), 21 (2012)

35. Richstone, L., Schwartz, M.J., Seideman, C., Cadeddu, J., Marshall, S., Kavoussi, L.R.: Eye metrics as an objective assessment of surgical skill. Ann. Surg. **252**(1), 177–182 (2010)

36. Rougier, C., Meunier, J., St-Arnaud, A., Rousseau, J.: Monocular 3D head tracking to detect falls of elderly people. In: 2006 International Conference of the IEEE Engineering in Medicine and Biology Society, pp. 6384–6387, August 2006

37. Rougier, C., Meunier, J., St-Arnaud, A., Rousseau, J.: Robust video surveillance for fall detection based on human shape deformation. IEEE Trans. Circuits Syst. Video Technol. **21**(5), 611–622 (2011)

38. Sehairi, K., Chouireb, F., Meunier, J.: Elderly fall detection system based on multiple shape features and motion analysis. In: 2018 International Conference on Intelligent Systems and Computer Vision (ISCV), pp. 1–8, April 2018

39. Stauffer, C., Grimson, W.E.L.: Adaptive background mixture models for real-time tracking. In: IEEE Computer Society Conference on Computer Vision and Pattern Recognition, 1999, vol. 2, pp. 246–252, Los Alamitos, CA, USA, August 1999

40. Xiu, Y., Li, J., Wang, H., Fang, Y., Lu, C.: Pose Flow: efficient online pose tracking. In: BMVC (2018)

41. Yao, L., Min, W., Lu, K.: A new approach to fall detection based on the human torso motion model. Appl. Sci. **7**(10) (2017)

42. Yu, M., Yu, Y., Rhuma, A., Naqvi, S.M.R., Wang, L., Chambers, J.A.: An online one class support vector machine-based person-specific fall detection system for monitoring an elderly individual in a room environment. IEEE J. Biomed. Health Inform. **17**(6), 1002–1014 (2013)

43. Zigel, Y., Litvak, D., Gannot*, I.: A method for automatic fall detection of elderly people using floor vibrations and sound-proof of concept on human mimicking doll falls. IEEE Trans. Biomed. Eng. **56**(12), 2858–2867 (2009)

GradXcepUNet: Explainable AI Based Medical Image Segmentation

Amandeep Kaur[1]([✉]), Guanfang Dong[2], and Anup Basu[2]

[1] Indira Gandhi Delhi Technical University for Women, Delhi, India
amandeep022btit18@igdtuw.ac.in
[2] University of Alberta, Edmonton, Canada
{guanfang,basu}@ualberta.ca

Abstract. Medical images segmentation is an important research area. Physicians and radiologists can diagnose diseases in their patients by observing the visual features using various imaging methods like CT, MRI, X-ray, and Ultrasound. AI-based medical image segmentation models can help radiologists and experts analyze various ailments. However, the prediction results are only trustworthy when the results can be interpreted by a doctor. Our work utilizes "Explainable AI(XAI)." We propose GradXcepUNet, an XAI-based medical image segmentation model, that couples the segmentation power of U-Net and explainability features of the Xception classification network by Grad-CAM. The Grad-CAM trained images highlight the critical regions for the Xception classification network. Then, as the guidance, the visualized results for critical regions are combined with an existing segmentation model (U-Net) to produce the final segmentation results. With the assistance of XAI analysis and visualization, our GradXcepUNet outperforms the original U-Net and many state-of-the-art methods. The evaluation results show that we can reach a Dice coefficient of 97.73% and an Intersection over Union (IoU) score of 78.86% on the 3D-IRCADb-01 database.

Keywords: Image processing · Multimedia and health-care

1 Introduction

Liver cancer is the third most prevalent reason for cancer-induced deaths in 2020, according to the World Health Organization (WHO) [1]. Liver segmentation is crucial in determining the medicaments for liver diseases like Hepatocellular carcinoma (HCC), intrahepatic cholangiocarcinoma, and hepatoblastoma. The liver segmentation task is demanding since the liver shape is non-rigid. Also, pathologies can affect the surface of the liver. Most of the time, clearly defined edges are not visible in the liver, resulting in incorrect segmentations. Thus, accurate and efficient segmentation would assist the detection of cancerous cells present in patients and lead to better treatments of liver diseases.

Supplementary Information The online version contains supplementary material available at https://doi.org/10.1007/978-3-031-22061-6_13.

With advancements in computational power and the availability of datasets, the field of deep learning holds great potential. Convolution Neural Networks (CNNs) have shown promise in dealing with complex noise, image reconstruction, and segmentation. High-quality enhanced images can help in tasks like tracking, segmentation, and detection. Due to its frequent overfitting property and "Black box" effect, deep learning-based approaches are not yet viable on a large scale in the medical domain, where doctors may not be able to interpret the results provided by a model even if it can achieve high accuracy.

The human body is a complex adaptive system with dynamically changing characteristics. The information that benefits medical images segmentation may be hidden in high dimensional spaces [2]. To detect this information, a neural network can perform high-dimensional convolution and find many kernels which are suitable for training. Thus, for a well-trained model, the model's parameters contain the explanation for the relationships and patterns of the targeted task. However, it is very challenging to find the explanatory information within the model and reuse it appropriately through a reasonable algorithm. Our research is targeted at developing solutions to address these difficulties in the context of medical image segmentation.

Our present work concentrates on the Explainability of AI so that its potential is significantly utilized while inculcating confidence in the architecture. We propose GradXcepUNet, an XAI-based medical image segmentation model that couples the power of U-Net and Xception (XcepU-Net) with explainable features through grad-CAM. The model inputs the original 2D slices of patients. First, images are labeled by a well-trained Xception classification network. With the assistance of Grad-CAM, the critical regions of the targeted label are highlighted. Then, the critical regions are reorganized for guidance and input into the U-Net classification network. Together with the original 2D images, U-Net achieves better segmentation results. Our GradXcepUNet model outperforms many state-of-the-art methods. The final segmented prediction achieves a Dice coefficient of 97.73% and an Intersection over Union (IoU) score of 78.76%.

Following are the novel contributions of our work.

- The proposed GradXcepUNet model is transparent and adaptable to the dataset used and the knowledge of human experts. In the 3D-IRCADb-01 database, our GradXcepUNet model not only achieves a higher segmentation accuracy than the original U-net and other similar models, but the saliency map matches the region of interest with ground truth from human experts.
- Our XAI-based medical image segmentation model balances interpretability and accuracy. In addition, interpretable data can help traditional deep neural networks to be significantly more accurate in segmentation prediction. In the future, our architecture is expected to be applied to different medical imaging tasks and existing models.

The organization of this paper is as follows. Related work based on image segmentation and XAI-based approaches is presented in Sect. 2. In Sect. 3 we propose our approach GradXcepUNet. Section 4 highlights the experimental procedure and results. Finally, we conclude by stating our plans and extensions of the proposed method in Sect. 5.

2   Related Work

2.1   Related Work in Image Segmentation

Extremely unbalanced segmentation is common in medical image analysis since the target region of interest is smaller in magnitude than the background. In the biomedical image processing domain, each pixel needs to be associated with a class label. Ronneberger et al. introduced the revolutionary architecture of U-Net, which highlights yet another task, namely the separation of adjacent organs of the same class in the image [4]. For this task, they proposed the use of weighted loss. As a result, the labels between adjacent cells are given more significant weight in the loss function. Milletari et al. proposed V-Net as a volume-based, 3D image segmentation method using FCN. They also introduced a novel loss function based on Dice coefficient maximization [6]. The features captured by the deep layers help in perceiving the whole anatomy of interest at once. Liang-Chieh et al. proposed DeepLabv3, which uses Atrous convolution to capture dense feature maps [7]. Since Atrous convolutions are used, the field of view is increased considerably. Their cascaded module increases the Atrous rates by a factor of 2. The proposed Atrous spatial pyramid pooling augments image features with features of filters in the practical field of views. "DeepLabv3" improves on their previous implementations with results comparable to other state-of-the-art models. Ma et al. proposed a Robust-PCA-based method to deal with traditional segmentation tasks [3]. Their results are refined by the superpixel method. Hoel et al. proposed boundary loss, which works on contours, unlike regions [8]. The loss reduces the difficulty of unbalanced or skewed data as it uses integral over the intersection between different areas. It is crucial to train a network with better recall than precision for tasks like lesion detection in a skewed dataset. Thus, Salehi et al. used a loss function using the Tversky index to mitigate the problem of highly imbalanced data while gaining a good compensation between precision and recall on the U-Net architecture [9]. Ken et al. proposed an exponential logarithmic loss that tries to balance labels by segmentation difficulties [10]. The dice coefficient obtained was 82%, and the processing time was reduced to 0.4s compared to V-net, which takes more time.

Liver segmentation is challenging because of the absence of evident edges on most liver borders and other organs appearing nearby. To obtain an accurate segmentation result, Moghbel et al. used a random walker-based structure that segments the contrast-enhanced CT images with better speed and efficiency [11]. The area of the right lung lobe is used to detect the lung lobe. Furthermore, computational requirements are minimized by focusing on the rib-cage area. By using UNet as the base model, Chlebus et al. proved that cascading 2D FCN operating at a voxel level and trained on hand-crafted features reduces false detetions while simultaneously improving the segmentation quality [14].

By using cascaded fully convolutional neural networks (CFCNs), Christ et al. automatically segmented the liver and lesions in the 3D-IRCADb-01 database [15]. First, an FCN has trained to segment the liver as the ROI, which was input to the second FCN to segment lesions within the predicted liver ROI. Thus, the method trains and cascades two FCNs for the segmentation task. Jiang et al. designed the Attention Hybrid Connection Network (AHCnet) that combines the soft-hard attention mechanism and long-short skip links [19]. The joint dice loss function trains the liver localization network to achieve precise 3D liver bounding boxes and produce liver segmentation results. By utilizing the benefits of combining state-of-the-art models, better insights into the problem are proposed. Efremova et al. combined LinkNet-34, with ResNet-34 as the feature encoder and U-Net to present the segmentation results [17]. Jin et al. proposed a novel 3D residual attention aware network, RA-UNet [18]. The model can extract 3D structures in a pixel-to-pixel fashion. Attention modules get attention-aware features. Following this, residual blocks make the architecture deep and mitigate the gradient vanishing problem. Finally, the UNet model is deployed to blend low-level characteristics with high-level features. Spatial information is not fully leveraged with the third dimension in 2D convolutions. While computational expense and GPU memory use are high in 3D convolutions, Li et al. proposed the Hybrid Densely connected UNet (H-DenseUNet) to address these issues [12]. A 2D DenseUNet is introduced for extracting intra-slice features from the images, while the 3D convolutions aggregate volumetric contexts. The training method of H-DenseUNet is done in an end-to-end fashion, with a hybrid feature fusion (HFF) layer optimizing the intra-slice features and inter-slice characteristics.

2.2   Related Work Based on XAI Methods

Deep CNNs have demonstrated impressive classification predictions on benchmark datasets like ImageNet. However, it is not clear 'why' they function well. Zeiler et al. proposed a novel visualization technique called DeconvNet, which returns the feature activations to the input pixel space [21]. Dosovitskiy et al. proposed Guided BackPropagation, known as a guided saliency map, which serves as a variant of the above-mentioned approaches on Deconvolution [22]. Guided Backpropagation replaces the max-pooling layers with a convolution layer with increased stride. This method was tested on various image recognition benchmarks and shown to have no loss in accuracy. The local interpretable model-agnostic explanations (LIME) proposed by Ribeiro et al. generates interpretations for arbitrary sampled data around the input [23]. New predictions are produced and weighted by their closeness to the input example using the information from the neighboring examples. Then a simple interpretable model is trained on this dataset. By understanding this 'local' model, the original model gets visualized. LIME is the most commonly used XAI method in the research field.

Zhou et al. introduced Class Activation Maps (CAMs), which focus on the discriminative regions of an image [25]. This leads to the final predictions for the categorization of the image. The feature vector is generated by combining the average activation of the feature map found in the last convolutional layer. By back-projecting the weights of the output layer onto the feature maps, regions that strongly influence the final forecast are identified. The significant drawbacks of this approach are that CAM requires neural networks to have a well-defined structure in the last layers; otherwise, the model needs to be trained again. In addition, it is unable to interpret other layers except for the last one, thereby constraining visualizations. Selvaraju et al. proposed Grad-CAM which can produce visualizations for any convolution neural network, combating the previous drawbacks of CAM [26]. Grad-CAM produces a course localization map of the essential regions by having the class-specific gradient data flowing into the final convolution layer based on gradients. This leads to more transparent model predictions.

Following the benefits of the CAM-based approach, studies like Wenwu et al. introduced Probabilistic-CAM (PCAM) pooling, which can leverage the localization ability of CAM for training [27]. This work depicts clear and sharp boundaries in the region of interest in the ChestX-ray14 dataset in comparison to the standard CAMs. Maloca et al. introduced the Traceable Relevance Explainability (T-REX) technique which enhances the predictions for optical coherence tomography image segmentation [28]. The prediction process was transparent and led to optimized applications. The method is based on ground truth annotation, calculating Hamming distances among annotators, and a machine learning algorithm. For lesion detection on a fundus image, Hongyang et al. used Grad-CAM to precisely detect lesion regions [29]. Experiments demonstrated the method's effectiveness. Also, Dong et al. proposed feature points in medical images for portable ultrasound devices by an XAI method. They employed Laplace Pyramid Fusion with feature points to achieve better denoising results [5].

3   Proposed Method

Overall, our approach consists of two components. First, we find the areas of interest in a liver by the XAI-based method. Then, we deploy XAI methods with state-of-the-art deep learning models. The reason for deploying the XAI approach is that the results can be regarded as prior knowledge for deep learning models. As Diligenti et al. showed, integrating valid prior knowledge can restrict the space to search for optimal parameters, which leads to a better trained medical image segmentation model [16]. Figure 1 shows our model structure.

3.1   Xception-Grad-CAM

Grad-CAM can produce visualizations for a convolution neural network [26]. If a base network is classification-oriented (the final result is the label of the input), Grad-CAM produces a coarse localization map of the essential regions using the class-specific gradient data. Specifically, the gradient data flows into the final convolution layer, which leads to a transparent visual explanation for model predictions. Usually, deeper layers of the CNN architecture are efficient in capturing the high-level heuristic features. These heuristic features often contribute significantly to the prediction. The last layer of the architecture (just before the

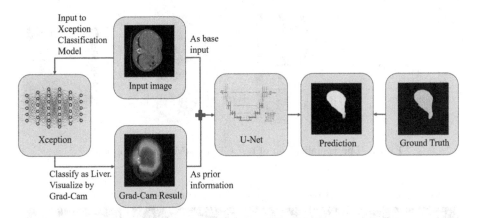

Fig. 1. Our GradXcepUNet framework: The sliced 2D image of a patient is feed into Xception to receive Grad-Cam visualization result. Then, Grad-CAM trained saliency maps together with original images are used as the U-Net input to train the segmentation model.

fully connected layer) is expected to contain both high-level interpretation and detailed spatial knowledge. Thus, Grad-CAM utilizes the gradient learning that flows into the last convolution layer to assign an importance value to each neuron for a specific class. As the mathematical representation, let $A^k = [a_{i_j}]^k$ denote a feature map in the final convolution layer for input x. Then, the value of the feature map A^k to the score $z_c(x)$ is given by:

$$\alpha_k^c = \frac{1}{Z} \sum_{ij} \frac{\partial z_c}{\partial a_{i_j}^k} \tag{1}$$

where Z indicates the representation of pixels in A^k denoting average Global Pooling. The Grad-CAM heatmap can be described as follows:

$$L_{Grad-CAM}^C = ReLU(\sum_k \alpha_k^c A^k) \tag{2}$$

ReLU function is deployed to extract only those characteristics or attributes that positively impact the class prediction. Grad-CAM visualizations outperform not only the interpretability but also the faithfulness to the original model.

For Grad-CAM results, we deployed Xception as the base model to obtain the visualization results. Xception is binary trained as $\{C_{original}, C_{liver\_only}\}$. Then, for input images, we only extract the visualization result for C_{liver_only}. It is worth mentioning that we actually trained two CNN architectures: VGG16-Grad-CAM and Xception-Grad-CAM. We found Xception-Grad-CAM empirically achieves better visualization results. We speculate that this is due to the complexity of Xception-Grad-CAM being greater than VGG16-Grad-CAM. Figure 3 shows the visualization comparison.

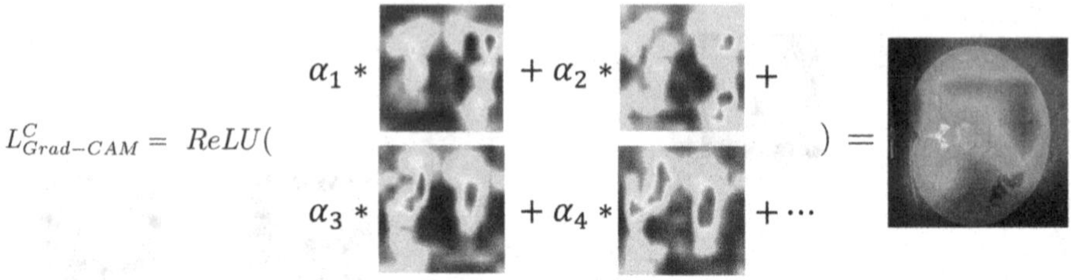

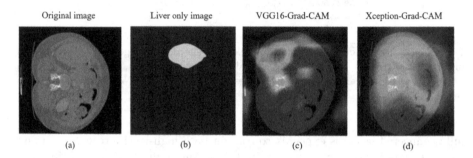

$$L^C_{Grad-CAM} = ReLU(\; \alpha_1 * [\;] + \alpha_2 * [\;] + \alpha_3 * [\;] + \alpha_4 * [\;] + \cdots \;) = [\;]$$

Fig. 2. The visualization of our Grad-CAM implementation.

| Original image | Liver only image | VGG16-Grad-CAM | Xception-Grad-CAM |

(a)   (b)   (c)   (d)

Fig. 3. Grad-CAM predictions for a sample image: (a) The original 2D slice image. (b) The segmented mask of liver. (c) Grad-CAM prediction with VGG 16 as the base model; and (d) Grad-CAM prediction with Xception as the base model.

3.2 Xception

Here we introduce our Xception classification model [30]. Xception utilizes the properties of cross-channel and spatial correlations. Thus, based on the above properties, the feature maps of CNN can be decoupled. According to this, Xception is designed to replace inception modules with depthwise separable convolutions in deep learning networks. Our Grad-CAM visualization explanation shows its power in feature detection as well as interpretability, as described by Kaiser et al. [24]. Mathematically, let

$$Conv(W, y)_{(i,j)} = \sum_{r,s,t}^{R,S,T} W_{(r,s,t)} \cdot y_{(i+r,j+s,t)} \tag{3}$$

be defined as the convolution layer, where Ws are trainable weights. Similar to spatial separable convolution, the depthwise separable convolution splits the kernel into two parts as Depthwise and Pointwise convolutions. They can be described as:

$$DepthwiseConv(W, y)_{(i,j)} = \sum_{r,s}^{R,S} W_{(r,s)} \odot y_{(i+r,j+s)} \tag{4}$$

In the depthwise convolution, the input image is given a convolution without changing depth. This is done by using kernels with dimensions $(k * k * 1)$. Each $(k * k * 1)$ kernel repeats one channel of the image, obtaining the scalar products of every K pixel group, produces a $(k + 3 * k + 3 * 1)$ image. Accumulating these images creates a $(k + 3 * k + 3 * 3)$ image.

$$PointwiseConv(W, y)_{(i,j)} = \sum_{t}^{T} W_t \cdot y_{(i,j,t)} \tag{5}$$

In the pointwise convolution, a $1 * 1$ kernel is used which has a depth of the input image's channel. In our images, 3 $(1 * 1 * 3)$ kernels iterate through the $(k + 3 * k + 3 * 3)$ image to get a $(k + 3 * k + 3 * 1)$ image.

Finally, the depthwise separable convolution layer is given by:

$$SepConv(W_p, W_d, y)_{(i,j)} = PointwiseConv_{(i,j)}(W_p, DepthwiseConv_{(i,j)}(W_d, y)) \tag{6}$$

The significant difference between the normal convolution and depthwise convolution is that, in the former, the images are transformed 256 times; but, in the latter, the transformation is done only once, and the image is simply elongated to 256 channels. We use this property of Xception in our model, reducing computational cost.

3.3 U-Net

The U-Net model [4] is a revolutionary approach in medical image segmentation with each pixel assigned a class label. As mentioned by Isensee et al. using the U-Net as a benchmark necessitates adapting the strategy to the novel problem [13]. The U-Net model proposes applying weighted loss, separating background markers between connected cells, and seizing a significant weight in the loss function. The model consists of a contracting part constituting the downsampling path and an expanding path, which is the upsampling path. Thus, the expanding path and contracting path are symmetrical, forming a U-shaped architecture. The input image is extrapolated in a mirror directly to forecast pixels in the border region. The sampled output is then merged with high-resolution features from the contraction path for feature localization. The architecture has several feature channels in the upsampling process. The previous convolution layer eventually learns to assemble a more precise result based on this data. The U-Net is used to achieve multi-scale attention learning and can unite low-level characteristics with high-level attributes. The energy function of U-Net is given by:

$$EF = \sum w(x) \log \left(p_{k(x)}(x) \right) \tag{7}$$

with p_k being the pixel-wise SoftMax function applied over the final feature map, defined as:

$$p_k = \exp \left(a_k(x) \right) / \sum_{k'=1}^{K} \exp \left(a_k(x)' \right) \tag{8}$$

with a_k denoting the activation in channel k.

Due to its context-based learning, U-net is faster to train than most other segmentation models. U-net is incredibly valuable due to its incredible modularity and mutability. U-net has considerable development potential because its modular design allows it to continue expanding by incorporating new unique concepts within itself, Siddique et al. [20].

3.4   GradXcepUNet

Combining the power of U-Net and Xception, with explainability features through Grad-CAM, we propose the architecture of the GradXcepUNet model. The model takes the original 2D image slices of patients as input, and the Grad-CAM generated saliency map. The whole GradXcepUNet can be represented as $Z(X, W, G)$. Where,

$$Z(X, W, G) = f_{u-net}(W \oplus X)/\Delta G \qquad (9)$$

Here, X corresponds to the original 2D image slices, W corresponds to the Grad-CAM generated saliency map and G corresponds to the ground truth mask for the image, are input into the XcepU-Net model.

The Xception model uses the Binary cross-entropy function as the loss function defined as:

$$H_p(q) = -\frac{1}{N} \sum_{i=1}^{N} y_i.log(p(y_i)) + (1 - y_i).log(1 - p(y_i)) \qquad (10)$$

where N is the output size, y is the label and p(y) is the predicted probability of the output belonging to a specific class.

We define an epoch as the iteration over 4 training batches. The parameters of the network are trained by back-propagating the derivatives of loss throughout the network and updating the parameters using the Adam optimizer with an initial learning rate of 1 x 10 $^{-3}$ for all experiments. By combining Gradcam and Xception, our result surpasses the original segmentation results from classical U-Net, and other methods as described in Table 1.

4   Experimental Results

4.1   Dataset

The dataset used for this experiment is the 3D-IRCADb-01 database from a medical research center in France. It comprises of 3D CT scans of 20 venous phase CT patients - ten men and ten women, with hepatic tumors in 75% of the cases, from multiple European hospitals with varying CT scanners and involving 120 liver tumors of varying sizes. The data is available in DICOM and NifTi formats. We use the NifTi format for our experiments. NifTi files are frequently used in neuroscience and neuroradiology imaging research tasks. The dataset is split into train and test. Thirteen patients are kept in the training (and validation) set, while seven patients are for the test. Both the sets are independent, with no patient data repeating in either. This data repository contains 20 medical examinations in 3D format, with the source images and the segmenting masks of the liver. Each segmentation is done by a radiologist, manually outlining the liver contours for all images. Since it consists of only 20 data points, we augment it by converting the 3D images to their 2D slices, making 2823 images in total. Of 2823, 1888 are used for training and validation from the training dataset. From the 7 patients in the test dataset, 935 images are extracted for testing. In addition to the training data mentioned above, we added 1888 images from our Xception-trained Grad-CAM model, corresponding to each original 2D slice of the medical image. Thus, we had 3776 training images and 935 testing images, approximately 75% in training and 25% in testing. Training images were normalized and restricted to the gray-scale channel. Figure 4 describes the data split and preparation methodology applied for creating the final cohorts of the train and test datasets.

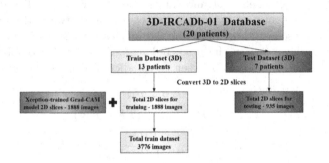

Fig. 4. Data set preparation methodology.

4.2 Experimental Setup

The experiments were performed on Keras with Tensorflow v2.5 backend. The programming language used for implementation is python version 3.7.11. The Grad-CAM model trains the 2D slices obtained from the above step to predict the essential regions that have resulted in the final prediction. These images with the original images are given as input to the model. Of the 3776 images in the training set, 1800 images were randomly selected for validation. A series of experiments were performed for the hyperparameter tuning for the parameters - learning rate, batch size, and number of epochs. The final network (GradXcepUNet) is trained with a batch size of 4 and a learning rate of 10^{-3} for 30 epochs.

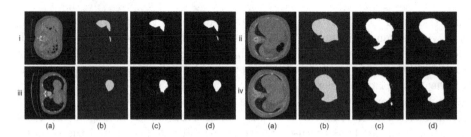

Fig. 5. Examples of segmentation results by GradXcepUNet on the Test dataset with (a) Original Slice; (b) Ground Truth; (c) Prediction with U-Net; and (d) Prediction with GradXcepUNet.

4.3   Results and Evaluation

The model was tested on 935 images, the results were converted to their corresponding image format for better visualization. For evaluation, we deploy two metrics, namely Dice coefficient and Intersection Over Union. The dice coefficient computes the ratio of twice the area of overlap between the two images to the total representation of pixels in the images. Intersection over Union is the ratio of overlay between the predicted segment and the ground truth segmentation with the union area between the predicted segment and ground truth segmentation.

$$Dice = \frac{2 \times TP}{(TP + FP) + (TP + FN)} \tag{11}$$

$$IoU = \frac{TP}{(TP + FP + FN)} \tag{12}$$

Here, TP: True Positives; FP: False Positives; FN: False Negatives

Figure 5 shows our results compared with the classical U-Net model, and our proposed approach GradXcepUNet outperforms the former model. More results can be viewed in Fig. 6. We take the pixel value 50 as our threshold. In the prediction result, all pixel points greater than 50 will be considered pixel points of the liver. Thus, we can generate a binary mask to indicate where the liver is segmented.

As a comparison of the results of the ablation experiments, we found that the liver segmentation results after receiving a priori knowledge analyzed by explainable AI methods were better in terms of edge smoothing, noise removal, and detail retention than the U-Net-only method. In Fig. 5, if we look closely we can see that there is a certain amount of noise in the segmentation results of the U-Net. We suspect that this is due to a lack of guidance on the location of the liver features. In group i, the U-Net method misses some discontinuous regions. In groups ii and iv, U-Net added many regions that should not be livers. For group iii, the U-Net method performed similarly to our method. Overall, because

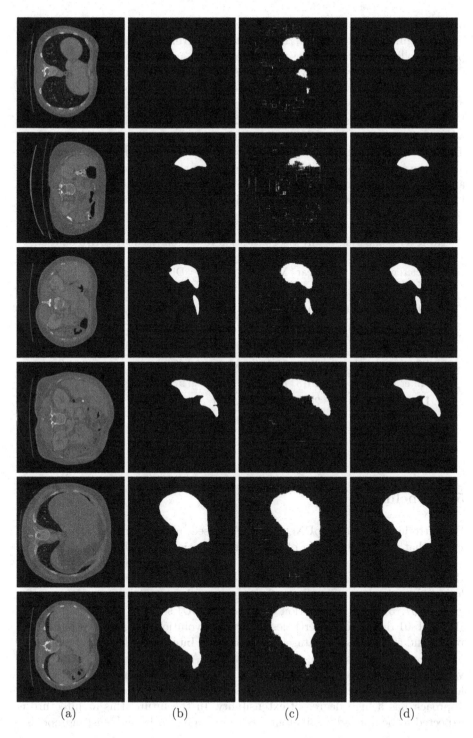

Fig. 6. (a) Original Slice (b) Ground Truth (c) Prediction with U-Net (d) Prediction with GradXcepUNet.

our method had less noise and smoother edges, our method is substantially ahead of the original U-Net method. The ablation experiments also demonstrate that prior knowledge extraction based on explainable artificial intelligence by Grad-Cam can significantly improve the segmentation accuracy of existing methods.

Table 1 illustrates that GradXcepUNet can achieve comparable state-of-the-art results and score second highest in the category. The Dice coefficient obtained by GradXcepUNet is 97.73%, and the IoU score is 78.76%. The efficacy and adaptability of GradXcepUNet validate the encouraging results achieved. From Table 1 we infer, that our proposed approach is better than most of the models, and is the second top in the test data category at the global level, bested only by a model that is not explainable. Li et al. uses both 3D as well as the 2D dataset for their consideration, while our method uses only the 2D slices of patients [12].

Table 1. Comparing the liver segmentation results on the 3DIRCADb dataset.

Reference	Year	Model	Dice coefficient(%)	Type
Moghbel et al. [11]	2016	–	91.1	Liver
Chlebus et al. [14]	2018	U-Net	92.3	Liver
Christ et al. [15]	2017	–	94.3	Liver
Jiang et al. [19]	2019	AHCNet	94.5	Liver
Efremova et al. [17]	2019	LinkNet-34	96.2	Liver
Jin et al. [18]	2018	RA-UNet	97.70	Liver
Li et al. [12]	2017	H-DenseU-Net	98.2 ± 1	Liver
Ours	2021	**GradXcepUNet**	**97.73**	Liver

5   Conclusion

We introduced GradXcepUNet, an XAI-based medical image segmentation model that couples the power of U-Net and Xception (XcepU-Net) with explainability of features through Grad-CAM. The Grad-CAM trained images highlight the critical regions for the XcepU-Net model to focus on, and the combined model learns from these to produce the final segmented prediction with a Dice coefficient of 97.73% and Intersection over Union (IoU) score of 78.86% on the 3D-IRCADb-01 database. Nevertheless, with the evolution of better computational power and an increase in dataset size, Artificial Intelligence has the potential to outperform other techniques. We believe that AI-based image segmentation in various domains, especially healthcare, has the potential of increasing efficiency, lowering costs, leading to better diagnoses and health outcomes. The proposed approach has a high degree of extensibility. In the future, this architecture is expected to be applied to different medical imaging tasks and existing models.

References

1. Cancer. World Health Organization, World Health Organization, 21 Sept 2021. https://who.int/news-room/fact-sheets/detail/cancer
2. Holzinger, A., et al.: Towards the augmented pathologist: challenges of explainable-AI in digital pathology. arXiv preprint arXiv:1712.06657 (2017)
3. Ma, Y., Dong, G., Zhao, C., Basu, A., Wu, Z.: Background subtraction based on principal motion for a freely moving camera. In: McDaniel, T., Berretti, S., Curcio, I.D.D., Basu, A. (eds.) ICSM 2019. LNCS, vol. 12015, pp. 67–78. Springer, Cham (2020). https://doi.org/10.1007/978-3-030-54407-2_6
4. Ronneberger, O.: Invited talk: u-net convolutional networks for biomedical image segmentation. In: Bildverarbeitung für die Medizin 2017. I, p. 3. Springer, Heidelberg (2017). https://doi.org/10.1007/978-3-662-54345-0_3
5. Dong, G., Ma, Y., Basu, A.: Feature-guided CNN for denoising images from portable ultrasound devices. IEEE Access 9, 28272–28281 (2021)
6. Fausto, M., Navab, N., Ahmadi, S.-A.: V-Net: fully convolutional neural networks for volumetric medical image segmentation. In: 2016 fourth international conference on 3D vision (3DV). IEEE, (2016)
7. Chen, L.-C., et al.: Rethinking atrous convolution for semantic image segmentation. arXiv preprint arXiv:1706.05587 (2017)
8. Kervadec, H., et al.: Boundary loss for highly unbalanced segmentation. In: International Conference on Medical Imaging with Deep Learning, vol. 102, pp. 285-296. PMLR (2019)
9. Salehi, S.S.M., Erdogmus, D., Gholipour, A.: Tversky loss function for image segmentation using 3D fully convolutional deep networks. In: Wang, Q., Shi, Y., Suk, H.-I., Suzuki, K. (eds.) MLMI 2017. LNCS, vol. 10541, pp. 379–387. Springer, Cham (2017). https://doi.org/10.1007/978-3-319-67389-9_44
10. Wong, K.C.L., Moradi, M., Tang, H., Syeda-Mahmood, T.: 3D segmentation with exponential logarithmic loss for highly unbalanced object sizes. In: Frangi, A.F., Schnabel, J.A., Davatzikos, C., Alberola-López, C., Fichtinger, G. (eds.) MICCAI 2018. LNCS, vol. 11072, pp. 612–619. Springer, Cham (2018). https://doi.org/10.1007/978-3-030-00931-1_70
11. Moghbel, M., et al.: Automatic liver segmentation on computed tomography using random walkers for treatment planning. EXCLI J. 15, 500 (2016)
12. Li, X., et al.: H-DenseUNet: hybrid densely connected UNet for liver and tumor segmentation from CT volumes. IEEE Trans. Med. Imaging 37(12), 2663–2674 (2018)
13. Isensee, F., et al.: nnU-Net: self-adapting framework for U-Net-based medical image segmentation. arXiv preprint arXiv:1809.10486 (2018)
14. Chlebus, G., et al.: Automatic liver tumor segmentation in CT with fully convolutional neural networks and object-based postprocessing. Sci. Rep. 8(1), 1–7 (2018)
15. Christ, P.F., et al.: Automatic liver and tumor segmentation of CT and MRI volumes using cascaded fully convolutional neural networks. arXiv preprint arXiv:1702.05970 (2017)
16. Diligenti, M., Roychowdhury, S., Gori, M.: Integrating prior knowledge into deep learning. In: 2017 16th IEEE International Conference on Machine Learning and applications (ICMLA), pp. 920-923. IEEE (2017)
17. Efremova, D.B., et al.: Automatic segmentation of kidney and liver tumors in CT images. arXiv preprint arXiv:1908.01279 (2019)

18. Jin, Q., et al.: RA-UNet: a hybrid deep attention-aware network to extract liver and tumor in CT scans. Front. Bioeng. Biotechnol. **8**, 1471 (2020)
19. Jiang, H., et al.: AHCNET: an application of attention mechanism and hybrid connection for liver tumor segmentation in CT volumes. IEEE Access **7**, 24898–24909 (2019)
20. Siddique, N., et al.: U-Net and its variants for medical image segmentation: a review of theory and applications. IEEE Access **9**, 82031-82057 (2021)
21. Zeiler, M.D., Fergus, R.: Visualizing and Understanding Convolutional Networks. In: Fleet, D., Pajdla, T., Schiele, B., Tuytelaars, T. (eds.) ECCV 2014. LNCS, vol. 8689, pp. 818–833. Springer, Cham (2014). https://doi.org/10.1007/978-3-319-10590-1_53
22. Springenberg, J.T., et al.: Striving for simplicity: the all convolutional Net. arXiv preprint arXiv:1412.6806 (2014)
23. Ribeiro, M.T., Singh, S., Guestrin, C.: Why should i trust you?" explaining the predictions of any classifier. In: Proceedings of the 22nd ACM SIGKDD International Conference on Knowledge Discovery and Data Mining (2016)
24. Kaiser, L., Gomez, A.N., Chollet, F.: Depthwise separable convolutions for neural machine translation. arXiv preprint arXiv:1706.03059 (2017)
25. Zhou, B., et al.: Learning deep features for discriminative localization. In: Proceedings of the IEEE Conference on Computer Vision and Pattern Recognition (2016)
26. Selvaraju, R.R., et al.: Grad-CAM: visual explanations from deep networks via gradient-based localization. In: Proceedings of the IEEE International Conference on Computer Vision (2017)
27. Ye, W., et al.: Weakly supervised lesion localization with probabilistic-CAM pooling. arXiv preprint arXiv:2005.14480 (2020)
28. Maloca, P.M., et al.: Unraveling the deep learning gearbox in optical coherence tomography image segmentation towards explainable artificial intelligence. Commun. Biol. **4**(1), 1–12 (2021)
29. Jiang, H., et al.: A multi-label deep learning model with interpretable grad-CAM for diabetic retinopathy classification. In: 2020 42nd Annual International Conference of the IEEE Engineering in Medicine & Biology Society (EMBC), pp. 1560-1563. IEEE (2020)
30. Chollet, F.: Xception: deep learning with depthwise separable convolutions. In: Proceedings of the IEEE Conference on Computer Vision and Pattern Recognition (2017)

Non-invasive Anemia Detection
from Conjunctival Images

Rahatara Ferdousi[1]($\boxtimes$)(iD), Nabila Mabruba[1], Fedwa Laamarti[1,2](iD),
Abdulmotaleb El Saddik[1,2](iD), and Chunsheng Yang[3](iD)

[1] University of Ottawa, Ottawa, ON, Canada
{rferd068,nmabr089,flaamart,elsaddik}@uottawa.ca,
{fedwa.laamarti,a.elsaddik}@mbzuai.ac.ae
[2] Mohamed bin Zayed University of Artificial Intelligence, Abu Dhabi, UAE
[3] National Research Council Canada, Ottawa, ON, Canada
chunsheng.yang@nrc.gc.ca

Abstract. Anemia is a worldwide health issue. To diagnose anemia,
blood must be drawn to examine the hemoglobin level. The procedure is
time-consuming and labor-intensive. The existing Artificial Intelligence
(AI)-based anemia detection methods in literature have shortcomings,
including, i) specially designed data collection device, ii) manual feature
extraction, iii) small data size for training the model, and iv)user's trust
in AI prediction. In this paper, we aim to provide a non-invasive model
of anemia detection from visible signs. We trained a CNN model on
eye-membrane image data collected from real patients and open image
sources. Our model predicts anemic patients with good accuracy at 98%.
In addition, we proposed the explainable AI method as a part of the
non-invasive diagnosis to enhance the user's trust in the CNN model's
prediction.

Keywords: Anemia · AI · Computer vision · CNN · Deep learning ·
Explainable AI · Non-invasive

1 Introduction

Artificial Intelligence is used in a variety of areas of health and well-being. Applications range from real-time feedback on physical activities [5] to digital twin
systems for health [7]. One of the important areas of health that can benefit from
the use of AI is Anemia detection. Anemia is a disease in which the number of
red blood cells and their oxygen-carrying capacity are insufficient to the body's
tissues [8]. Nowadays, it is a major global public health issue; around 1.62 billion people worldwide (24.8% of the total population) suffer from this disease.
Children under the age of five (47.4%) and pregnant women(41.8%) are at high
risk of this disease [3]. Early detection of anemia and proper treatment can
reduce the adversarial affects (e.g., death during childbirth) of severed anemia
[10]. However, to detect anemia at the right time, frequent screening is necessary.

© The Author(s), under exclusive license to Springer Nature Switzerland AG 2022
S. Berretti and G.-M. Su (Eds.): ICSM 2022, LNCS 13497, pp. 189–201, 2022.
https://doi.org/10.1007/978-3-031-22061-6_14

Therefore, the anemia diagnosis needs to be non-invasive and cost-free (e.g., no blood drawn or in-person doctor appointment) [4]. Non-invasive AI based anemia detection from visual signs such as-conjunctival pallor, fingernail spooning would be beneficial for instant and cost-free anemia detection. The traditional anemia detection approach to determine hemoglobin level from complete blood count is invasive, costly and require expertise and lab-setting [6].

Image processing [10] (e.g., RGB color analysis, segmentation); machine learning [12] (e..g., K-means, Support Vector Machine (SVM), K-nearest algorithm); deep learning [11] (e.g., Multilayer perceptron, Convolution neural network) have been frequently proposed to automate the anemia diagnosis process through AI. However, the main problems of recent techniques are- i) additional device requirement, ii) manual feature extraction, and iii)limited data samples. The issues with the additional devices are manufacturing costs as well as expertise to use those devices. The traditional machine learning approaches (e.g., K-NN, SVM) requires another level of image processing, such as segmentation, color analysis, etc. to extract features for training the anemia prediction model. The challenge of small data persists due to the unavailability of authorized and relevant images of visual symptoms.

In this study, we propose a computer vision-based non-invasive anemia detection model using convolutional neural network (CNN). The proposed method uses smartphone to collect patient's image. The use of smartphone for data collection does not require any specific cost or expertise. We trained a deep learning CNN model which only requires the image with labels for the training. The features are decided by the neural network by convolution. Similar to the existing approaches, we also encountered the data size problem. Although the data used in our model is comparatively larger than the existing work, it is still a small data size to train a CNN model. Therefore, we used data augmentation to avoid data bias and overfitting issues. The key contributions of this research are two folds. First, we created an anemia prediction dataset by collecting conjuctival pallor images from real patients, google open images and Kaggle public dataset.[1]Please click this link to access the full dataset. Second, we developed a CNN model with comparatively better accuracy and explained the prediction using LIME.

The rest of the paper is organized as follows. In Sect. 2, we discuss the anemia prediction models proposed literature. In Sect. 3, we describe the high-level structure of our proposed anemia prediction approach. In Sect. 4, we present the experiment and results of anemia detection. We compare the results with existing work and explain the prediction. Finally, we conclude this study in Sect. 5 including the limitations and future scope.

2   Related Work

This section summarizes contemporary studies for non-invasive anemia detection. The goal of this section is to summarize the key contributions of existing

---

[1] https://drive.google.com/drive/folders/1qR3mTTj7N-Law6ylR_YnI5JeUI27Pt8H?
usp=sharing.

literature. We first categorize the AI methodologies proposed in existing work into two groups as follows, then we compare the existing methodologies in a comparative table later in this section.

2.1   Color Analysis-Based Method

1. In [4], the authors present a data acquisition device based on color clustering to eliminate the issues of improper light. The macro-lens captures a closer view of the images, highlighting more detail of the data. The spacer containing LED lights helps to focus the eye areas and eliminates the issue with ambient light. Such detailed capturing of the data provided better prediction accuracy. For the classification of the anemia, they used various machine learning algorithms like Naïve Bayes, Decision tree, and a balanced dataset to obtain better accuracy.

2. In [10], the authors aim at developing a non-invasive system by color analysis using AI technologies. The model consists of data collection, image processing, augmentation, segmentation, feature extraction, and modeling. For each of the steps, the author proposed existing machine learning algorithms, The prediction was regression based.

3. In [8], the authors use a K-means algorithm to cluster the RGB pixel values to segment the dominant color(R, G, or B) in the image. The under-eye images are segmented by finding the closest centroid of the pixels. The Euclidean distance was measured to find the closest centroid of the cluster. The proposed K-means color clustering is used to find the similar images in patient's gallery to decide whether it's a sign of anemic or not.

4. Similar to [8], the method in [12], also involves a color analysis algorithm that measures hemoglobin through an image color analysis algorithm. The users need to select the areas on fingernails to detect the hemoglobin level. The images taken in low-light were ignored to improve the accuracy. The authors narrated that smartphone based anemia detection application can replace the blood count based and in-person anemia detection.

5. The system in [9], used a Raspberry Pi Zero with a 5M pixel 1080P camera to capture a real-time image of the user's nail and then used colour classification and image classification to detect anaemia levels. They used existing image data set to train this system that produces more accurate data. The key difference between this study and other studies is that their system performs real-time image analysis.

6. The authors in [2], obtained the conjunctiva images with a camera and then calculated the ROI (region of interest). Following the extraction of the ROI, the intensity values of the image pixels' R and G components were measured and compared to the ROI. The model predicted less red eye-membranes as pallor.

2.2   Neural Network-Based Method

1. Sickle cells in the human body may be recognized with great precision using data mining techniques such as the multilayer perceptron classifier algorithm [13]. The authors proposed an MLP (Multi-Layer Perceptron) classification algorithm that classifies Sickle Cell Anemia (SCA) into three groups- Normal (N), Sickle Cells (S), and Thalassemia (T). The authors obtained 99% prediction accuracy for classifying sickle cell anemia.
2. Similar to [13], the authors in [1] also use blood cell images to train their model. The pre-trained model is updated in such a way that the first layer learns low-level features such as colors and edges, while the last layer learns task-specific features. The last layers are then replaced with new layers to learn target task-specific features for training a fine-tuned model on the target dataset.
3. In [11], the proposed system takes images from the smartphone camera and analyses them upon servers in order to predict anemia, along with the results delivered to the mobile phone. A CNN model was developed to predict anemia. Although the underlying technology of the model is CNN, the authors used R and G color analysis and segmentation for understanding the pale eye membrane area.

We summarize and compare the data and technology used in recent anemia detection AI models. The comparison is presented in Table 1. Here, source represents the data collection source. Authorized indicates that the data are labeled and approved by a medical practitioner. Sample type defines the type of symptoms considered as data samples. Size reflects the amount of data used to train the prediction model. Access denotes the availability of data for usage.

From Table 1, we can summarize the existing work as follows.

– In [4], the authors used a specially designed device to capture the high-quality image of eye membrane. Although this approach is non-invasive, it requires an additional cost and expertise to use the device.
– The sickle-cell anemia prediction models in [1,13] is trained with more data ranging from 200–1387. However, the blood cell images are collected by surveying blood test reports in invasive anemia detection. As this procedure requires blood to be drawn each time, in practice it is not feasible to apply frequently. To address this issue smartphone has been chosen by some research to capture visible symptoms for non-invasive anemia detection.
– The AI models trained with visual signs like conjunctival pallor fingernails spooning has been used a small amount of data ranging from 36–102.
  From the above discussion and table, it can be observed that the data used in the non-invasive anemia detection is smaller and restricted to access. This indicates that accessing real patients' sign symptoms is challenging. Moreover, most of the images are not labelled by medical practitioners.

Table 1. Comparison of existing anemia detection AI models.

Authors	Source	Authorized	Sample Type	Size	Access
Diamauro [4]	Device	No	Conjunctival pallor image	102	Private
Nrayan [10]	Smartphone camera	No	Conjunctival pallor image, Blood test report	–	Private
Mannino [8]	Survey		Conjunctival pallor image	36	Private
Sevani [12]	Smartphone camera	No	Fingernails image	100	Private
Yeruva [13]	Medical institute	Yes	Blood cell image	1387	Private
Ghosal [6]	Smartphone camera	Yes	Conjunctival pallor image	60	Private
Saldivar [11]	Smartphone camera	Yes	Conjunctival pallor image	300	Public
Alzubaidi [1]	–	Yes	Blood cell image	200-626	Private
Mitra [9]	Medical institute	No	Fingernails image	–	Private
Bauskar [2]	–	No	Conjunctival pallor image	99	Public

3   Method

In this section, first we present the high-level structure of our proposed method. Then we elaborate the detail of each steps in the structure.

Fig. 1. Block diagram of the high-level structure of the proposed method.

3.1   High-Level Structure of the Proposed Method

The high-level structure of the proposed anemia detection approach is listed as follows. The high-level structure was illustrated in Fig. 1. The description of the steps is given below.

- **Data Collection:** The first step is to collect and aggregate conjunctival image data.

- **Data Pre-processing:** The second step involves cleaning, augmentation, and domain expert validation.
- **Model Tuning and Training:** In the third step, we tune parameters such as epochs, layers, batch size, kernel size, etc., to obtain the final model that we will use for anemia prediction. After determining the parameters we trained a CNN model with our anemia dataset.
- **Evaluation:** In the fourth stage, the performance of the anemia prediction model is evaluated.
- **Explanation:** Finally, the prediction is explained using the LIME algorithm to highlight the area that caused a specific output (e.g., anemic, non anemic).

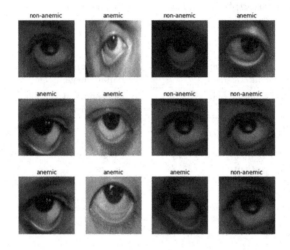

Fig. 2. Example of samples in the anemia detection dataset. Link to access the dataset

Data Collection. We required under eye membrane's image data of both-anemic and non-anemic to build the classifier. We collected the data as follows.

- First, We used an open-source dataset from Kaggle. That data was not authorized by a medical doctor.
- We then collected data from a doctor's office in Bangladesh. These data contain conjunctival images of patients who have been tested as anemic or non-anemic as per their blood test report.
- After that, we collected some data from the Google's open source image database.
- Ultimately we aggregated data from these three sources and the medical practitioner scrutinized those data as anemic and non-anemic.

One of the reasons of selecting conjunctiva images as data is that the eye-membrane color doesn't vary significantly for different skin tones. Nevertheless, we considered a combination of samples from three different sources. An example of the data is demonstrated in Fig. 2.

Data Pre-Processing. The collected data is pre-processed in the following ways.

- First, we cleaned the google open source images, as mostly the images were blur and noisy.
- Then, we resized all the images into 128×128 size.
- After that, images were normalized to scale down the pixel values to 0 to 1.
- Though we collected data from three sources, this data was not good enough to train a CNN model. So we performed data augmentation by flipping and rotating all the training images to enhance the dataset.
- The clean dataset contains: 150 publicly available anemia dataset from Kaggle, 50 Real patient image, 72 Google open images. Data distribution per class: 158 Anemic 114 Non-anemic.

Data Partitioning. We coded a function to split our dataset into three parts: training 60%, testing 20%, and validation 20%. As our dataset is small, to avoid empty batches in each split we considered 20% data in the validation and test set. The training set is used to fit the models; the validation set is used to estimate prediction error for model selection; the test set is used for assessment of the generalization error of the final chosen model. Validation set is used to avoid over-fitting problem.

Model Tuning and Training. In this step, we determined various parameters to customize the CNN architecture for our target dataset. After several trial and error, we determined the following selections to build the model.

- Batch Size = 32. We tried 8, 16 and 32 too.
- IMAGE SIZE = 128×128. We also tried 64×64 and 256×256.
- Channels = 3, because we need color image to understand the paleness. So R, G, and B channels were determined.
- EPOCHS = 30, we considered, 5, 10, 15, 20, 25, and 30.
- Metrics = Categorical Cross Entropy. Because this function estimates loss optimally for the categorical classification.
- Number of hidden layers (Convolutional and Pooling layer pairs) four. Here we also considered 2,3,5, and 6 hidden layers
- Kernel Size of Convolution = 3×3, Kernel Size of Maxpooling = 2×2. We also tried Convolution = 2×3, Kernel Size of Maxpooling = 1×1.
- For activating the hidden layer Rectified Linear Unit (RelU) function is used.
- As we performed a binary classification, to activate the output layer we used softmax function that provides probability score (1 or 0) for an image.

The values of the parameters that we considered did not provide better accuracy or resulted into significant validation loss. Finally we trained the model by compiling it with the selected parameters mentioned above. The architecture of the CNN model used in this work is illustrated in Fig. 3.

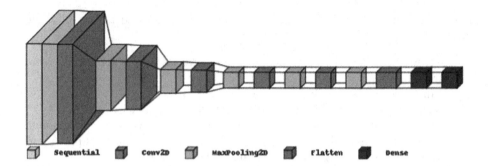

Fig. 3. Architecture of the CNN model.

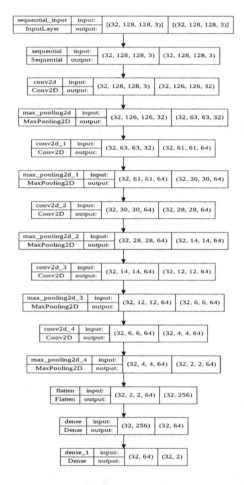

Fig. 4. Summary of the CNN model trained for the prediction.

It can be observed that the neural network has a sequential input layer, followed by four pairs of Convolutional layer and Maxpooling layer. The convolutional layers are gradually squeezed and flattened and connected to the output layer. The summary of this model is illustarted in Fig. 4.

To understand the summary in Fig. 4,

- let us first consider the input layer. The size of each image is (128, 128, 3), as per the dimension we defined for image size and channel.
- The keras then appends an extra dimension for processing 32 batches. By processing we mean training multiple images in every step of a single epoch.
- Convolving a (128, 128) image with a (3, 3) filter, with strides and dilation rate of 1, and 'valid' padding, results in an output of size $(128 - 3 + 1, 128 - 3 + 1) = (126, 126)$. Since, this layer have 32 such filters, the output shape becomes (126, 126, 32).
- We used default MaxPooling kernel having a shape of (2, 2) and strides of (2, 2). Applying that to a (126, 126) image results in an image of shape $(((126 - 2)//2) + 1, ((126 - 2)//2) + 1)) = (63, 63)$.
- This pattern is extended to all the following Conv2D and MaxPooling layers with 64 filters.
- The Flatten layer takes all pixels along all channels and creates a 1D vector (not considering batch size). Therefore, an input of (2, 2, 64) is flattened to $(2 * 2 * 64) = 256$ values.
  The number of parameters for a Conv2D layer is given by: (kernel height * kernel width * input channels * output channels) + (output channels if bias is used). In this way, the model obtains 146,754 trainable parameters.

4 Evaluation

First, we evaluated the model's performance. We trained the model at 30 EPOCHS and observed the training and validation loss in parallel. Second, we performed experiment on individual and batch images. Finally, we compared the performance with existing work.

4.1 Model Performance

Figure 5, demonstrates that demonstrates that after epoch 5 the accuracy gradually increases and the loss gradually decreases. In addition, there is a very subtle difference between the validation and training loss. This result indicates that the model may not suffering from any significant overfitting issue. Near Epoch 25, there is a peak in the loss, which may be due to presence of noisy data in the dataset.

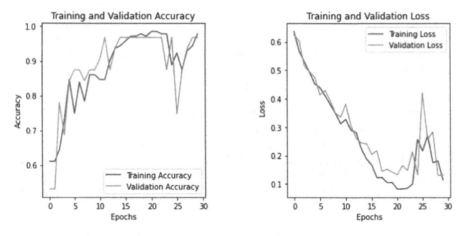

Fig. 5. Comparison of training and validation accuracy (left) and loss (right).

4.2   Prediction Accuracy

We performed the prediction for individual and one batch of images and found that overall the confidence score ranges between 86% – 99%. The confidence score of prediction for one batch is depicted in Fig. 6. From this result we found that the model predicted anemia with good confidence score. A few images were predicted with 100% confidence score due to minor overfitting issues. As we mentioned before, that to mitigate the overfitting issues we used data augmentation to subdue the overfitting issues due to the small size of the dataset.

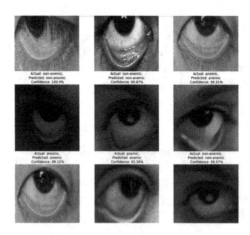

Fig. 6. Confidence score of prediction on images from a specific batch.

4.3 Comparison with Existing Work

In this section, we discuss the relative investigation of previous existing method-ologies. We consider a comparative analysis based on a few parameters which are mentioned in Table 2.

Table 2. Comparative analysis of existing anemia detection approach.

Author	Model Name	Model Type	Accuracy (%)
Dimauro[4]	Logistic Regression, Naïve Bayes, KNN, Decision Tree	Classification	95
Narayan[10]	SLIC Superpixelation	Regrassion	–
Sevani [12]	K-means algorithm	Classification	90
Mannino [8]	Color Analysis algorithm	Classification	97
Yeruva [13]	Multilayer Perceptron (MLP)	Classification	99
Ghosal [6]	Image Processing	Classification	89
Saldivar [11]	Deep Learning CNN	Classification	77.58
Alzubaidi[1]	Transfer Learning	Classification	98.87
Mitra[9]	Modified Deep Learning	Classification	–
Bauskar[2]	Machine Learning	Regression	93
Proposed	Convolutional Neutral Netowrk	Deep learning	98

From the above table we can observe that studies [1, 13] demonstrate higher accuracy. However, they use invasive, whereas our accuracy is high, and we use non-invasive detection.

4.4 Explanation

The CNN is a black-box model and it only provides us with the decision that whether a patient is anemic or non-anemic. However, to calibrate the trust of the users in our AI model we used LIME algorithm to identify the region responsible for providing the output. The LIME tool worked as follows.

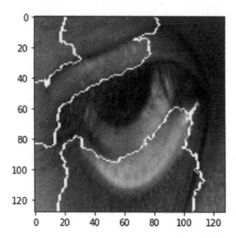

Fig. 7. Explanation of the prediction by the CNN model using LIME tool.

- To explain the prediction, we applied the LIME algorithm after the output layer. The algorithm modifies single data points based on the feature values.
- The LIME algorithm split the image outputs into superpixels, where patches of the part images have similar visual features: color or brightness. To explain the prediction of the CNN model, first, it splits an output image into K number of superpixels. These superpixels represent similar image patches.
- Then the algorithm randomly creates new images by changing the superpixel values to 1 and 0. Then it searches for the original prediction class (anemic or non-anemic) and fits the CNN model to a surrogate model (e.g. regression model) for a specific class preserving the superpixel values to 1 and 0.

As shown in Fig. 7, we can find that the LIME model identified an image as anemic due to under eye-membrane and parabolic shape. It can be observed that the model considers the eye-membrane area, however, it has also taken some other regions in the eye, which we did not expect.

5   Conclusions

In this paper, we propose a CNN model-based anemia detection using real patients and open source data. In addition, the trained model demonstrates good performance. There are scopes to improve and extend this research. CNN is a data-hungry model as it needs arguably 10,000 images to provide better performance. Although the dataset used in this work is very limited according to the requirement of the CNN model, comparatively it was higher than other existing works in the literature. The transfer learning model like You only look once(YOLO) can be trained with this data to mitigate the unavailability of the data. In addition a bounding box labelling can be applied to the existing dataset. It can exclude the other regions of the image from the eye-membrane. Such a dataset can be used for real-time prediction.

Acknowledgements. This research is supported in part by collaborative research funding from the National Program Office under National Research Council of Canada's Artificial Intelligence for Logistics Program.
Also, we cordially thank Dr.Nusrat Jahan, MBBS,BCS (health), PDS code : 141946, Assistant Surgeon at government of the people's republic of Bangladesh; for supporting the medical background and data scrutinizing of this research .

References

1. Alzubaidi, L., Fadhel, M.A., Al-Shamma, O., Zhang, J., Duan, Y.: Deep learning models for classification of red blood cells in microscopy images to aid in sickle cell anemia diagnosis. Electronics **9**(3), 427 (2020)
2. Bauskar, S., Jain, P., Gyanchandani, M.: A noninvasive computerized technique to detect anemia using images of eye conjunctiva. Pattern Recogn. Image Anal. **29**(3), 438–446 (2019)
3. De Benoist, B., Cogswell, M., Egli, I., McLean, E.: Worldwide prevalence of Anaemia 1993–2005; who global database of Anaemia (2008)
4. Dimauro, G., Guarini, A., Caivano, D., Girardi, F., Pasciolla, C., Iacobazzi, A.: Detecting clinical signs of Anaemia from digital images of the palpebral conjunctiva. IEEE Access **7**, 113488–113498 (2019)
5. Díaz, R.G., Laamarti, F., El Saddik, A.: DTCoach: your digital twin coach on the edge during COVID-19 and beyond. IEEE Instrum. Measur. Mag. **24**, 22–28 (2021)
6. Ghosal, S., Das, D., Udutalapally, V., Talukder, A.K., Misra, S.: sHEMO: Smartphone spectroscopy for blood hemoglobin level monitoring in smart anemia-care. IEEE Sensors J. **21**(6), 8520–8529 (2020)
7. Laamarti, F., Badawi, H.F., Ding, Y., Arafsha, F., Hafidh, B., El Saddik, A.: An ISO/IEEE 11073 standardized digital twin framework for health and well-being in smart cities. IEEE Access **8**, 105950–105961 (2020)
8. Mannino, R.G., et al.: Smartphone app for non-invasive detection of anemia using only patient-sourced photos. Nature commun. **9**(1), 1–10 (2018)
9. Mitra, S., Rathore, S., Gupta, S.K.: A novel non-invasive algorithm using the concept of nail based anemia disease detection. J. Univ. Shanghai Sci. Technol. **23**(2), 265–273 (2021)
10. Narayan, S.S., et al.: A smartphone based multi input workflow for non-invasive estimation of Haemoglobin levels using machine learning techniques. arXiv preprint arXiv:2011.14370 (2020)
11. Saldivar-Espinoza, B., et al.: Portable system for the prediction of anemia based on the ocular conjunctiva using artificial intelligence. arXiv preprint arXiv:1910.12399 (2019)
12. Sevani, N., et al.: Detection anemia based on conjunctiva pallor level using k-means algorithm. In: IOP Conf. Ser. Mater. Sci. Eng. **420**, 012101 (2018). IOP Publishing (2018)
13. Yeruva, S., Varalakshmi, M.S., Gowtham, B.P., Chandana, Y.H., Prasad, P.K.: Identification of sickle cell anemia using deep neural networks. Emerg. Sci. J. **5**(2), 200–210 (2021)

3D Segmentation and Visualization of Human Brain CT Images for Surgical Training - A VTK Approach

Palak[1]($\boxtimes$), Benjamin Delbos[2], Rémi Chalard[2], Richard Moreau[2], Arnaud Lelevé[2], and Irene Cheng[1]

[1] Department of Computing Science, University of Alberta, Edmonton, Canada
{ptiwary,locheng}@ualberta.ca
[2] Univ Lyon, INSA Lyon, Villeurbanne, France
{benjamin.delbos,remi.chalard,richard.moreau,
arnaud.leleve}@insa-lyon.fr

Abstract. Simulated surgical planning and training has been proved to be effective in enhancing the performance of surgical operations. 3D medical image modeling and visualization is therefore gaining increasing attention in the research community as the navigation using 3D DICOM data provides a more realistic planning and training environment. However, medical applications often have specific targets, e.g. TB, cancer and tumor. Algorithms developed for the respective applications are designed based on the characteristics, like shape and intensity, of the target. In order to analyze the target for symptom diagnostic or monitoring purposes, the target region needs to be segmented out from its background so that the algorithm output will not be adversely affected by irrelevant signals close to the target. In this paper, we focus on segmentation and visualization, with a use case of developing a 3D environment for Ventricular puncture operation planning and training. The difference between our work and other segmentation techniques is that we need to segment not only one target, but also the path along the surgical tool inserted into the brain. This creates challenges to the algorithm design because a set of segmentation parameters may be effective for one region, but may not be effective for another due to the different data region contrasts, densities, shapes and so on. Segmentation is only an initial step but is necessary in order to conduct the actual surgical training. While many researchers or clinicians waste effort in generating segmentation results, our contribution lies in our VTK approach, which is fast to implement so that the users can focus on the core process. Our experimental results demonstrate the feasibility of providing a realistic 3D visualization and interactive environment for surgical planning and training.

Keywords: Surgical training · 3D brain modeling · VTK · CT imaging · Hounsfield · Catheter simulation · Haptic feedback

S. Berretti and G.-M. Su (Eds.): ICSM 2022, LNCS 13497, pp. 202–212, 2022.
https://doi.org/10.1007/978-3-031-22061-6_15

1   Introduction

A CT (Computed tomography) scan is an x-ray procedure that creates a stack of cross-sectional images with the help of computer technology. CT images are more detailed than conventional x-ray images and can reveal bones as well as soft tissues and organs. DICOM image slices that CT scans produce are in serial form and present 3D visualization of the human body.

They can reveal abnormal structures and help the physician plan and monitor treatments. 3D visualization plays an important role in diagnostic healthcare, medical research and education. A head CT can be used to evaluate various structures of the brain and look for abnormalities, areas of bleeding, stroke, tumor, etc. The Hounsfield scale is a quantitative measure of radiodensity [11]. Hounsfield Unit (HU) is computed from the values of Rescale Intercept and Rescale Slope available in DICOM images. Pixel values in a CT image are displayed in terms of relative radiodensity. It is displayed according to the mean attenuation of the tissue that it corresponds to on a scale from -1024 to over 3000 on the Hounsfield scale.

HU can be calculated from pixel data as follows [12]:

$$HU = PixelValue * RescaleSlope + RescaleIntercept$$

Figure 1 [3] shows the general HU values of a human body. In this paper, we present techniques that can segment and visualize different ranges of HU index using the VTK (Visualization Toolkit) library for DICOM images.

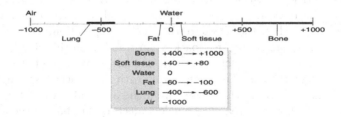

Fig. 1. Hounsfield units of a human body [3]

Segmentation and visualization are two fundamental and necessary steps in medical image analysis. Segmentation aims to separate the different regions in the brain, e.g. bones, white matter, gray matter, tissues and so on, so that the target object, e.g., ventricles or tumor, can be extracted for further detailed analysis. Segmentation provides a navigation map (visualization) for the deployment of surgical tools. In recent years, there is an increasing research effort to develop simulated tools for surgical training and planning. The objective of our work is to provide insight into the feasibility of combing haptic-device feedback with software visualization to simulate surgical operations for clinical training and planning, with the use case of developing a 3D environment for ventricular

puncture. Ventricular drainage is commonly performed in neurosurgery depart-
ments or in the emergency room. It consists of inserting a catheter into the brain,
using a needle, until it reaches the frontal horn to drain cerebrospinal fluid for
therapeutic or diagnostic purposes [14]. Figure 2 shows our skull phantom. The
long rod is a catheter used by surgeons. They have different sizes of catheter.
We used this one because it is quite rigid and thus it does not deform during
insertion, whereas smaller catheter are deformable. Once the catheter is in place
they can insert through it another tool to drain the ventricule.

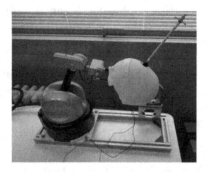

Fig. 2. Left is our haptic-device, and a brain skull phantom with a catheter inserted
is displayed on the right.

Nowadays, the learning of this gesture is only performed by companionship:
there is no effective simulator for training in this type of surgery. Motivated by
this need, we aim to design a haptic-enhanced visual simulator that renders the
various layers crossed by the needle during such an operation, so that surgeons
get a realistic visual-and-touch feedback when training on this simulator before
operating on a patient. The challenge is to model the volume of the brain, the
ventricle, and the virtual needle synchronized with the haptic-device handled by
the trainee. The location of the needle tip in the 3D model will activate haptic
cues specific to each brain area. The whole trajectory will be recorded for rapid
objective evaluation purposes [15]. The 3D model will have to be calibrated with
the haptic-device to synchronize the device position with the virtual needle in
real-time. In this paper, we focus on the segmentation and visualization steps.

2   Related Work

3D visualization and reconstruction can be achieved using surface rendering
and volume rendering techniques. In [1], 3D Visualization of brain MR DICOM
images was achieved using these techniques through VTK platform and free
plane clipping method was applied to achieve virtual dissection. In 2016, a VTK-
based medical image 3D visualization system was built [7] to render the results
of 3D reconstruction using two different techniques - the Marching Cubes algorithm and Ray Casting algorithm. A similar 3D rendering software was designed

for MR images using vtkFixedPointVolumeRayCastMapper and Microsoft.NET framework [9]. These algorithms have been used widely in medical 3D reconstruction and have proven to be very efficient [2].

Marching Cubes is a high resolution 3D surface reconstruction algorithm. It generates a 3D surface by processing the 3D medical data in scan-line order. The algorithm requires the user to specify a threshold value which is the density value corresponding to the surface that is desired to be visualized. [8] Surfaces corresponding to the threshold are then located and triangles are created. Finally, a visualizable image is constructed by creating normals to the surface at each vertex of the triangle. Ray Casting algorithm is used for rendering 3D models where opacity and color details are collected along the ray traversing path through the entire volume. VTK provides users with functions for both the algorithms mentioned above.

3D visualization plays an important role in detecting abnormalities in human bodies, such as tumors, cancers, blockages, etc. It contributes towards timely detection and treatment. In 2018, the marching cube algorithm was applied to a segmented brain tumor to generate the vertices and the faces to form a 3D model. This information was stored in a stl file [5]. Region-growing algorithms have been used for automatic segmentation of lungs from chest CT scans. Image processing functions from the ITK library and volume rendering methods from the VTK library were also used [10]. In 2018, a brain haemorrhage region detection technique was proposed on DICOM images using the HU-based approach. The segmentation was based on the HU values in the range of 40–50 that corresponds to the haemorrhages [6]. The above mentioned techniques have been proven to be efficient for segmentation of one target structure, e.g., the lung or abnormalities like a brain tumor, cancer, or haemorrhage. In this paper, we aim at segmenting and visualizing multiple anatomical regions of the human head along the path of a surgical tool. These structures may be the brain, bones, ventricles, and blockages, having very different contrasts and densities. Algorithms designed to target a particular structure characteristics is unlikely to segment effectively other brain regions composed of different material types.

3   Methodology

CT data is used as input and reconstructed in order to prepare for segmentation. 3D DICOM Reconstruction can be implemented using two techniques: surface rendering and volume rendering. In this paper, we present the results of HU-based surface rendering and volume rendering of brain CT DICOM images (Fig. 3) using VTK libraries (vtkContourFilter, vtkGPUVolumeRayCastMapper and vtkMarchingCubes).

The challenge of this step is to identify isosurface values, which need to be defined as thresholds in the algorithms in order to visualize the respective surfaces. We first convert raw pixel values to HU values using rescale intercept and rescale slope, that are stored in the DICOM header, and then plot its histogram.

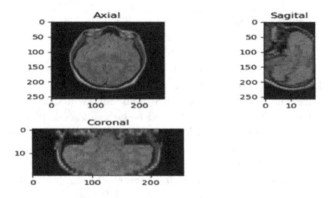

Fig. 3. CT brain DICOM images from different views

As illustrated in Fig. 4, there are lots of air, white matter, grey matter, blood, parts of brain and some bones, corresponding to different HU values.

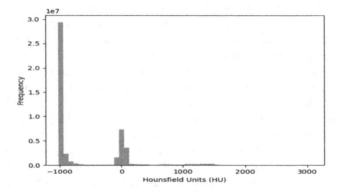

Fig. 4. Histogram of hounsfield units

Since we know that air exhibits HU value of -1000, we apply thresholding to mask out air from the image. We then move onto surface extraction using the VTK functions mentioned above. Finally, we render the generated 3D surfaces as described below.

3.1 Volume Rendering

Volume rendering can be achieved using vtkVolumeRayCastMapper, which maps the volume data from vtkImageData using vtkVolumeRayCastFunction to calculate the pixels and display them in the render window when signalled by vtkRenderer. This mapper uses vtk transfer function to create a 1-D lookup texture that the shader program can use to perform composite ray casting [13]. This transfer function serves as the input to the vtkVolumeProperty. The transfer function consists of a piece-wise function called vtkPiecewiseFunction for the scalar opacity and a function called vtkColorTransferFunction for the color.

3.2   Surface Rendering

Surface rendering can be achieved using vtkContourFilter, which creates an isosurface through the data at a specific threshold value (HU value). This function finds points in the volume data where the scalar value corresponds to the threshold value given by the user. It creates isosurfaces by scanning through the data volume and connecting points of the same value. The threshold value provided might occur several times in the data volume. This leads to multiple isocontours being returned by the algorithm.

Marching-cubes is one of the most widely used surface rendering methods. vtkMarchingCubes implements marching-cubes technique and is more efficient than vtkContourFilter and vtkVolumeRayCastMapper methods because surface rendering can be achieved with more resolution and higher precision. vtkMarchingCubes is a filter that takes volumetric image data as its input and generates an output of one or more isosurfaces. One or more iso values must be specified to generate the isosurfaces. A min/max scalar range and number of contours can also be specified to generate a series of evenly spaced contour values.

4   Experimental Result

Our 3D Brain CT DICOM images used for visualization were provided by a surgeon at Lyon. Figure 5 shows the properties of these DICOM images. Pixel data is usually given in axial orientation in a high resolution grid of 512×512 pixels.

Properties	Value
Slice thickness (mm)	1.25
Pixel Spacing (row, col)(mm)	0.488281,0.488281
Rescale Intercept	-1024
Rescale Slope	1
Number of slices	188
Dimensions	512x512

Fig. 5. Properties of DICOM images

These images were thresholded to different patches based on the HU values to remove air and generate surfaces for skin, bone and soft tissues. The result of vtkMarchingCubes is shown in Fig. 6 and Fig. 7, where different colours represent different meshes. HU values of +500 to +3000 represent bone in CT images. Figure 8 shows the result of vtkContourFilter when a HU threshold value of

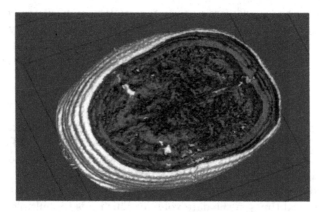

Fig. 6. Surface rendering of skin, brain and soft tissues with marching cubes

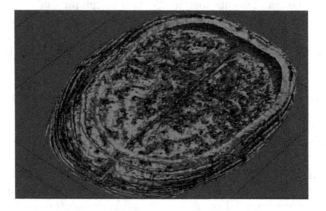

Fig. 7. Surface rendering of brain and soft tissues with marching cubes

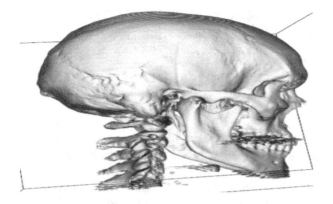

Fig. 8. Surface rendering of bone structure using vtkContourFilter

Fig. 9. Surface rendering of brain and soft tissues with vtkContourFilter

500 was defined in order to extract a surface for bone or bone-like density. Soft tissues and skin were rendered in a similar manner as shown in Fig. 9.

The DICOM images were thresholded using a HU value of -1000 to remove all air and then vtkGPUVolumeRayCastMapper was used to create a 3D volume of the remaining structures as shown in Fig. 10 and Fig. 11.

Fig. 10. Result of volume rendering using vtkGPUVolumeRayCastMapper

5   Discussion

Without segmentation, clinicians and surgeons can only rely on visual clues to carry out surgical planning and training. For example, in our use case of ventricular puncture operation, Fig. 12 shows how the CT data can be displayed from multiple views along the three major axes to provide visual guidance to the

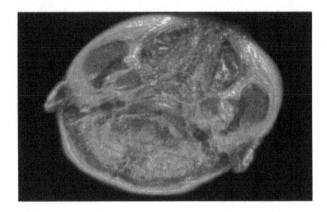

Fig. 11. Result of volume rendering using vtkGPUVolumeRayCastMapper with more isosurfaces

users. The planned trajectory of the surgical tool (catheter) is projected onto the display. The advantage of 3D segmentation and modeling is to allow the surgical experts to define the relative material values in each segmented region or layer. These material values are mapped onto a feedback force table associated with the haptic-device. Therefore, in addition to the visual guide, users navigating the catheter can also feel a force simulating the pressure generated from the anatomical structures, e.g., bones and soft tissues, when the catheter is inserted into the human brain.

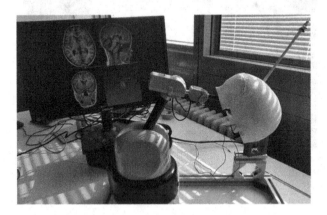

Fig. 12. The screen illustrates visualization of CT data without 3D segmentation and modeling, with our haptic-device and head phantom and catheter inserted.

We observed that it can be challenging when it comes to segment more finely-divided individual structures when these structures have similar HU vales or fall under the same HU value window. VTK is not very efficient when it comes to segment structures with small HU ranges. One solution is to apply progressive interpolation between two surfaces so that the gradually changing feedback forces detected can supplement the visualization clue (visual-touch perception) for the clinician or surgeons during the planning/training process. In future work, we will explore this research further.

6 Conclusion

3D reconstruction, modeling and visualization of medical data has been proven to be useful to clinicians and surgeons in identifying abnormalities in the human body and helping them in surgical planning and training, which results in better surgical performance. However, an initial but necessary step is segmentation. While algorithms are often designed for specific targets, e.g., lung and tumor, they are unlikely to be effective on multiple targets, as different regions or layers like bones, arteries and soft tissues have their unique contrasts, material densities and so on, which require specific set of parameter values in the algorithms in order to define the patch boundaries. Different from related work, which focus on a target structure, we aim to segment regions (layers) along the path of a surgical tool inserted into the brain, with ventricular puncture as our use case. In this paper, we presented results of segmentation and visualization of Brain CT DICOM data. Visualization of different surfaces composed of bone, skin and soft tissue was implemented using VTK. Our approach is fast and provides a realistic 3D environment for interactive navigation of the surgical tool. Our objective is to help the users to focus on the core process of surgical planning and training, without wasting too much efforts on the initial segmentation task.

Acknowledgment. The financial supports from NSERC and the Mitacs Globalink Program, Canada, are gratefully appreciated.

References

1. Du, R., Lee, H.J.: A Visualization System of Brain MR image based on VTK (2019)
2. Zhao, K., Sun, Q., Liu, Z.: 3D Reconstruction of Human Head CT Images Based on VTK, pp. 16–20 (2020). https://doi.org/10.1109/ICARM49381.2020.9195291
3. Lima, R.F., Marengoni, M.: Visualization 3D Reconstruction - Volume Rendering of Mucus into Paranasal Sinuses. VISAPP (2015)
4. http://www.intl.elsevierhealth.com/e-books/pdf/940.pdf
5. Al-Rei, M.: Automated 3D Visualization of Brain Cancer (2017)
6. Phan, A.C., Võ, V.Q., Phan, T.C.: A Hounsfield value-based approach for automatic recognition of brain haemorrhage. J. Inf. Telecommun. **3**, 1–14 (2018). https://doi.org/10.1080/24751839.2018.1547951
7. Tan, J., Chen, J., Wang, Y., Li, L., Bao, Y.: Design of 3D visualization system based on VTK utilizing marching cubes and ray casting algorithm. In: 8th International Conference on Intelligent Human-Machine Systems and Cybernetics (IHMSC), pp. 192–197 (2016). https://doi.org/10.1109/IHMSC.2016.153

8. Lorensen, W.E., Cline, H.E.: Marching cubes: a high resolution 3D surface construction algorithm. SIGGRAPH Comput. Graph. **21**, 163–169 (1987). https://doi.org/10.1145/37402.37422

9. Madusanka, N., Zaben, N.A., Shidaifat, A.A., Choi, H.-K.: 3D rendering of magnetic resonance images using visualization toolkit and Microsoft.NET framework. J. Multimedia Inf. Syst. **2**(2), 207–214 (2015). https://doi.org/10.9717/JMIS.2015.2.2.207

10. da Nobrega, R.V.M., Rodrigues, M.B., Reboucas Filho, P.P.: Segmentation and Visualization of the Lungs in Three Dimensions Using 3D Region Growing and Visualization Toolkit in CT Examinations of the Chest (2017). https://doi.org/10.1109/CBMS.2017.23

11. Hounsfield scale, 14 March 2018. https://en.wikipedia.org/wiki/Hounsfield$_$scale. Accessed 21 March 2018

12. Surface extraction: creating a mesh from pixel-data using Python and VTK, 26 October 2014. https://pyscience.wordpress.com/2014/09/11/surface-extraction-creating-amesh-from-pixel-data-using-python-and-vtk/. Accessed 5 March 2022

13. Real-time volume rendering with hardware-accelerated raycasting, 18 April 2006. https://citeseerx.ist.psu.edu/viewdoc/download?doi=10.1.1.121.304&rep=rep1&type=pdf. Accessed 13 March 2022

14. Toma, A.K., Camp, S., Watkins, L.D., Grieve, J., Kitchen, N.D.: External ventricular drain insertion accuracy: is there a need for change in practice? Neurosurgery **65**(6), 1197–200 (2009). https://doi.org/10.1227/01.NEU.0000356973.39913.0B

15. Raabe, C., Fichtner, J., Beck, J., Gralla, J., Raabe, A.: Revisiting the rules for freehand ventriculostomy: a virtual reality analysis. J. Neurosurg. JNS, **128**(4), 1250–1257 (2018). https://thejns.org/view/journals/j-neurosurg/128/4/article-p1250.xml. Accessed 9 Dec 2020

Smart Homes

The Energy 4.0 Concept and Its Relationship with the S³ Framework

Pedro Ponce[1](✉) ⓘD, Omar Mata[1] ⓘD, Sergio Castellanos[2] ⓘD, Arturo Molina[1] ⓘD,
Troy McDaniel[3] ⓘD, and Juana Isabel Mendez[1] ⓘD

[1] Institute of Advanced Materials for Sustainable Manufacturing, Tecnologico de Monterrey,
64849 Monterrey, Nuevo Leon, Mexico
{pedro.ponce,omar.mata,armolina,isabelmendez}@tec.mx
[2] Cockrell School of Engineering, The University of Texas at Austin, Austin, USA
sergioc@utexas.edu
[3] The Polytechnic School Ira A. Fulton Schools of Engineering, Arizona State University,
Tempe, AZ, USA
troy.mcdaniel@asu.edu

Abstract. The integration of cyber-physical systems into industrial production systems is called Industry 4.0, referring to the fourth industrial revolution. On the other hand, a cyber-physical system is understood as incorporating sensing, computation, control, and networking into physical objects, linking them and the Internet. Currently, the technologies that compose cyber-physical systems have a fast development attracting the attention of the industry and academy. As a result, comments, discussions, and analysis to applying the industry 4.0 concept and elements that are integrated have emerged in several fields such as medicine, engineering, economics, energy, etc. Mainly, the energy sector is focusing on the digitalization of the energy industry, where the information and communications technology, energy management, and intelligent buildings are considered. Moreover, the concept of energy 4.0 is not clear enough since it could be considered only to integrate technology to the network. However, this paper coins a novel definition in which energy 4.0 integrates green, digital and intelligent management of energy according with the end-user social and economic requirements. This new structure of the energy sector has been called Energy 4.0. In this paper, the S3 (innovative, sensing, and sustainably) Framework is applied as an alternative for product innovation in Energy sector. Furthermore, two examples of Energy 4.0 systems are shown.

Keywords: Energy 4.0 · S3 product · Renewable energy · Energy management systems · Industry automation

1 Introduction

The first mention of the term "Industry 4.0" was at the Hannover Fair 2011 [1, 2]. Industry 4.0 is understood as incorporating cyber-physical systems (CPS) into industrial production systems, pursuing a higher level of automation brought by the included interconnection capabilities. Then, sensors, communications, control systems, and the Internet should interact to guarantee the production requirements following objectives such as flexibility and integration of components [3].

Considering CPS as an enabler of Industry 4.0, the following four fields of application—up to 2025—have been identified [4]: (i) Energy, the application of CPS to add intelligence into the electrical grid; (ii) Mobility, refers to not-fixed systems, i.e., the intelligent transportation system (ITS) and vehicular networking systems; (iii) Health, involving telemedicine and remote diagnostic and; (iv) Industry, dealing with automated production.

Then, the conventional energy systems responsible for providing power for light and heat are merged with communications, sensors, control algorithms, and the Internet, working as one entity to supply the energy demand—the synergy between the concepts of Energy and Industry 4.0 results in the notion of Energy 4.0. However, the definition of the term Energy 4.0 currently lacks technical precision. It seems to describe a potential field of applications rather than a prevailing reality, hindering the development and understanding of Energy 4.0 systems. The following features and drivers of Energy 4.0 have been identified [5, 6] and in the report of PWC [7]: (i) Features: efficiency, smart, integrated, and decentralized; (ii) Drivers: stakeholders, client, renewables, and communications.

On the other hand, the methodologies for product development in the context of Energy 4.0 do not exist. The strategies to be applied for Energy 4.0 systems must consider a sensing, smart and sustainable (S3) viewpoint [8].

The Energy 4.0 concept is discussed in this paper, and a definition for Energy 4.0 is presented. Considering the framework S3 (sensing, smart and sustainable), a path for product development that boosts Energy 4.0 systems is given. Furthermore, two examples of Energy 4.0 systems are presented to illustrate the application of the proposed Energy 4.0 definition and S3 Framework.

The structure of the paper is as follows: Sect. 2 deals with the conventional approach to energy 4.0, Sect. 3 presents the concept of Sensing, Smart, and Sustainable (S3) products, Sect. 4 presents the energy 4.0 definition, and Sect. 5 shows some examples of Energy 4.0 products, finally Sect. 6 shows the conclusions.

2 The Conventional Approach to Energy 4.0 and Its Challenges

The formulation of the "Industry 4.0" concept has not only served as a framework for the identification of production methodologies and the creation of technological products. Still, it has also triggered reflection, in various areas, about the possible paradigmatic changes that have occurred in each one, affecting the current way of approaching them. Thus, the addition of the suffix "4.0" to a specific area seeks to represent a similar change, regularly focused on the integration of technological elements in procedures or results. In this way, it has been taken as a general guideline that a particular "area 4.0" includes integrating cyber-physical systems, which consider the existence of sensors, communication systems, control systems, and management systems interacting as a unit.

In this way, the technological advances that are increasingly permeating all areas of knowledge have led to the identification of elements like those mentioned above, resulting in the proposal of paradigmatic frameworks such as Agriculture 4.0 and Education 4.0. Following a similar scheme, it is proposed a partial definition of the concept "energy 4.0" that includes the following elements [9]:

- Decentralized intelligence: machine-to-machine communication
- Cyber-physical production systems
- Smart sensors for decision-making
- Modular addition of features
- Wireless communication and interconnected operation
- Real-time monitoring and creation of prospects
- Human machine interfaces
- Energy management
- Digital twins
- Storage systems
- Renewable energy
- Power electronics
- Distribution and generation systems

Although there is no explicit definition for the previous iterations of Energy 4.0, they can be described in relation with the previous concepts of industry:

- Energy 1.0: operator-based decisions. The system has no sensors; therefore, the experience-based knowledge of the operator is used for making decisions. There are not strict regulations.
- Energy 2.0: sensor-based decisions. Sensors are being incorporated into the system. The operator is still responsible of the decision making but he has more information about the status of the system.
- Energy 3.0: software-based decisions. With the implementation of more sensors, the grid is being monitored for fail detection and the demand. Smart meters start to appear increasing the penetration of alternative energy sources.

In the same article [9], the speed of technological changes and the technical challenges they entail are discussed, such as variability in electricity generation, local networks, and control of storage systems. Ultimately, the above approach recognizes that the Smart-Grid is a crucial conceptual model for the Energy 4.0 approach.

However, there is a fundamental problem with such a definition. It had been said that 4.0, as a suffix, applied to identifiable paradigmatic changes in the practice of some area and it is necessary to recognize that, although electrical engineering has applied various advanced technologies and now offers a different overall context, some factors endanger the incipient concept "energy 4.0" as defined above:

- The traditional process of generation, transmission, distribution, and consumption is still prevalent worldwide. However, more effecting generation and distribution systems have been implemented in the last years such as HVDC in distribution lines.

- The primary energy sources are combustion-based, and novel alternative generation devices still need a reliable backbone to synchronize to conventional power plants. Overall, new technologies cannot deal with black-start or reactive power support as effectively as their traditional counterparts.
- Changes to the electric grid have been gradual: the modern power grid comprises the successive incorporation of differentiable technologies applied to solve local problems. Traditional and new devices coexist depending on the need at hand. Incorporating CPS' potentials is not beneficial in a general sense.
- Many new generation and storage technologies are being incorporated without adequately assessing their technical or environmental impact. Their usage may lead to hindrances.
- The incorporation of measurement interconnected devices typically follows an analytical or preventive aim and are generally understood as costs unrelated to a higher utility.
- The Smart-grid concept goes radically beyond any current electric project or deployed grid. Moreover, the smart grid concept is difficult to achieve when there are conventional electric grids that are transformed to a digital topology. Smart grid concept usually could be a purely conceptual approach. Although there are examples that enjoy a high technological level, there is no application that achieves all the desired characteristics for the Smart-grid. Attaching the "energy 4.0" concept to the Smart-Grid, as done in [9], relegates it to a conceptual reference model, not a paradigm shifts as expected.
- The definition provided does not escape a mechatronics approach to the technologies involved. It does not go beyond the "super-automation". It is more applicable to identify challenges than to talk about the current state of the area.

3 Design and Development of Sensing, Smart and Sustainable Energy Products

The concept of "S3" is used to design technology-based products/services that provide sensing, smart and sustainable solutions. By applying the S3 concept for the design and development of new products, the designers can carry out a systematic design process and a conscious design process where sensing, smart and sustainable features must be considered during all the design processes. To create S3 technologies, the authors use the "S3 product development reference framework" [10]. This reference framework guides designers to carry out a systematic design process where the activities to be performed are given and the tools/techniques to be used are recommended. The four defined stages that consider this reference framework are (i) product ideation, (ii) concept design and target specification, (iii) detailed design, and (iv) prototyping.

The reference framework provides guidelines for the S3 product design through the sensing, smart and sustainable domains. The sensing domain refers to the necessary activities to design a sensing system (e.g., sensors selection, electronic/electrical design, and mechanical design) considering environmental and economic factors and technical requirements. The smart domain is focused on providing the necessary guidelines to design smart solutions to be implemented in the product to be developed. This domain is focused on the design of the three core elements of a smart product (i) the design of the physical components, (ii) the design of the smart components and (iii) the design of the connectivity system. Finally, the sustainable domain focuses on provides a toolbox that includes activities, techniques, methods, tools, among others to obtain sustainable technologies. This domain considers the triple bottom line (3BL) of the sustainability. Therefore, the technologies to be developed will pursue reduced environmental impacts through the efficient use of materials, energy, and manufacturing processes. The social aspect is related to the contribution of a product to people's quality of life, considering health, education, culture, and housing, at the employee, customer, and community level. Finally, the economic aspect is related to productivity, the development of low-cost products, the birth of enterprises, employment generation, and profitability.

In the next section the definition of Energy 4.0 and the application of the S3 Framework to the development of Energy 4.0 products is presented.

4   Energy 4.0 Definition

Focusing on the growth of CPS, the concept of Industry 4.0 emerges, marking a pattern in the features that any system must have to be efficient and autonomous and provide overall enhancement in human, social and economic development indicators [7].

Hence, the Energy 4.0 concept does not mean just an energy system composed of CPS to power delivery, but it means a system that is affordable and clean/green. Also, Energy 4.0 involves heat generation, not only is electric generation. Therefore, the definition of Energy 4.0 is the next:

Definition 1: Energy 4.0 is understood as the energy given from physical or chemical resources, which effectively integrates Internet, advanced sensors, computing, and networks technologies, to guarantee a sustainable and affordable operation.

Then, any system fitting with the definition 1 is called Energy 4.0 system. Note that the framework S3 (see Sect. 2) can be applied to provide a methodology for product development in the context of Energy 4.0, as is shown in the following subsection.

In this paper is presented below a definition of energy 4.0 for electrical energy. Energy 4.0 is an advanced network between the generation sector to the end-users that allows for transmission, monitoring, and adjusting the energy according to the end-users' economic and technological needs and the electric companies. This network also provides more efficiency and sustainable energy production. Also, all the devices and systems included in the network include intelligent systems that can offer a fast and easy way to be interconnected in each layer of the energy generation like economic and demanding needs, operation and control, supervision and control, sensing and protection. Besides, the end-user also can be integrated as a part of the generation system.

4.1   Sensing, Smart, and Sustainable Products for Energy 4.0 Systems

Applying the S3 product concept to the vision of Energy 4.0 for the creation of technologies that address the current challenges of Energy 4.0 development, the following steps must be considered [10]:

1. Product ideation: execute a search to select and develop a promising product that fits the Energy 4.0 definition.
2. Concept design and target specification: compile the customer requirements and physical information such as constraints, parametric and geometric relations.
3. Detailed design: establish the form, dimensions, material properties of each individual component; and produce the drawings, manufacturing documents, and testing reports.
4. Prototyping: check any remaining mistake in the product and execute the final test to verify the functionality and possible modifications.

4.2 Possible Benefits of Using Energy 4.0

Industry 4.0 combines embedded production system technologies with intelligent production processes, paving the way for intelligent energy management as part of an energy 4.0 system [11]. This energy management system should not be seen as just a technological solution but rather a comprehensive set of functions that can achieve sustainability. Some of these functions are the following:

1. Improved production management. Waste reduction, production efficiency, quality control, and other capabilities can be achieved by the implementation of production monitoring that can lead to energy efficiency and sustainability [12].
2. Improved methods of production. The reduction of CO_2 emissions, improvement of productivity and energy use optimization, in addition of renewable energy as a consequence of the implementation of digital technologies [13].
3. Informed decision making. The implementation of digital twins, and added capabilities of data mining, data processing, and data analytics can enable a better-informed decision-making about products and services [14].
4. Energy sector transformation. Sustainable energy systems as the smart grid offer new energy management capabilities and at the same time involving the consumer in the energy management process [15].

Furthermore, the smart grid, in the context of industry 4.0, allows not only energy utilities to monitor and control the power generation, transmission and distribution processes but also the users to have more participation in those processes. Hence, the energy 4.0 can benefit some of the challenges presented in [16]:

- Demand response. Scheduling electricity usage helps managing peak power conditions reducing plant stress and inefficient operation.
- Advance metering infrastructure. The implementation of smart meters may help keep informed the consumers about their load information, billing details, etc.
- Substation automation. To provide real-time analysis, monitoring, protection, fault management, etc., at a local level of the smart grid.
- Home energy management. Comprised of smart appliances, control systems, smart meters, and in home displays [17], those permit the customer to monitor, control and manage their power consumption.

- Transmission line monitoring. Detection of anomalies and failures for maintaining the quality of service.
- Plug-in hybrid and electric vehicles. With the increasing popularity of these types of vehicles, a dedicated energy management system is needed to supply the necessary energy on demand.

Moreover, to evaluate the effectiveness of the smart grid, some reliability indices based on IEEE standard were presented on [18, 19]. Therefore, the energy 4.0 could adopt the same metrics:

- SAIDI (System Average Interruption Duration): indicates the total duration of interruption for the average customer over a period of time.
- SAIFI (System Average Interrupt Frequency): indicates how often the average customer experiences an interruption over a period of time.
- CAIDI (Customer Average Interruption Duration): it is the division between SAIDI and SAIFI.
- MAIGI (Momentary Average Interruption Frequency): indicates the average frequency of momentary interruptions.

In the next section some examples of Energy 4.0 systems, where S3 Framework can be applied, are illustrated.

5   Examples of Energy 4.0 Systems

5.1   Microgrid

A microgrid is understood as a set of loads, distributed generators, and energy storage systems working to satisfy the electrical demand, with a unique point of connection with the electrical grid, which is named point of standard coupling (PCC) [20].

The microgrid is capable to work in two modes: island and grid connected. In the island mode, the microgrid is not tied to the electrical grid working independently to satisfy the demand of electricity. The grid-connected mode refers to the microgrid working together with the electrical grid. A schematic of a microgrid is presented in Fig. 1, it is observed the PCC, the distributed generators represented as photovoltaic panels and wind turbines, the bulbs refer to the loads, and the storage systems as the batteries.

On the other hand, the microgrid is capable to work in a decentralized approach, where the main objective is the energy management in the microgrid considering the autonomy of its elements such as distributed generators and loads. Also, the interconnection between microgrids is possible, then microgrids can be coordinated considering several energy sellers and the energy market.

Considering the features mentioned above, the microgrid is an example of Energy 4.0 system because microgrid provides energy considering: sustainable operation since the primary sources of microgrids are clean energy systems; affordable operation because a set of microgrids work in coordination to consider the energy market where the main objective is to found what type of energy is cheaper at a determined time; the integration

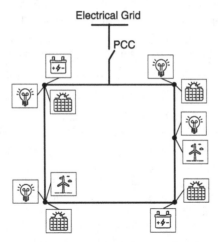

Fig. 1. Schematic representation of a microgrid.

of Internet, sensors, computing, and networks is clear because, in both island and grid-connected mode, the microgrids require advance technologies to be coordinated or to coordinate its sources of energy.

5.2 Smart Thermostats in HVAC Systems

A smart thermostat can present controllers and indicators employing software interfaces that run on mobile devices, i.e., cell phones and tablets, which can remotely interact with the devices using the web [21]. The information about the energy consumed by the thermostat is constantly updated to the user. Furthermore, advanced sensors are integrated into the thermostat to detect sleep and occupancy patterns to be turned off automatically to save energy. The definition of an intelligent thermostat was complemented considering the user's expectation using usability analysis [22].

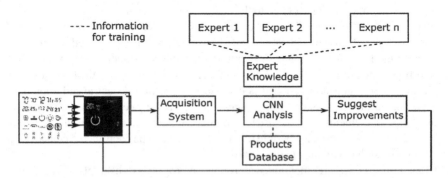

Fig. 2. Application of artificial intelligence in the product design process

Figure 2 presents the block diagram of the implementation of artificial intelligence techniques in thermostats to make it smart. In addition to the communications, sensors, and electronic device interfaces, the smart thermostat has the capacity to learn the preferences of its operators. Then, the compiled information is used to find the preference set of buttons, icons, and displays, of the user, which increment the usability of the product [22].

From Definition 1, the smart thermostats in HVAC systems can be considered an example of Energy 4.0 because these systems provide energy (heat) using advanced sensors, computing, networks, and the Internet. The sustainability and affordability of smart thermostats lie in their function to save energy, which reduces cost and environmental impacts.

5.3 Cyber-Physical Systems in Microgrids and Smart Grids

The smart grid (SG) is described as a convergence of information technology and communication technology with power systems engineering. But as stated before, this concept as complex as it is, it has stayed purely conceptual. Hence the actual smart grid concept comprises the reinforcement of the conventional power infrastructure with information and communication technology, introducing novel products and services inherent to this hybridization [23].

The cyber-physical systems (CPS) integrate computing, communication and storage capabilities focusing on the intersection of the physical and cyber world, see Fig. 3. The CPS provide advantages in sensing, communication, and processing of information locally and globally [24].

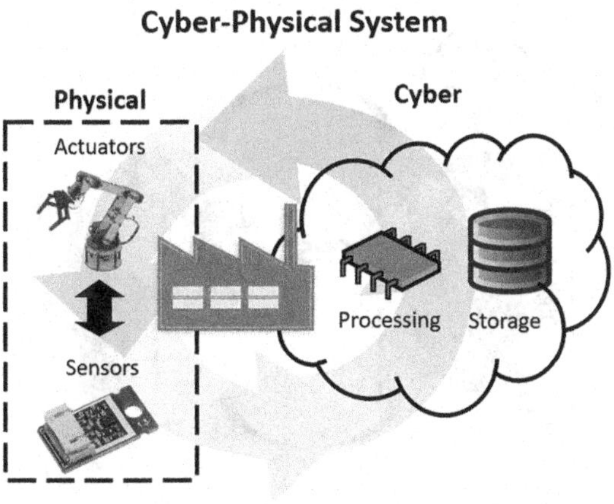

Fig. 3. Cyber-physical system diagram

As the CPS performs extensive computational operations on the data received from the physical devices, analyses and interprets the data, and finally takes actions in real-time, the SG can be considered a CPS consisting of physical devices. Moreover, those devices can be design using the S3 framework in which the Sustainable part is inherent of the smart grid objective to deliver sustainable energy to communities using local energy resources [25] and to save energy. The sensing part is focused on the physical world with all the devices that gather information in the SG domains: electricity, water, gas and oil, transportation, human activity, security, and others [26]. Finally, the smart part can be found on how all the gathered information is processed within the CPS using effective data fusion techniques [26] or even enhancing the cybersecurity that becomes essential for implementing advance applications [25, 26].

5.4   Real Time Simulation and Smart Grid

Digital twin is a technology to achieve the physical and virtual world fusion. It refers to a comprehensive physical and functional description with all available data of a system, or component. The defined characteristics of a digital twin are [27]:

- Must include operation and engineering data and behaviour descriptions via simulation models.
- It evolves along with the real system or component along the whole life cycle.
- It is not only used to describe the behaviour but rather find solutions for the real system.

As Industry 4.0 involve the physical world and the cyber world, the digital twins are the ideal tool to create a digital simulation model that updates and changes as it receives information from their physical analogue as shown in Fig. 4.

Fig. 4. Digital twin example

Therefore, this technology can also be applied to the smart grid concept and take advantage of the information created during the design and engineering to be available for evaluation during the operation of the grid. Since so far, a complete smart grid has not been implemented, the first approach has been real time simulation-based solutions for optimized operations and failure predictions. The real time simulations consist in big number of calculations that are made in parallel, to numerically simulate a network or a specific system as close as it would in real life. Either these simulations are from a complete smart grid [28] or specific subsystems inside the smart grid [29] or even the behavior of agents affecting the smart grid [30].

6 Conclusions

The notion of energy 4.0 has been discussed in several forums, conferences, and reports, without establishing a concert definition. The lack of meaning for Energy 4.0 affects the development of new technologies to face the necessities of the current world in matters of energy. In this paper, a specific definition of Energy 4.0 is presented with the objective of starting a path for the development of Energy 4.0. Furthermore, the application of the S3 Framework in the context of Energy 4.0 offers a method for product innovation. The microgrid and smart thermostats are only two examples of Energy 4.0 systems that normally are seen because of Industry 4.0 due to the missed definition of Energy 4.0 systems.

References

1. Drath, R., Horch, A.: Industrie 4.0: hit or hype? [Industry forum]. IEEE Ind. Electron. Mag. **8**(2), 56–58 (2014). https://doi.org/10.1109/MIE.2014.2312079
2. Kagermann, H., Lukas, W.D., Wahlster, W.: Industrie 4.0: Mit dem Internet der Dinge auf dem Weg zur 4. industriellen Revolution. VDI Nachrichten, (13) (2011). http://www.vdi-nac hrichten.com/Technik-Gesellschaft/Industrie-40-Mit-Internet-Dinge-Weg-4-industriellen-Revolution
3. Pfeiffer, T., Hellmers, J., Schön E., Thomaschewski, J.: Empowering user interfaces for Industrie 4.0. Proc. IEEE **104**(5), 986–996 (2016). https://doi.org/10.1109/JPROC.2015.250 8640
4. MacDougall, W.: Industrie 4.0—Smart manufacturing for the future. Germany Trade & Invest (2014). https://www.manufacturing-policy.eng.cam.ac.uk/documents-folder/policies/germany-industrie-4-0-smart-manufacturing-for-the-future-gtai/view
5. Krasser, M.: Energy 4.0: Ensuring a reliable energy system for the future. Public hearing European Economic and Social Committee: New energy market - New deal for energy consumers, December 2015. https://www.eesc.europa.eu/en/news-media/presentations/energy-40-ensuring-reliable-energy-system-future
6. Misiunas, D.: The future in now: Energy 4.0 (2017). https://futurepreneurs.eu/wp-content/upl oads/2017/11/2.D.pdf
7. PWC: Energy 4.0: Energy transition towards 2030. 7th India Energy Congress, February 2018. http://www.indiaenergycongress.in/iec18/Downloads/background_papers/Final/Reaching%20the%20last%20mile.pdf
8. Miranda, J., Pérez-Rodríguez, R., Borja, V., Wright, P.K., Molina, A.: Sensing, smart and sustainable product development (S3 product) reference framework. Int. J. Prod. Res. **57**, 1–22 (2017)
9. Seixas, M., Melicio, R., Mendes, V.: Comparison of offshore and onshore wind systems with MPC five-level converter under energy 4.0. Electr. Power Compon. Syst. **46**(13), 1399–1415 (2018)

10. Miranda, J., Pérez-Rodríguez, R., Borja, V., Wright, P.K., Molina, A.: Integrated Product, Process and Manufacturing System Development Reference Model to develop CPS- The Sensing, Smart and Sustainable Micro-factory Case. Elsevier Enhanced Reader, 50-1, pp. 13065–13071 (2017)
11. Medojevic, M., Díaz, P., Cosic, I., Rikalovic, A., Sremčev, N., Lazarevic, M.: Energy Management in Industry 4.0 Ecosystem: A Review on Possibilities and Concerns, pp. 0674–0680 (2018). https://doi.org/10.2507/29th.daaam.proceedings.097
12. Ghobakhloo, M., Fathi, M.: Corporate survival in Industry 4.0 era: the enabling role of lean-digitized manufacturing. J. Manuf. Technol. Manag. 31(1), 1–30 (2019). https://doi.org/10.1108/JMTM-11-2018-0417
13. Ghobakhloo, M.: Determinants of information and digital technology implementation for smart manufacturing. Int. J. Prod. Res. 58(8), 2384–2405 (2020). https://doi.org/10.1080/00207543.2019.1630775
14. Dalenogare, L.S., Benitez, G.B., Ayala, N.F., Frank, A.G.: The expected contribution of Industry 4.0 technologies for industrial performance. Int. J. Prod. Econ. 204, 383–394 (2018). https://doi.org/10.1016/j.ijpe.2018.08.019
15. Di Silvestre, M.L., Favuzza, S., Riva Sanseverino, E., Zizzo, G.: How decarbonization, digitalization and decentralization are changing key power infrastructures. Renew. Sustain. Energy Rev. 93, 483–498 (2018). https://doi.org/10.1016/j.rser.2018.05.068
16. Faheem, M., et al.: Smart grid communication and information technologies in the perspective of Industry 4.0: opportunities and challenges. Comput. Sci. Rev. 30, 1–30 (2018). https://doi.org/10.1016/j.cosrev.2018.08.001
17. Méndez, J.I., Peffer, T., Ponce, P., Meier, A., Molina, A.: Empowering saving energy at home through serious games on thermostat interfaces. Energy Build. 263, 112026 (2022). https://doi.org/10.1016/j.enbuild.2022.112026
18. Foba, V.J., Boum, A.T., Mbey, C.F.: Optimal reliability of a smart grid. Int. J. Smart Grid - ijSmartGrid 5(2), 74–82 (2021)
19. Kyrylenko, O.V., Strzelecki, R., Denysiuk, S.P., Derevianko, D.G.: Main Features of the Stability and Reliability Enhancement of Electricity Grid with DG in Ukraine Based on IEEE Standards (2013). http://dspace.nbuv.gov.ua/xmlui/handle/123456789/100756
20. Olivares, D.E. et al.: Trends in microgrid control. IEEE Trans. Smart Grid 5(4), 1905–1919 (2014). https://doi.org/10.1109/TSG.2013.2295514
21. Ponce, P., Peffer, T., Molina, A.: Framework for communicating with consumers using an expectation interface in smart thermostats. Energy Build. 145, 44–56 (2017). ISSN 0378-7788, https://doi.org/10.1016/j.enbuild.2017.03.065
22. Ponce, P., Balderas, D., Peffer, T., Molina, A.: Deep learning for automatic usability evaluations based on images: a case study of the usability heuristics of thermostats. Energy Build. 163, 111–120 (2018). ISSN 0378-7788, https://doi.org/10.1016/j.enbuild.2017.12.043
23. Brandt, T., Feuerriegel, S., Neumann, D.: Modeling interferences in information systems design for cyberphysical systems: insights from a smart grid application. Eur. J. Inf. Syst. 27(2), 207–220 (2018)
24. Mata, O., Ponce, P., McDaniel, T., Méndez, J.I., Peffer, T., Molina, A.: Smart city concept based on cyber-physical social systems with hierarchical ethical agents approach. In: Antona, M., Stephanidis, C. (eds.) Universal Access in Human-Computer Interaction Access to Media, Learning and Assistive Environments, vol. 12769, pp. 424–437. Springer, Cham (2021). https://doi.org/10.1007/978-3-030-78095-1_31
25. Kuruvila, A.P., Zografopoulos, I., Basu, K., Konstantinou, C.: Hardware-assisted detection of firmware attacks in inverter-based cyberphysical microgrids. Int. J. Electr. Power Energy Syst. 132, 107150 (2021)
26. Lu, X., Chen, B., Chen, C., Wang, J.: Coupled cyber and physical systems: embracing smart cities with multistream data flow. IEEE Electrif. Mag. 6(2), 73–83 (2018)

27. Boschert, S., Heinrich, C., Rosen, R.: Next generation digital twin. In: Proceedings of TMCE, vol. 2018, pp. 7–11. Las Palmas de Gran Canaria, Spain (2018)
28. Guo, F., et al.: Comprehensive real-time simulation of the smart grid. IEEE Trans. Ind. Appl. **49**(2), 899–908 (2013)
29. Brenna, M., et al.: Real time simulation of smart grids for interface protection test and analysis. In: Proceedings of 14th International Conference on Harmonics and Quality of Power-ICHQP 2010, pp. 1–6. IEEE (2010)
30. Deilami, S., Masoum, A.S., Moses, P.S., Masoum, M.A.: Real-time coordination of plug-in electric vehicle charging in smart grids to minimize power losses and improve voltage profile. IEEE Trans. Smart Grid **2**(3), 456–467 (2011)

A Real-Time Adaptive Thermal Comfort Model for Sustainable Energy in Interactive Smart Homes: Part I

Juana Isabel Méndez[1]([✉]) [iD], Adán Medina[1] [iD], Pedro Ponce[1] [iD], Therese Peffer[2] [iD], Alan Meier[3] [iD], and Arturo Molina[1] [iD]

[1] Institute of Advanced Materials for Sustainable Manufacturing, Tecnologico de Monterrey, Monterrey, MX 64849, USA
{isabelmendez,A01331840,pedro.ponce,armolina}@tec.mx
[2] Institute for Energy and Environment, University of California, Berkeley, CA 94720, USA
tpeffer@berkeley.edu
[3] Energy and Efficiency Institute, University of California, Davis, CA 95616, USA
akmeier@ucdavis.edu

Abstract. A successful smart city implementation needs to efficiently use natural and human resources. This can be achieved by dividing the smart city into smaller modules, such as a smart community, and even smaller such as a smart home, to allow energy management systems to monitor the city's behavior. The electricity end-user sector is often divided into the residential, commercial, and public transport, industrial, and agricultural sectors. On the other hand, HVAC systems constitute from 40% up to 60% of energy consumption in buildings. Nevertheless, householders do not entirely accept connected devices due to complex interfaces, lack of interest, or acquired habits of thermostat usage that affect thermal comfort, hence, usability and behavioral problems. Thermal comfort is widely defined as that state of mind which conveys satisfaction with the thermal surroundings. This paper obtains an adaptive comfort model for measuring these three features through energy simulations to compare them during the year. This paper analyzes three energy model scenarios to review the adaptive behavior of a community of twelve houses. Three energy models were simulated for Mexico City, Concord (California), and Ontario (Canada) and later deployed into an interactive online platform to determine what further actions are required to improve the quality of life of householders without losing thermal comfort and maximizing energy savings. Besides, this platform allows worldwide users to interact with the platform and learn how clothing insulation, activity, and location affect energy consumption and thermal comfort.

Keywords: Energy simulations · Interactive platform · Thermal comfort · Gamification · Communities

1 Introduction

Energy measures the electricity consumption needed to manage generation, consumption, and electricity conservation. The electricity end-user sector is often divided into the residential, commercial, and public transport, industrial, and agricultural sectors. For instance, in the US, consumption was about 27.238 trillion kWh, with 22%, 18%, 28%, and 35%, respectively. According to the last record in SENER, in 2019, the Mexican electricity consumption was about 284 billion kWh, with 22.7%, 12.3%, 0.4%, 60%, and 4.6% respectively [1]. In other countries such as Canada, electricity consumption is much higher, reaching 3606.38 billion kWh, with 12%, 8%, 20%, 28%, and 2%, respectively [2]. Therefore, reducing electrical consumption without being invasive or losing quality of life is challenging. Moreover, citizens represent the linkage, the main actor, and the sensor to the smart city as they interact with it.

Consequently, users' interaction plays a primary role in understanding and knowing the city to reduce electrical consumption. However, teaching them how to reduce electrical consumption is a challenge as sometimes they have other interests than lowering their consumption. In that regard, a smart home environment seems the ideal place, to begin with, as they provide the interaction between the user and the city to become energy aware.

Gamification, or the use of game elements in real contexts or for real purposes applied to platforms or interfaces, teaches, motivates, and engages the householders in reducing energy consumption or becoming energy aware [3–7]. In that sense, research suggests including ludic elements, for instance, within interfaces or through platforms to teach individuals how energy consumption affects [3–5]. Energy awareness campaigns or comparisons between individuals or homes sensitize citizens about how their consumption affects the environment or community [6, 8–10]. In [9], they suggest including interactive and visual models to engage citizens to adapt to learn new topics.

2 Thermal Comfort

The standard definition of thermal comfort is *"That condition of mind which expresses satisfaction with the thermal environment"* [11, 12]. In 1970, Fanger [13] proposed a comfort model considering physically based determinism and comfort equations. This model is commonly known as the Predicted Mean Vote/ Predicted Percentage of Dissatisfied (PMV/PPD) and considers six mandatory parameters [11]:

- Metabolic rate → related to the occupant
- Clothing insulation → related to the occupant
- Air temperature → related to the surrounding environment
- Radiant temperature → related to the surrounding environment
- Air speed → related to the surrounding environment
- Humidity → related to the surrounding environment

In 1998, Brager and de Dear [14] proposed the adaptive method based on human behavior through three categories of thermal adaptation:

- Behavioral adaptation → related to personal, technological, and cultural responses.
- Clothing insulation → related to genetic and acclimatization
- Air temperature → related to expectations and experiences derived from self-perception and sensory information.

Thus, thermal comfort not only implicates energy saving but also has a personal benefit not only in the perception of well-being but also profitable in a working environment. Therefore, its benefits go beyond a simple energy-saving thing. Moreover, the World Health Organization (WHO) proposed a healthy indoor temperature range between 18 °C and 24 °C [15]. Moreover, increasing 1 °C in a thermostat setpoint during summer can save at least 5% of the electricity [16].

2.1 Metabolic Rate

It is the rate of energy transformation into heat and mechanical work expressed in met units. The metabolic rate depends on the surface skin. Typically, the metabolic rate considers 1.8 m^2 of surface skin. This surface belonged to a man with a height of 1.70 m and 68 kg. Table 1 shows some of the household activities.

Table 1. Typical activities at home.

Activity	Met = 58.1 W/m^2	W/m^2	W/person
Sleeping	0.7	40	72
Sitting	1	60	108
Typing	1.1	65	117
Standing	1.2	70	126
Cooking	1.8	105	189

2.2 Clothing Insulation

As mentioned before, human thermal comfort depends on several factors, such as climate factors and physical activity or clothes worn. These last two are part of the human body's heat loss; since the human being converts food into work and heat. The amount of this heat lost depends intensely on the amount of physical activity being performed and the cloth insulation that affects how much heat is produced by the human body is transferred into the ambient [17].

The main factor that the cloth insulation helps control is the air velocity across the body and controls the exchange of heat between the body and the ambient. Most exchanges are when the body transfers heat into the ambient. However, it can also be the other way around, and the ambient can transfer heat to the body; clothing insulation controls both transfers.

The thermal insulation factor is determined in standards ASHRAE 55 and others like ISO 7730 with the use of thermal manikins to assess the clothing insulation value of each garment, with a measuring unit called clo or m^2K/W when using SI units, and then produce tables such as Table 2 [18]. This table shows different clothing insulation values for various clothing ensembles, and these values are obtained by adding up all the clothing insulation values of each garment.

Table 2. Clothing insulation values for everyday clothing ensembles

Clothing ensemble	Icl (clo)	Rcl (m²K/W)
Warm socks, briefs, shoes, woven shirt, cool trousers	0.42	0.065
Cool socks, briefs, undershirt, shoes, woven shirt, cool trousers	0.51	0.113
Cool socks, briefs, undershirt, shoes, woven shirt, warm jacket, warm trousers	0.77	0.119
Warm dress, pantyhose, bra and panties, shoes	0.49	0.076
Warm sweater, warm skirt, warm blouse, pantyhose, bra and panties, shoes	0.64	0.099
Warm sweater, warm slacks, warm blouse, pantyhose, bra, and panties, shoes	0.77	0.119

2.3 Behavior

Adaptive behavior requires considering the conditions to which an individual is exposed. These conditions include clothing insulation or activities that increase or reduce heat production. For instance, the individual should require turning on the fans or the Heating, Ventilation, and Air-Conditioning (HVAC) system to feel comfortable [12]. A higher rate of metabolic heat or insulation could rely on indoor thermal comfort. Householders use several strategies to achieve thermal comfort; these strategies lead to choices as [19]:

- Building site: areas shaded by trees, near a pond, or can shelter from wind or dust.
- Architectural design and construction materials: shape, orientation, thermal insulation, glazed areas, open areas, floors.
- Rooms' occupancy is based on the season.
- Type of HVAC system
- Type of controls: programmable or connected thermostat, ceiling fans, openable windows.
- Garments are based on the climate, season, indoor temperature, and fashion style.
- Different activities or postures based on the season.
- Attitude toward the indoor operative temperature: if the individual prefers to save or spend money by accepting wider ranges of indoor temperatures.

Behavioral measures provide insights into individuals' thermal comfort; however, it is complex to measure their satisfaction because their comfort is related to perception [18, 20, 21]. It is possible to identify adaptive opportunities by, for instance, measuring through cameras or voice individuals' reactions [22–24], habits, and garments can help identify adaptive opportunities that can impact comfort while saving energy.

2.4 Determining Acceptable Thermal Comfort in Occupied Spaces

The Graphical Comfort Zone Method considers representative households with metabolic rates between 1.0 and 1.3 met (sitting, typing, and standing activities) and clothing insulation Icl between 0.5 and 1.0 clo who are not exposed to direct-beam solar radiation and average air speed up to 0.2 m/s. Moreover, this method is limited to a humidity ratio at or below 0.012 kg*H2O/kg dry air, the water pressure of 1.91 kPa at standard pressure, or a dew-point temperature (tdp) of 16.8 °C. Equation 1 describes the comfort zone values for intermediate values of Icl. These values are determined by linear interpolation between the limits for 0.5 and 1.0 clo using the following criteria:

$$t_{min, I_{cl, active}} = \big[(I_{cl} - 0.5\ clo) t_{min, 1.0\ clo} + (1.0\ clo - I_{cl}) t_{min, 0.5\ clo} \big] / 0.5\ clo \quad (1)$$

$$t_{max, I_{cl, active}} = \big[(I_{cl} - 0.5\ clo) t_{max, 1.0\ clo} + (1.0\ clo - I_{cl}) t_{max, 0.5\ clo} \big] / 0.5\ clo$$

where:

t_{min}, $I_{cl, active}$ = lower operative temperature t0 limit for clothing insulation I_{cl}.
t_{max}, $I_{cl, active}$ = upper operative temperature t0 limit for clothing insulation I_{cl}.
Icl = thermal insulation of the clothing in clo.

2.5 Energy Models for Analyzing Thermal Comfort

Energy model simulators provide information about thermal comfort and energy savings. One of the most common energy simulators is EnergyPlus [25]. This software simulates and predicts the overall energy consumption in the building. Some of the studies performed for thermal comfort analysis include:

- Thermal comfort models in African households improved comfort by 76%.
- Effects on thermal sensation when individuals have a higher level of perceived control and allow energy savings up to 9% [26].
- Improve thermal performance and comfort by replacing glazings or insulation [27].

This paper analyses the adaptive behavior in communities based on clothing insulation, metabolic rate, and behavioral action to promote energy reductions while achieving thermal comfort. The thermal comfort is depicted using the Graphical Comfort Zone Method for 0.5 clo and 1.0 clo during summer and winter, and a metabolic rate up to 1.2 met. These variables are only considered because both the PMV/PPD model and the adaptive model have in common. Besides, these variables are related to the occupant. Hence, this paper proposes feedback to the householder about their clothes or activities and how they relate to electrical consumption. It compares their house with other householders to engage them and promote energy reductions.

3 Proposal

This paper proposes three energy model scenarios to analyze the adaptive behavior of a community of twelve houses. Three energy models were simulated for Mexico City, Concord (California), and Ontario (Canada). Besides, as a result of the energy models, an interactive platform with thermal comfort analysis, HVAC usage, and indoor and outdoor temperature is deployed on the Genial.ly webpage [28].

Figure 1 depicts the community distribution plan view created in Rhinoceros v7 + Grasshopper. The energy model was developed using LadybugTools 1.3.0 from Grasshopper 1.0.0007 [29]. Thus, Table 3 describes the characteristics considered for the energy model input values. These values come from the EnergyPlus templates for the Mid-Rise Apartment Building Program with the following considerations [30]:

- People per area: 0.028309 people/m^2
- Equipment loads per area: 6.7 W/m^2
- Lighting density: 6.5 W/m^2
- Number of people per area: 0.025 ppl/m^2
- Setpoint: Heating: 21.7 °C; Cooling: 24.4 °C

3.1 Graphical Comfort Zone Method Analysis

Once the energy simulation results were obtained, the Graphical Comfort Zone method was interpreted considering 1.0 and 0.5 clo with a metabolic rate of 1.0 to determine the thermal comfort of these communities. Then, based on the thermal comfort, further examination was performed by analyzing the clo and metabolic rate to determine which activities and clothing insulation deliver a higher percentage of thermal comfort.

3.2 Energy Consumption by Community and Household

The energy consumption by community and household was analyzed to determine which location and whether the heating or ventilation wastes more energy than the other. Besides, the surface was obtained to determine the relationship between the area, thermal comfort, and energy depletion.

Table 3. Households' characteristics

Building characteristics	Description
Location	Mexico City, Mexico (ASHRAE Climate zone 3) Concord, California (ASHRAE Climate zone 3) Ontario, Canada (ASHRAE Climate zone 6)
Construction Type	Wood Framed
% Occupancy	9 pm to 6 am: 100% 7 am: 85% 8 am: 39% 9 am a 3 pm: 25% 4 pm: 30% 5 pm: 52% 6 pm to 8 pm: 87%

(*continued*)

Table 3. (*continued*)

Building characteristics	Description
Metabolic rate	95 W/m^2
Construction materials Climate Zone 3	
External wall mass: Typical Insulated Wood Framed Exterior Wall-R12	Construction layers: 25 mm Stucco, 5/8 in. Gypsum Board, Typical Insulation-R10, 5/8 in. Gypsum Board
External window	U 0.42 SHGC 0.25 Simple Glazing Window,
Exterior Roof: Typical Wood Joist Attic Floor-R38	Construction layers: 5/8 in. Gypsum Board, Typical Insulation-R37;
Floor: Typical Insulated Carpeted 8in Slab Floor	Construction layers: Typical Insulation, 8 in. Normal weight Concrete Floor, Typical Carpet Pad
Construction materials Climate Zone 6	
External wall mass: Typical Insulated Wood Framed Exterior Wall-R20	Construction layers: 25mm Stucco, 5/8 in. Gypsum Board, Typical Insulation-R19, 5/8 in. Gypsum Board
External window	U 0.34 SHGC 0.38 Simple Glazing Window,
Exterior Roof: Typical Wood Joist Attic Floor-R48	Construction layers: 5/8 in. Gypsum Board, Typical Insulation-R47
Floor: Typical Insulated Carpeted 8in Slab Floor-R5	Construction layers: Typical Insulation-R4, 8 in. Normal weight Concrete Floor, Typical Carpet Pad

4 Results

Figure 2 depicts the Graphical Comfort Zone Method for 0.5 clo, and 1.0 clo with a metabolic rate of 1 met or for a sitting activity at Mexico City, Concord, and Ontario indoors for Home 1. The three graphics indicate that with 1.0 clo, the individuals feel comfortable during a significant part of the year; on the contrary, with 0.5 clo, there is lesser comfort than with the other clo. The Ontario community reflects that it requires more insulation than 1.0 to feel comfortable. As a complement, Fig. 3 depicted the percentage of comfortable time for the three communities during the year and displayed it by home. It considers the following clo values from Table 2: 0.51, 0.77, and includes a 1.0 clo value. The activities considered were sitting (1 met), typing (1.1 met), and standing (1.2 met). The other values were not considered due to being lower than 0.5, and the Graphical Comfort Zone Method indicates that lower than 0.5 will not be comfortable.

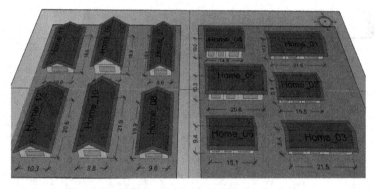

Fig. 1. Community distribution of the twelve houses.

Table 4 displays the area of each household and the percentage of annual HVAC energy consumption compared with the other households. Homes 5, 9, and 10 consumed more than the other homes, although home 1 had more area than these homes. Figure 4 shows the HVAC annual breakdown consumption by home and location. The significant differences relied on the heating; as Ontario is in a cold zone, it required more heating than the other locations located in warmer zones. On the other hand, homes 5, 9, and 10 required more heating than the other homes. However, to propose, for instance, energy reduction alternatives to these types of homes, thermal comfort analysis must be performed. Thus, analyzing Fig. 3 revealed that the householders were most comfortable with 1.0 during the typing activity (1.1 met) for the case of Mexico and the USA. For Ontario, it showed that standing activities (1.2 met) were more comfortable than the other activities; however, it did not increase more than 74% due to Ontario requiring higher insulation of clothes or higher metabolic rates. These results confirm what the Graphical Comfort Zone Method exposed in Fig. 2.

Table 4. Square meters and percentage of annual energy consumption by households.

Household	1	2	3	4	**5**	6
Square meters (area)	222.4	144.2	180.6	148	**212.2**	141.9
% HVAC consumption (kWh)	7%	5%	6%	9%	**11%**	6%
Household	7	8	**9**	**10**	11	12
Square meters (area)	167.7	191	**198.5**	**214.6**	202.7	212.2
% HVAC consumption (kWh)	7%	8%	**11%**	**12%**	8%	10%

clo	Home	Sitting (1.0 met)			Typing (1.1 met)			Standing (1.2 met)		
		Mexico City	Concord	Ontario	Mexico City	Concord	Ontario	Mexico City	Concord	Ontario
0.51 clo	1	1.1%	5.4%	15.7%	20.0%	18.5%	25.5%	32.3%	27.2%	31.3%
	2	2.4%	6.6%	16.2%	21.3%	19.5%	25.2%	33.0%	27.6%	30.8%
	3	3.1%	7.2%	17.7%	23.0%	21.5%	26.5%	35.4%	29.8%	32.1%
	4	11.7%	12.9%	19.1%	26.8%	23.4%	25.4%	36.6%	30.7%	29.7%
	5	6.0%	9.3%	17.3%	23.3%	20.9%	24.3%	33.9%	28.4%	29.2%
	6	5.6%	8.9%	17.9%	24.4%	22.1%	25.7%	35.3%	30.1%	31.2%
	7	4.7%	8.2%	16.6%	23.2%	21.3%	24.5%	33.5%	29.2%	29.7%
	8	3.2%	7.2%	16.4%	22.1%	20.2%	24.3%	33.0%	28.4%	29.5%
	9	8.6%	10.5%	16.7%	24.1%	21.6%	23.6%	33.8%	28.8%	28.0%
	10	7.7%	9.7%	16.5%	23.5%	21.2%	23.5%	33.0%	28.9%	28.4%
	11	1.4%	6.0%	15.7%	20.6%	19.0%	24.1%	32.1%	27.4%	29.7%
	12	5.1%	8.3%	16.6%	22.9%	20.6%	24.0%	33.2%	28.7%	29.0%
0.77 clo	1	33.1%	27.6%	31.7%	47.6%	38.5%	37.3%	79.0%	68.9%	47.6%
	2	33.7%	27.9%	31.1%	47.4%	37.5%	35.1%	77.8%	64.8%	44.5%
	3	36.0%	30.3%	32.2%	48.7%	39.8%	34.9%	77.5%	65.3%	44.7%
	4	36.8%	30.9%	28.6%	47.3%	38.2%	29.2%	65.8%	52.1%	37.5%
	5	34.5%	28.7%	28.7%	46.2%	37.8%	30.7%	70.0%	56.1%	39.1%
	6	35.8%	30.4%	31.1%	47.9%	39.7%	32.9%	74.1%	61.2%	41.9%
	7	34.1%	29.5%	29.9%	46.0%	38.8%	33.8%	73.4%	60.8%	43.3%
	8	33.5%	28.7%	29.7%	45.8%	38.4%	33.9%	74.1%	62.1%	43.9%
	9	34.2%	28.9%	28.1%	45.1%	37.9%	30.6%	68.4%	54.3%	40.1%
	10	33.4%	29.0%	28.2%	44.6%	38.1%	31.1%	69.1%	55.7%	40.2%
	11	32.7%	27.8%	30.1%	45.6%	37.8%	34.7%	75.6%	64.2%	46.2%
	12	33.6%	29.0%	29.0%	45.4%	38.4%	32.3%	72.2%	59.1%	42.5%
1.0 clo	1	61.4%	49.8%	38.9%	92.9%	84.5%	58.3%	80.1%	81.4%	72.8%
	2	60.5%	46.8%	36.7%	90.2%	82.1%	53.5%	79.1%	80.4%	72.3%
	3	60.9%	48.4%	36.7%	88.6%	80.5%	53.7%	77.2%	78.6%	71.2%
	4	53.5%	40.6%	30.6%	75.9%	68.6%	44.3%	73.8%	76.8%	67.7%
	5	55.6%	42.5%	32.2%	83.1%	73.8%	46.4%	77.3%	79.4%	70.3%
	6	58.8%	46.1%	34.6%	84.5%	76.9%	49.9%	75.9%	78.0%	70.7%
	7	57.2%	46.0%	35.5%	85.1%	77.2%	51.2%	77.1%	79.1%	72.2%
	8	57.7%	46.6%	36.0%	87.2%	78.9%	52.1%	78.3%	80.1%	72.6%
	9	53.9%	41.9%	32.4%	79.4%	72.1%	47.5%	76.3%	78.8%	70.5%
	10	54.0%	43.2%	32.9%	80.4%	72.9%	47.6%	77.1%	79.3%	70.7%
	11	58.3%	47.4%	37.5%	90.3%	81.6%	55.5%	79.6%	81.1%	73.6%
	12	56.0%	45.1%	34.9%	84.1%	76.4%	50.2%	77.7%	79.7%	72.2%

Fig. 2. Percentage of comfortable time for three activities and three clothing insulations. Clo = 0.51, 0.77, and 1.0. Met = 1.0, 1.1, 1.2 (sitting, typing, and standing postures).

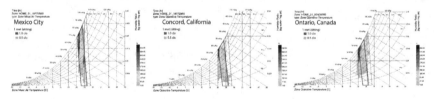

Fig. 3. Graphical Comfort Zone Method for 0.5 clo and 1.0 clo, with a metabolic rate of 1.0 (sitting activity).

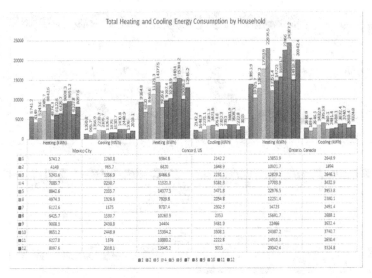

Total Heating and Cooling Energy Consumption by Household

	Mexico City		Concord, US		Ontario, Canada	
	Heating (kWh)	Cooling (kWh)	Heating (kWh)	Cooling (kWh)	Heating (kWh)	Cooling (kWh)
1	5741.2	1260.8	9364.8	2142.2	13853.9	2648.9
2	4149	965.7	6420	1648.9	10821.7	1894
3	5243.6	1356.9	8466.6	2231.1	12829.2	2646.1
4	7085.7	2230.7	11321.3	3181.3	17703.9	3432.9
5	8842.6	2103.7	14377.5	3471.8	22976.5	3953.8
6	4974.3	1926.6	7929.8	2054.8	12251.4	2380.1
7	6122.6	1575	9707.4	2302.5	14723	2491.4
8	6425.7	1530.7	10263.9	2353	15691.7	2688.1
9	9008.3	2430.8	14404	3481.9	22466	3632.4
10	9651.2	2448.9	15394.2	3508.1	24387.2	3740.7
11	6227.8	1376	10000.2	2222.8	14010.3	2650.4
12	8097.6	2018.1	12945.2	3015	20042.4	3124.8

■1 ■2 ■3 ■4 ■5 ■6 ■7 ■8 ■9 ■10 ■11 ■12

Fig. 4. HVAC Annual breakdown consumption by home and location.

4.1 Interactive Community Platform

Figure 5 depicts the community interactive platform and the number of homes available at Genial.ly [28]. This platform shows three views and a 360 view. Each 360 view deploys a central view from Mexico City, MEX, Concord, CA, and Ontario, CAN. The objective of each view is to show the player or householder how the location affects and how the shadow changes, having the same characteristics in terms of date and hour.

Figure 6 depicts the energy consumption, the indoor and outdoor temperature of the community, and the comparison between locations. The information came from Table 2, Figs. 2 and 4. The interactive views allow the community members can review each consumption, compare the HVAC usage between homes, and how the thermal comfort changes depending on the clothing, activity, and location. Figure 6(a) shows the energy use intensity by home and location. Thus, Figs. 6(b) to (e) show the case for home 1. Figure 6(b) depicts the indoor and outdoor temperature monthly graph for each location and the monthly HVAC consumption. Figures 6(c) to (e) illustrate the percentage of thermal comfort for three activities and the clothing insulation.

238 J. I. Méndez et al.

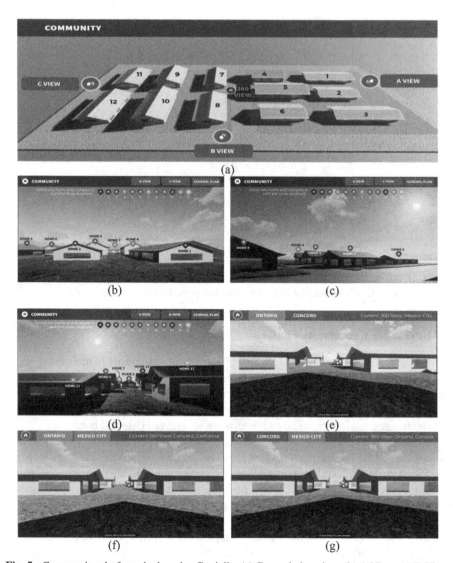

Fig. 5. Community platform deployed at Genially. (a) General plan view; (b) A View; (c) B View; (d) C View; (e) 360 Mexico City View's community; (f) 360 Concord View's community; (g) 360 Ontario View's community.

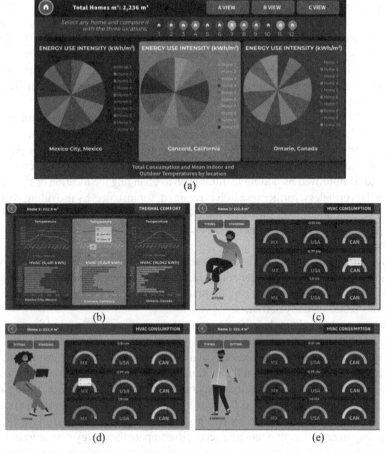

Fig. 6. Community platform deployed at genially. (a) Overall Energy Use Intensity by home and location; home 1 platforms; (b) indoor and outdoor temperature and HVAC usage; thermal comfort and different clothing for these activities; (c) sitting; (d) typing; and (e) standing.

5   Discussion

The results showed that three homes consumed more energy during winter periods and that the thermal comfort was like the other homes that consumed less energy. Therefore, in terms of comfort, if a proposal regarding saving energy is made, there should not be a major problem for homes 5, 9, and 10, as they are within the range of comfort. However, an interesting finding is that, for instance, home 4 consumes less energy than the other homes and is less comfortable than the other homes; thus, proposals that include higher metabolic rates or have more cloth insulation should be suggested.

This leads to a system proposal that evaluates the user's thermal comfort and shows them in interactive platforms or interfaces how to become energy aware and compare their consumption with the other homes. Future work includes personalizing the platforms. For instance, Méndez et al. [3, 16, 31] proposed to employ gamification elements based on personality traits, energy end-user segments, and gamified user types to show tailored interfaces that engage householders. Moreover, suggestions such as clothing element changes can help the user to maintain a thermal comfort sensation that helps not only the energy consumption of his environment but also helps him stay in a healthier environment that reduces stress and negative emotions. For instance, an interface proposal could include the thermal sensation scale, where the end-user can vote based on their preference.

This paper proposes an interactive online platform where worldwide users can play with the platform, analyze the difference between each location, and get their conclusions based on this usage. Besides, thermal comfort awareness is deployed by providing graphics that show how clothing and activity affect thermal comfort. However, this proposal could be improved by adding features such as changing the location or the number of homes. Nevertheless, further research is required, for instance, by providing Artificial Intelligence techniques that, based on the location, predict the energy consumption to speed up the process in which EnergyPlus is required. Therefore, more time is needed to run the simulation.

Conventional energy model simulators such as Ladybug Tools [29] allow personalization in terms of energy usage, schedules, activities, and clothes; however, this personalization has its limits when analyzing thermal comfort in real-time to provide strategies or solutions based on daily activities.

Thus, research should include, for instance, the use of computer vision for predicting clothing insulation and continuously updating the thermal comfort ranges in real-time. Moreover, thermal comfort methods associate activities with clothing insulation; thus, computer vision could even predict the thermal comfort based on the clothing insulation and then infer the type of activity that the householders will do. Besides, this research can be extended and implemented for every home by providing this information, for instance, in thermostat interfaces or smart homes interfaces [3, 22].

Providing interactive interfaces and access to compare the energy consumption of the other householders lets the community and the citizens or individuals know more about their environment and how their decisions may affect the environment, their spending money, or their thermal comfort.

6 Conclusion

This paper compared three communities in Mexico City, Concord, California, and Ontario, Canada, to analyze thermal comfort and HVAC energy consumption. The community comprises twelve homes in different areas; besides measuring the energy impacts and thermal comfort between cities, the occupancy remained the same, and the construction type and metabolic rate. The changes occurred in the climate zone. Mexico and Concord are warmer, whereas Ontario is in a colder climate.

Graphical Comfort Zone Method provides insights about the thermal comfort ranges within the home; if using clothing values of 1.0 or 0.5, the people would feel comfortable. Then, based on that analysis, propose decisions that improve the quality of life of householders and consider their thermal preferences. These communities showed that three homes consumed greater amounts of heating and remained within the thermal comfort ranges. In addition, homes like the fourth had lesser thermal comfort and lower energy consumption than the other three homes. Homes 5, 9, and 10 represented 34% of the total energy consumption in the community; thus, efforts require understanding their habits and how small changes could benefit the total amount of consumption and that it would not affect their thermal comfort.

Nevertheless, due to the increase of dynamic models, adaptive methods that predict, for instance, clothing properties are needed to understand how an individual could adapt to indoor environments. Moreover, those predictions could go further by predicting environmental impacts that lead to energy waste; thus, strategies must be addressed to promote energy reduction while achieving thermal comfort.

Therefore, efforts require focusing on how they can increase thermal comfort without increasing energy consumption and, if possible, decrease consumption. Thus, attempts can be targeted into clothing insulation, and real-time feedback can help householders improve their comfort without wasting more energy or money.

Acknowledgments. Research Project supported by Tecnologico de Monterrey and CITRIS under the collaboration ITESM-CITRIS Smart thermostat, deep learning, and gamification project (https://citris-uc.org/2019-itesm-seed-funding/).

References

1. SENER I Sistema de Información Energética I Usuarios de energía eléctrica por entidad federativa, https://sie.energia.gob.mx/bdiController.do?action=cuadro&subAction=applyOptions. Accessed 20 June 2021
2. Natural Resources Canada: Energy Fact Book 2020–2021. Natural Resources Canada (2020)
3. Méndez, J.I., Peffer, T., Ponce, P., Meier, A., Molina, A.: Empowering saving energy at home through serious games on thermostat interfaces. Energy Build. **263**, 112026 (2022). https://doi.org/10.1016/j.enbuild.2022.112026
4. Ponce, P., Meier, A., Mendez, J., Peffer, T., Molina, A., Mata, O.: Tailored gamification and serious game framework based on fuzzy logic for saving energy in smart thermostats. J. Clean. Prod. **262**, 121167 (2020). https://doi.org/10.1016/j.jclepro.2020.121167
5. Chou, Y.: Actionable Gamification: Beyond Points, Badges, and Leaderboards. Packt Publishing Ltd, Fremont, CA (2019)
6. Schiele, K.: Utilizing gamification to promote sustainable practices. In: Marques, J. (ed.) Handbook of Engaged Sustainability, pp. 427–444. Springer, Cham (2018). https://doi.org/10.1007/978-3-319-71312-0_16
7. AlSkaif, T., Lampropoulos, I., van den Broek, M., van Sark, W.: Gamification-based framework for engagement of residential customers in energy applications. Energy Res. Soc. Sci. **44**, 187–195 (2018). https://doi.org/10.1016/j.erss.2018.04.043
8. Mendez, J.I., Ponce, P., Medina, A., Peffer, T., Meier, A., Molina, A.: A smooth and accepted transition to the future of cities based on the Standard ISO 37120, Artificial Intelligence, and gamification constructors. In: 2021 IEEE European Technology and Engineering Management Summit (E-TEMS), pp. 65–71, IEEE, Dortmund, Germany (2021). https://doi.org/10.1109/E-TEMS51171.2021.9524900
9. Méndez, J.I., Ponce, P., Peffer, T., Meier, A., Molina, A.: A gamified HMI as a response for implementing a smart-sustainable university campus. In: Camarinha-Matos, L.M., Boucher, X., Afsarmanesh, H. (eds.) PRO-VE 2021. IAICT, vol. 629, pp. 683–691. Springer, Cham (2021). https://doi.org/10.1007/978-3-030-85969-5_64

10. Fraternali, P., et al.: A socio-technical system based on gamification towards energy savings. In: 2018 IEEE International Conference on Pervasive Computing and Communications Workshops (PerCom Workshops), pp. 59–64, IEEE, Athens (2018). https://doi.org/10.1109/PERCOMW.2018.8480405
11. Jenkins, M.: Thermal Comfort Basics: What is ASHRAE 55? I SimScale Blog. https://www.simscale.com/blog/2019/08/what-is-ashrae-55-thermal-comfort/. Accessed 3 June 2021
12. Parsons, K.C.: Human Thermal Comfort. CRC Press, Boca Raton, FL (2020)
13. Fanger, P.O., et al.: Thermal Comfort. Analysis and Applications in Environmental Engineering (1970)
14. de Dear, R.J., Brager, G.S.: Developing an Adaptive Model of Thermal Comfort and Preference. Center for the Built Environment, Berkeley (1998)
15. High Indoor Temperatures. World Health Organization (2018)
16. Méndez, J.I., et al.: Designing a consumer framework for social products within a gamified smart home context. In: Antona, M., Stephanidis, C. (eds.) HCII 2021. LNCS, vol. 12768, pp. 429–443. Springer, Cham (2021). https://doi.org/10.1007/978-3-030-78092-0_29
17. Nicol, F., Humphreys, M.A., Roaf, S.: Adaptive Thermal Comfort: Principles and Practice. Routledge, London (2012)
18. Fergus Nicol, J., Humphreys, M.A.: Principles of adaptive behaviours. In: Kubota, T., Rijal, H.B., Takaguchi, H. (eds.) Sustainable Houses and Living in the Hot-Humid Climates of Asia, pp. 209–217. Springer, Singapore (2018). https://doi.org/10.1007/978-981-10-8465-2_20
19. Humphreys, M., Nicol, F., Roaf, S.: Adaptive Thermal Comfort: Foundations and Analysis. Routledge (2015). https://doi.org/10.4324/9781315765815
20. de Dear, R., Xiong, J., Kim, J., Cao, B.: A review of adaptive thermal comfort research since 1998. Energy Build. **214**, 109893 (2020). https://doi.org/10.1016/j.enbuild.2020.109893
21. Yang, L., Fu, R., He, W., He, Q., Liu, Y.: Adaptive thermal comfort and climate responsive building design strategies in dry–hot and dry–cold areas: Case study in Turpan, China. Energy Build. **209**, 109678 (2020). https://doi.org/10.1016/j.enbuild.2019.109678
22. Medina, A., Méndez, J.I., Ponce, P., Peffer, T., Meier, A., Molina, A.: Using deep learning in real-time for clothing classification with connected thermostats. Energies **15**, 1811 (2022). https://doi.org/10.3390/en15051811
23. Méndez, J.I., Mata, O., Ponce, P., Meier, A., Peffer, T., Molina, A.: Multi-sensor system, gamification, and artificial intelligence for benefit elderly people. In: Ponce, H., Martínez-Villaseñor, L., Brieva, J., Moya-Albor, E. (eds.) Challenges and Trends in Multimodal Fall Detection for Healthcare. SSDC, vol. 273, pp. 207–235. Springer, Cham (2020). https://doi.org/10.1007/978-3-030-38748-8_9
24. Méndez, J.I., et al.: Smart homes as enablers for depression pre-diagnosis using PHQ-9 on HMI through fuzzy logic decision system. Sensors **21**, 7864 (2021). https://doi.org/10.3390/s21237864
25. EnergyPlus I EnergyPlus. https://energyplus.net/. Accessed 3 June 2021
26. Yun, G.Y.: Influences of perceived control on thermal comfort and energy use in buildings. Energy Build. **158**, 822–830 (2018). https://doi.org/10.1016/j.enbuild.2017.10.044
27. Shabunko, V., Lim, C.M., Mathew, S.: EnergyPlus models for the benchmarking of residential buildings in Brunei Darussalam. Energy Build. **169**, 507–516 (2018). https://doi.org/10.1016/j.enbuild.2016.03.039
28. Méndez, J.I.: Know Your Community – Thermal Comfort. https://view.genial.ly/6286e8f6ec1be60019eec51d/interactive-content-community-thermal-comfort. Accessed 27 May 2022
29. Ladybug Tools: Ladybug Tools I Home Page. https://www.ladybug.tools/. Accessed 2 May 2021

30. U.S. Department of Energy: EnergyPlus^{TM} Version 9.5.0 Documentation: Input Output Reference (2021)
31. Méndez, J.I., et al.: A rapid HMI prototyping based on personality traits and AI for social connected thermostats. In: Batyrshin, I., Gelbukh, A., Sidorov, G. (eds.) MICAI 2021. LNCS (LNAI), vol. 13068, pp. 216–227. Springer, Cham (2021). https://doi.org/10.1007/978-3-030-89820-5_18

A Real-Time Adaptive Thermal Comfort Model for Sustainable Energy in Interactive Smart Homes: Part II

Adán Medina[1] ⓘ, Juana Isabel Méndez[1] ⓘ, Pedro Ponce[1](✉) ⓘ, Therese Peffer[2] ⓘ, Alan Meier[3] ⓘ, and Arturo Molina[1] ⓘ

[1] Institute of Advanced Materials for Sustainable Manufacturing, Tecnologico de Monterrey, Monterrey, MX 64849, USA
{A01331840isabelmendez,pedro.ponce,armolina}@tec.mx
[2] Institute for Energy and Environment, University of California, Berkeley, CA 94720, USA
tpeffer@berkeley.edu
[3] Energy and Efficiency Institute, University of California, Davis, CA 95616, USA
akmeier@ucdavis.edu

Abstract. Clothing garments directly affect the human body's thermal balance and thermal comfort. The ideal thermal balance is when the body's temperature remains neutral and the environment is not affecting it. Nevertheless, achieving that thermal balance is very unlikely due to other variables, such as humidity, that need consideration. Therefore, these variables affect the human body's perception of the environment's temperature leading to behavioral problems and a lack of thermal comfort. Besides, adaptive methods require integrating dynamic models that predict clothing properties to provide accurate thermal comfort to the householder and understand how an individual adapts to indoor environments rather than the conventional thermal comfort analysis. Therefore, a computer vision system integrated into camera recognition is needed to implement an online clothing insulation recognition system to get feedback on thermal comfort and provide information to the householders about how the clothes and activities affect their thermal comfort. Besides, this recognition needs to be considered in dynamic interfaces such as connected thermostat interfaces. Furthermore, this vision system needs to detect the clothing worn by the user and infer possible metabolic activities based on the clothes. Hence, this paper proposes classifying the garments through a Deep Neural Network (DNN) using the YOLOv3 in which available external sources, such as cameras, gather the householder's clothes and postures to classify the type of cloth and activity and provide information to the householder through a dynamic interface in order to continue their thermal comfort. Thus, a 24-h simulation is performed considering three scenarios: (1) typical 0.5 clo value and 1.0 metabolic rate; (2) dynamic clo values with activities; and (3) dynamic values adding the underwear clo values. Hence, thermal comfort analysis results are included in an interactively connected thermostat mock-up. This mock-up and interaction are available online.

Keywords: Clothing insulation · Online monitoring · Yolov3 · Behavioral clothing and activities · Connected thermostat mock-up · Online platform · Interactive interface

S. Berretti and G.-M. Su (Eds.): ICSM 2022, LNCS 13497, pp. 244–258, 2022.
https://doi.org/10.1007/978-3-031-22061-6_18

1 Introduction

The worn clothing influences the Human body's thermal balance in different temperature environments. According to the laws of thermodynamics, the heat exchange between the human body and the environment depends on which is warmer than the other [1–5]. Thus, if the environment is warmer than the human body, it transfers heat into the body, and if the human body is warmer than the environment, it transfers heat to the surroundings. In the case of equal temperature, there would be a thermal balance between the environment and the body; hence, there would be no heat transfer. However, achieving that thermal balance is very unlikely due to other variables, such as humidity, that need consideration. Therefore, these variables affect the human body's perception of the environment's temperature leading to behavioral problems and a lack of thermal comfort [6].

Clothing garments act as a thermal insulator that affects the heat transfer in two aspects regarding the human body's perception and the environment's temperature [7]:

It prevents the human body from exchanging heat with the environment.

It acts on the evaporative resistance of the skin, which influences how the body.

Standards such as ASHRAE 55 [1, 8] provide clothing insulation values (clo value) tables for both clothing ensembles and individual clothing garments; the measurements were obtained from a standing thermal manikin [8]. Other methods like direct measurement while the clothing garments are worn, employing tables using thermal manikin values, or through regression methods on the physical characteristics of clothes to determine the clothing insulation are mentioned by Lotens and Hanevith [9]. However, the most accepted method is the thermal manikin method [8, 10].

In addition, metabolic rate influences the blood flow in the body and determines the body's temperature affecting the thermal comfort and thermal sensation of the clothes worn [11, 12]. Thus, Table 1 displays a list of actions regarding clothing, garments, and activities responding to cold or heat [13, 14].

Table 1. List of actions as a response to warm or cool environments.

Heat responses	Cold responses
Adopt an open posture to increase the area available for heat loss	Increase muscle tension and shivering to generate more heat in the muscles
Take off some garments to increase heat loss	Curling up or cuddling up to reduce the surface area available for heat loss
Reduce the level of activity to decrease body heat production	Increase the activity level to promote body heat
Adopt the siesta routine to avoid more heat production	Add clothes that reduce the rate of heat loss per unit area
Going for a swim	Go to bed to increase the heat

Nevertheless, due to the increase in dynamic models, adaptive methods that predict clothing properties need to understand how an individual adapts to indoor environments.

Matsumoto, Iwai, and Ishiguro [15] used a computer vision system and a combination of Histogram of Oriented Objects (HOG) and Support Vector Machines (SVM) to recognize clothing garments. Bouskill and Havenith [16] used a thermal manikin to determine the relationship between clothing insulation and clothing ventilation with different activities known as metabolic rates. They concluded that clothing insulation affects less than the design and fabric of the clothing garment; thus, they recommended analyzing the clothing garments worn in specific places during specific activities to determine the best outfit that avoids colder or warmer thermal sensations.

Using computer vision to detect clothing may seem expensive when thinking about the implementation of the camera system and the computer needed to process the information and run the solution. However, suppose cameras are being spread to different uses such as telecare [10, 17–19] or personal assistants like Alexa [18, 20]. In that case, the idea that cameras are part of the smart home infrastructure needs to be considered. Thus, there would be no need to invest in a camera system and only think about the processing part of the problem. Figure 1. Displays some household devices that can track end-user activities, moods, and garments. For example, through smart TV, camera detection can monitor householder reactions or postures; voice detection can analyze the householders' speech to detect if any possible disease or affection may affect the individual. Méndez et al. [21] proposed using cameras and Alexa to track householders' moods and as a tool for depression pre-diagnosis. In [20], they proposed using Alexa and a camera to track seniors' moods and emotions to prevent social isolation and depression. Besides, Medina et al. presented a detailed example of using online analysis to detect clothing insulation and show messages to the householder to suggest how to save energy while keeping thermal comfort. Thus, this picture shows the integration of household appliances that can help track householders' daily activities and moods; therefore, this paper considers cameras to classify the garment and infer daily activities.

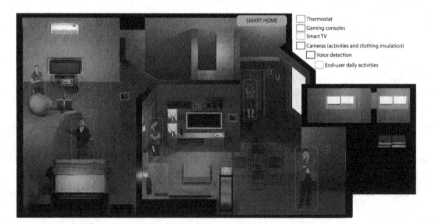

Fig. 1. Household devices integration in Smart Homes: cameras to classify householders' garments and activities.

1.1 Determining Acceptable Thermal Comfort in Occupied Spaces

ASHRAE 55 proposes three methods for determining thermal comfort during occupancy hours [22]:

- Graphic Comfort Zone Method: This method is limited to representative households with metabolic rates between 1.0 and 1.3 met (sitting, typing, and standing activities) and clothing insulation I_{cl} between 0.5 and 1.0 clo who are not exposed to direct-beam solar radiation and average air speed up to 0.2 m/s. Moreover, this method is limited to a humidity ratio at ro below 0.012 kg*H2O/kg dry air, a water pressure of 1.91 kPa at standard pressure or a dew-point temperature tdp of 16.8 °C.
- Analytical Comfort Zone Method: This method considers average metabolic rates between 1.0 and 2.0 met (sitting, typing, standing, and cooking activities) and clo up to 1.5. This method uses the ASHRAE thermal sensation scale and the PMV/PPD model. Acceptable indoor thermal comfort considers a 10% percentage of dissatisfied people in a range of predicted mean vote between −0.5 and 0.5.
- Elevated Air Speed Comfort Zone Method: this method considers activity levels with average metabolic rates between 1.0 and 2.0 met, clothing insulation up to 1.5, and average air speeds greater than 0.20 m/s. For instance, an increase of 0.1 clo or met reduces up to 0.8 °C the operative temperature, whereas a decrease of 0.1 clo or met increases the operative temperature to 0.8 °C or 0.5 °C

1.2 Interactive Connected Thermostat Interfaces

About 86% of the residential buildings have thermostats that control the Heating Ventilation and Air-Conditioning systems (HVAC) [6, 23]. Besides, studies revealed that connected thermostats reduce energy consumption by up to 35%, and with a behavioral change up to 5% can be reduced [6, 24]. Thus, these behavioral factors explain up to half of the variance of overall HVAC consumption. In that sense, in [6, 25] indicated that interfaces should teach, engage, and motivate the householder in performing activities that reduce energy consumption and ideally without losing thermal comfort [6].

Thus, they stated that an interface needs to provide information about their thermal comfort and not only the electricity bill or information about the energy consumption [6, 26]. This information can be provided through interactive platforms or through the use of game elements that engage the householder [6, 25, 26]. Moreover, in [27] they suggested using strategies that feedback to the user about how their garments affect their thermal comfort and possible activities that can improve their thermal comfort as exemplified in Table 1.

In that sense, a computer vision system integrated into camera recognition is needed to implement an online clothing insulation recognition system to get feedback on thermal comfort. Furthermore, this vision system needs to detect the clothing worn by the user and infer possible metabolic activities based on the body pose. Besides, using game elements in real contexts or for educational purposes help individuals understand better their products or how to take advantage of them. For instance, Méndez et al. [6] proposed a serious gamified interface for connected thermostats and proposed to use game elements as tips, feedback, rewards, or social communities to engage householders in energy activities [25]. Besides, Medina et al. [27] suggest including messages about the garments on the thermostat interfaces to show the householders how they can reduce energy while maintaining comfort by wearing the same clothes, lighter or heavier clothes.

Nevertheless, none of these proposals are available on online platforms or shows, for instance, how clothing and activities impact thermal comfort. Thus, this paper proposes to include an online platform as a result of the online analysis and show how the thermal comfort affects depending on the type of thermal comfort analysis, how it affects by analyzing in a conventional method or by using the methodology proposed in this research. In other words, the differences between the conventional model and online analysis.

2   Proposal

Figure 2 depicts the proposal divided into five steps. This proposal aims to determine if the householders are comfortable by comparing the conventional thermal comfort analysis and the DNN clothing and activity thermal comfort analysis. Thus, clothing recommendation feedback is provided if the householders are not comfortable.

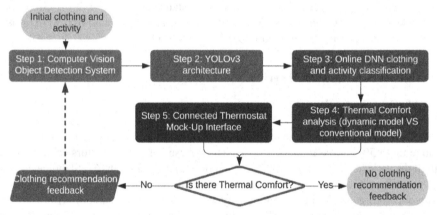

Fig. 2. General proposal for detecting clothing insulation and inferring activities using the YOLOv3.

- Step 1: it proposes a computer vision object detection system using a Deep Neural Network (DNN). The camera gathers the clothes and postures of the householder to classify the type of clothes.
- Step 2: The desired architecture needs to be fast due to the online complexity and accuracy; hence, Fig. 3 shows the YOLOv3 architecture chosen for this proposal. A single-stage object detection neural network presents results as the fastest algorithm. This accurate model is the You Only Look Once Version 3 algorithm (YOLOv3) [25].

  a. Since YOLOv3 uses a DarkNet53 backbone, the difference relies on the last three layers and eliminates the softmax connected layer with convolutional layers. The last one depends on the number of classes to be predicted (N); the architecture mentioned is listed in Table 2. This algorithm has a pre-trained model so we can apply transfer learning to be able to reuse the weights of the model trained for image classification and spend less time in the training steps, and have more free time to tweak the datasets used to train and test the model, specifically the labeled images that are discussed next.

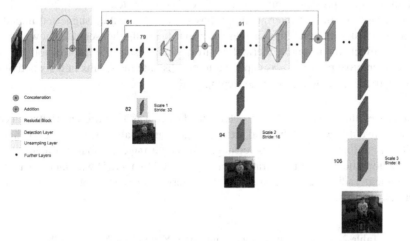

Fig. 3. DNN architecture: You only look once version 3 structure (YOLOv3).

- Step 3: The DNN was obtained, and an online analysis was performed. The DNN classifies the clothes based on the Modified National Institute of Standards and Technology Fashion database (FMNIST), containing 60,000 labeled images of different clothing garments. However, as this database was oriented toward the fashion industry, the images lack the background and people information; therefore, 1,000 images are manually labeled using VIA Image Annotator Software to cover this lack of information. These images were taken from free image databases and contained people with different clothing outfits in other locations and with an entire clothing ensemble instead of the individual clothing garment as presented in the FMNIST images. Table 3 describes the classes used for the data labeling; those classes were: Highly Insulating Jacket, Highly Insulating Shoes, Highly Insulating Trousers, Jacket, Trousers, Shoes, Socks, Hat, Gloves, Scarf, and Person so the network could recognize the individual wearing the jacket.
- Step 4: Once obtained from the DNN, the percentage of comfortable time was analyzed for three locations: Mexico City, Concord, and Ontario. Daily activity was analyzed to review the differences between static and dynamic thermal comfort. Thus, three scenarios were considered:

a. Conventional values from thermal comfort analysis: 0.5 clo and 1 met.

b. DNN clo value: A 24-h activity with four activities: sleeping, sitting, standing, reading, and typing. The DNN determined the clo values.
c. DNN + underwear: A 24-h activity considering four activities: sleeping, sitting, standing, reading, and typing. The clo values were determined by the DNN plus the underwear clo values.

- Step 5: A mock-up interface was deployed, considering a thermostat interface. Heating Ventilation and Air-Conditioning (HVAC) consume most of the energy at home, and thermostats are present in more than 86% of HVAC systems [6, 23, 26, 29]. Thus, a connected thermostat interface is proposed. In [6, 25], they suggested including in a thermostat interface tips or messages that teach the householder how specific actions affect, for instance, thermal comfort. Thus, using the information for step 4, an interactive platform was deployed to encapsulate the results into six activities: sitting, reading, standing, typing, sleeping, and reclining.

Moreover, this research aimed to determine if there was any difference between considering conventional clo values and metabolic rates and analyzing the end-users hourly activities in online. Besides, the percentage of time during each hour was analyzed for each location; the outdoor temperature was considered to not deal with indoor variables such as mean radiant temperature, building materials, or other heat sources that may affect this analysis.

Table 2. Deep neural network architecture (YOLOv3 architecture) [28].

Repetitions	Type	Filters	Size/stride	Output
	Convolutional	32	3 × 3	256 × 256
	Convolutional	64	3 × 3	112 × 112
1	Convolutional	32	1 × 1	
	Convolutional	64	3 × 3	
	Residual			128 × 128
	Convolutional	128	3 × 3/2	64 × 64
4	Convolutional	64	1 × 1	
	Convolutional	128	3 × 3	
	Residual			64 × 64
	Convolutional	256	3 × 3/2	32 × 32
8	Convolutional	128	1 × 1	
	Convolutional	256	3 × 3	
	Residual			32 × 32

(*continued*)

Table 2. (*continued*)

Repetitions	Type	Filters	Size/stride	Output
	Convolutional	512	3 × 3/2	16 × 16
8	Convolutional	256	1 × 1	
	Convolutional	512	3 × 3	
	Residual			16 × 16
	Convolutional	1024	3 × 3/2	8 × 8
8	Convolutional	512	1 × 1	
	Convolutional	1024	3 × 3	
	Residual			8 × 8
	Convolutional	1024	3 × 3	8 × 8
	Convolutional	1024	3 × 3	8 × 8
	Convolutional	1024	3 × 3	8 × 8
	Convolutional	N	1 × 1	8 × 8

Table 3. Clothing insulation values for the classes included in FMNIST and VIA images [6].

Class	Clo	$m^2 * °C/W$
Ankle boot	0.1	0.016
T-shirt/top	0.08	0.012
Dress	0.15	0.023
Pullover	0.28	0.043
Sneaker	0.04	0.006
Sandal	0.02	0.0032
Trouser	0.19	0.0304
Coat	0.6	0.096
Shirt	0.09	0.0144
Bag	0	0
Highly insulated jacket	0.4	0.064
Highly insulated trousers	0.35	0.056
Highly insulated shoes	0.1	0.016
Jacket	0.26	0.0416

(*continued*)

Table 3. (*continued*)

Class	Clo	m² * °C/W
Shoes	0.04	0.0064
Trousers	0.19	0.0304
Scarf	0.04	0.0064
Hat	0.04	0.0064
Gloves	0.05	0.008
Socks	0.02	0.0032

3   Results

This section presents the results of each step proposed in Fig. 2. Thus, the six activities with different clothing insulation were analyzed for the first step. Therefore, for the second step, the FMNIST dataset was considered. Figure 4(a) shows examples of the images in the FMNIST dataset; the main problem with this dataset is that it has no images containing human beings or background information; therefore, it could present problems with the implementation. Consequently, Fig. 4(b) displays a set of images gathered to create additional training examples by labeling the images with the VGG Image Annotator (VIA). Figure 5 shows the FMNIST dataset class samples. These classes were classified to follow ASHRAE 55 and recognize and validate the different clothing insulation values and their effects on thermal comfort calculations.

(a)   (b)

Fig. 4. FMNIST dataset (a); and (b) additional images using VGG Image Annotator labelling.

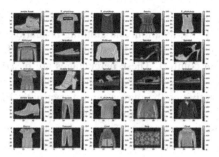

Fig. 5. FMNIST dataset classes.

Thus, Table 4 depicts the clothing insulation values considered for the analysis and the activity and metabolic rate for the third and fourth steps. The bold words were the garments that the DNN analyzed. Figure 6 depicts the results considering the three scenarios stated in the previous section. The first scenario shows that just for Mexico, there were hours where the conventional values had a higher percentage of comfortable time during each hour during sleeping activities or a metabolic rate of 0.8; for activities with 1.1 met, the percentage of comfortability remains the same in the three cases. The second scenario revealed that the comfortable percentage of time increased or remained the same as the conventional scenario; just for Mexico during 10 pm and 11 pm, it was comfortable with the first scenario. Finally, the third scenario was the one that had higher values where the householder had more percentage during each hour of comfortability. The difference relies on the underwear's value. The DNN cannot classify the underwear; it had lower clo values; nevertheless, it had a higher percentage of comfortable time than the conventional. However, it had a lower rate and no greater than 25% in terms of time comfortable because the analysis performed only considered the outdoor temperature.

Table 4. Clothing insulation values for common clothing ensembles

Clothing ensemble	DNN clo value	DNN + underwear	Activity	Met
Men's underwear, **shoes, trousers, t-shirt**	0.35	0.31	Sitting	1
Men's underwear, **jacket, socks, trousers, t-shirt**	0.51	0.47	Reclining	0.8
Men's underwear, **sandals, trousers, t-shirt**	0.33	0.29	Standing	1.2
Men's underwear, socks, **ankle boots, trousers, shirt**	0.44	0.38	Typing	1.1

The fifth step is explained in Fig. 7. This figure depicts a mock-up of a connected thermostat interface that mainly focus on the results presented in Fig. 6 for the three types of thermal comfort analysis: conventional, DNN clo value, and DNN + underwear clo value). The thermostat mock-up is available at [29]. Figure 7(a) shows the connected thermostat home view, and a message or tip is deployed as suggested in [6]. This message explains how the activity, location, and clothing insulation affect thermal comfort percentage. The messages invite the reader to interact with each activity and visualize the percentage of thermal comfort. These percentages are based on the results shown in Fig. 6 and consider just the clothing insulation scenarios with the associated activity. Figure (b) displays the activities the player can interact with. Figure 7(c) to (d) show the values for the sitting or reading, standing, typing, and sleeping or reclining activities with different clo values. It shows the differences between conventional analysis, the DNN clo value analysis, and underwear consideration for three locations (Mexico City, Concord, and Ontario).

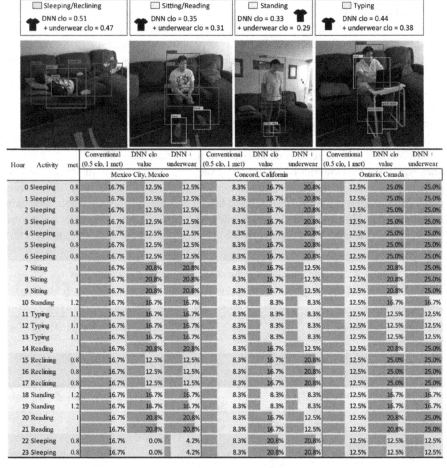

Hour	Activity	met	Conventional (0.5 clo, 1 met)	DNN clo value	DNN underwear	Conventional (0.5 clo, 1 met)	DNN clo value	DNN underwear	Conventional (0.5 clo, 1 met)	DNN clo value	DNN underwear
			Mexico City, Mexico			Concord, California			Ontario, Canada		
0	Sleeping	0.8	16.7%	12.5%	12.5%	8.3%	16.7%	20.8%	12.5%	25.0%	25.0%
1	Sleeping	0.8	16.7%	12.5%	12.5%	8.3%	16.7%	20.8%	12.5%	25.0%	25.0%
2	Sleeping	0.8	16.7%	12.5%	12.5%	8.3%	16.7%	20.8%	12.5%	25.0%	25.0%
3	Sleeping	0.8	16.7%	12.5%	12.5%	8.3%	16.7%	20.8%	12.5%	25.0%	25.0%
4	Sleeping	0.8	16.7%	12.5%	12.5%	8.3%	16.7%	20.8%	12.5%	25.0%	25.0%
5	Sleeping	0.8	16.7%	12.5%	12.5%	8.3%	16.7%	20.8%	12.5%	25.0%	25.0%
6	Sleeping	0.8	16.7%	12.5%	12.5%	8.3%	16.7%	20.8%	12.5%	25.0%	25.0%
7	Sitting	1	16.7%	20.8%	20.8%	8.3%	16.7%	12.5%	12.5%	20.8%	25.0%
8	Sitting	1	16.7%	20.8%	20.8%	8.3%	16.7%	12.5%	12.5%	20.8%	25.0%
9	Sitting	1	16.7%	20.8%	20.8%	8.3%	16.7%	12.5%	12.5%	20.8%	25.0%
10	Standing	1.2	16.7%	16.7%	16.7%	8.3%	8.3%	8.3%	12.5%	16.7%	16.7%
11	Typing	1.1	16.7%	16.7%	16.7%	8.3%	8.3%	8.3%	12.5%	12.5%	12.5%
12	Typing	1.1	16.7%	16.7%	16.7%	8.3%	8.3%	8.3%	12.5%	12.5%	12.5%
13	Typing	1.1	16.7%	16.7%	16.7%	8.3%	8.3%	8.3%	12.5%	12.5%	12.5%
14	Reading	1	16.7%	20.8%	20.8%	8.3%	16.7%	12.5%	12.5%	20.8%	25.0%
15	Reclining	0.8	16.7%	12.5%	12.5%	8.3%	16.7%	20.8%	12.5%	25.0%	25.0%
16	Reclining	0.8	16.7%	12.5%	12.5%	8.3%	16.7%	20.8%	12.5%	25.0%	25.0%
17	Reclining	0.8	16.7%	12.5%	12.5%	8.3%	16.7%	20.8%	12.5%	25.0%	25.0%
18	Standing	1.2	16.7%	16.7%	16.7%	8.3%	8.3%	8.3%	12.5%	16.7%	16.7%
19	Standing	1.2	16.7%	16.7%	16.7%	8.3%	8.3%	8.3%	12.5%	16.7%	16.7%
20	Reading	1	16.7%	20.8%	20.8%	8.3%	16.7%	12.5%	12.5%	20.8%	25.0%
21	Reading	1	16.7%	20.8%	20.8%	8.3%	16.7%	12.5%	12.5%	20.8%	25.0%
22	Sleeping	0.8	16.7%	0.0%	4.2%	8.3%	20.8%	20.8%	12.5%	12.5%	12.5%
23	Sleeping	0.8	16.7%	0.0%	4.2%	8.3%	20.8%	20.8%	12.5%	12.5%	12.5%

Fig. 6. Comparison between scenarios, clo and daily activities with traditional values.

Fig. 7. Connected thermostat mock-up deployed at genially. The thermal comfort depends on the clothing insulation, activities, and location: (a) Home view; (b) activities selection view; (c) sitting or reading activity; (d) standing activity; (e) typing activity; (f) sleeping or reclining activity.

4   Discussion

Part I of these two papers shows a conventional thermal comfort analysis for a community of twelve homes and depicts an online interactive platform. However, this part states that online implementation of clothing insulation recognition could reveal the end-user's comfort. State of the art found that Histogram of Oriented Objects (HOG) and color histogram with the help of a Support Vector Machine (SVM) was implemented to determine which class the cloth garment detected belongs to [8]. Unfortunately, this approach can be used only on static images and not on video or dynamic interfaces.

After the obtention of the clothing garments of the user obtained by the DNN, standards as ASHRAE 55 or ISO help as a guideline for getting the clo value, nevertheless, a problem arises, as it was presented for the second scenario, the DNN cannot consider the underwear clo values, as the camera cannot detect it. However, the underwear can be considered as a constant, so at the end of the garment analysis, an additional 0.04 clo value could be added. That affected the overall score of the clothing insulation value, as it was compared with the third scenario.

Online feedback and monitoring allow integrating these into dynamic interfaces where the end-user can interact directly with, for instance, gamified interfaces. Thus, as Table 1 indicates, the end-user can make recommendations to increase comfort. Besides, a connected thermostat mock-up was also deployed to show the differences between conventional thermal comfort analysis and online analysis.

Furthermore, this research provides insights into possible recommendations for including messages in the thermostat interface. For example, one of the recommendations that can be given inside the interface is the change of the clothing garments worn by the user. These recommendations can be either to change to garments with lower clothing insulation values as proposed by Medina et al. [27] or alternatives such as those proposed by Wang and Cao [8]. They concluded that it was better to have a higher insulation clothing garment in the lower part of the body to avoid feeling cold. Besides, people are expected to be more accustomed to using more clothing garments in the upper body and fewer in the lower part of the body, having an unbalanced clothing insulation value across the different body sections.

5 Conclusion

This paper proposes a system that detects and promotes thermal comfort considering the type of clothing and activity. This system uses a computer system with a deep neural network for online clothing insulation value detection. Once obtained the analysis, a connected thermostat mock-up is deployed.

Three communities in three different countries, Mexico City in Mexico, Concord in California, and Ontario in Canada, were used to demonstrate the difference between an online clothing insulation value and the current approach using constant values. As shown in Fig. 5, there are cases where the percentage varies as little as 8.3% and others that vary as much as 12.5% of the comfortable time, reaching a peak of 25.0% of time spent in the thermal comfort zone. In contrast, the constant value only reaches 16.7% of the time in the thermal comfort zone. However, the variation between the communities can be explained because the temperature used to perform the thermal comfort calculation was the outdoor temperature. Besides, communities located far north from the Equator, such as Ontario, can reach a thermal comfort time of a community closer to the Equator, such as the Mexico City community.

Another measurement is proposed in an indoor enclosure where all the other variables are kept constant to analyze better how the online clothing insulation value calculation affects the overall thermal comfort time. However since in our proposal we talk about a gamified interface where it can make suggestions to the user the presented results seem sufficient to denote the difference between constant values and online calculations and this allows the interface to be able to offer both short term solutions such as the change of a clothing garment in order to avoid entering a cycle of the user entering and leaving the thermal comfort zone, and long term solutions such as recommendations on change of wardrobe garments in order to have a balanced clothing insulation values across the entire body and even in further stages train the model for a specific season so the recommendations take into account if it's winter or summer and the location so there is no thermal stress from the changes between outdoor and indoor temperatures, also trying to make a more efficient use of an HVAC system by better controlling the clothing insulation and therefore reducing the need of alteration of settings in the HVAC system and finally achieving both a higher quality of life for the user and a more efficient energy use for the building.

Future work includes enhancing this platform by providing more scenarios to analyze and adding a feature where the user can scan their outfit so the platform displays the thermal insulation. Besides, depending on the user location, include the percentage of thermal comfort based on the location, considering that this percentage is based only on the outdoor temperature available on climate files and not on the indoor temperature. Thus, the ranges of thermal comfort may vary.

Acknowledgments. Research Project supported by Tecnologico de Monterrey and CITRIS under the collaboration ITESM-CITRIS Smart thermostat, deep learning, and gamification project (https://citris-uc.org/2019-itesm-seed-funding/).

References

1. Nagashima, K., Tokizawa, K., Marui, S.: Thermal comfort. In: Handbook of Clinical Neurology, pp. 249–260. Elsevier (2018). https://doi.org/10.1016/B978-0-444-63912-7.000 15-1
2. Romanovsky, A.A.: The thermoregulation system and how it works. In: Handbook of Clinical Neurology, pp. 3–43. Elsevier (2018). https://doi.org/10.1016/B978-0-444-63912-7.00001-1
3. Székely, M., Garai, J.: Thermoregulation and age. In: Handbook of Clinical Neurology, pp. 377–395. Elsevier (2018). https://doi.org/10.1016/B978-0-444-63912-7.00023-0
4. Te Lindert, B.H.W., Van Someren, E.J.W.: Skin temperature, sleep, and vigilance. In: Handbook of Clinical Neurology, pp. 353–365. Elsevier (2018). https://doi.org/10.1016/B978-0-444-63912-7.00021-7
5. Appenheimer, M.M., Evans, S.S.: Temperature and adaptive immunity. In: Handbook of Clinical Neurology, pp. 397–415. Elsevier (2018). https://doi.org/10.1016/B978-0-444-63912-7.00024-2
6. Méndez, J.I., Peffer, T., Ponce, P., Meier, A., Molina, A.: Empowering saving energy at home through serious games on thermostat interfaces. Energy Build. **263**, 112026 (2022). https://doi.org/10.1016/j.enbuild.2022.112026
7. Hudie, L.-A.: Ergonomics of the thermal environment: Determination of metabolic rate (2016)
8. Clothing Thermal Insulation During Sweating – Y.S. Chen, J. Fan, W. Zhang, 2003. https://journals.sagepub.com/doi/abs/https://doi.org/10.1177/004051750307300210. Accessed 2 Feb 2021
9. Lotens, W.A., Havenith, G.: Calculation of clothing insulation and vapour resistance. Ergonomics **34**, 233–254 (1991). https://doi.org/10.1080/00140139108967309
10. Caine, K.E., et al.: DigiSwitch: a device to allow older adults to monitor and direct the collection and transmission of health information collected at home. J Med Syst. **35**, 1181–1195 (2011). https://doi.org/10.1007/s10916-011-9722-1
11. Parsons, K.C.: Human Thermal Comfort. CRC Press, Boca Raton, FL (2020)
12. Luo, M., Wang, Z., Ke, K., Cao, B., Zhai, Y., Zhou, X.: Human metabolic rate and thermal comfort in buildings: The problem and challenge. Build. Environ. **131**, 44–52 (2018). https://doi.org/10.1016/j.buildenv.2018.01.005

13. Nicol, F., Humphreys, M.A., Roaf, S.: Adaptive Thermal Comfort: Principles and Practice. Routledge, London (2012)
14. de Dear, R.J., Brager, G.S.: Developing an Adaptive Model of Thermal Comfort and Preference. Center for the Built Environment, UC Berkeley (1998)
15. Matsumoto, H., Iwai, Y., Ishiguro, H.: Estimation of thermal comfort by measuring clo value without contact. In: MVA, pp. 491–494. Citeseer (2011)
16. Bouskill, L.M., Havenith, G., Kuklane, K., Parsons, K.C., Withey, W.R.: Relationship between clothing ventilation and thermal insulation. AIHA J. **63**, 262–268 (2002). https://doi.org/10.1080/15428110208984712
17. Solli, H., Hvalvik, S., Bjørk, I.T., Hellesø, R.: Characteristics of the relationship that develops from nurse-caregiver communication during telecare. J. Clin. Nurs. **24**, 1995–2004 (2015). https://doi.org/10.1111/jocn.12786
18. Sudharsan, B., Corcoran, P., Ali, M.I.: Smart speaker design and implementation with biometric authentication and advanced voice interaction capability, p. 12
19. Maharjan, R., Bækgaard, P., Bardram, J.E.: "Hear me out": Smart speaker based conversational agent to monitor symptoms in mental health. In: Adjunct proceedings of the 2019 ACM international joint conference on pervasive and ubiquitous computing and proceedings of the 2019 ACM international symposium on wearable computers, pp. 929–933. Association for Computing Machinery, New York, NY, USA (2019). https://doi.org/10.1145/3341162.3346270
20. Méndez, J.I., Mata, O., Ponce, P., Meier, A., Peffer, T., Molina, A.: Multi-sensor System, Gamification, and Artificial Intelligence for Benefit Elderly People. In: Ponce, H., Martínez-Villaseñor, L., Brieva, J., Moya-Albor, E. (eds.) Challenges and Trends in Multimodal Fall Detection for Healthcare. SSDC, vol. 273, pp. 207–235. Springer, Cham (2020). https://doi.org/10.1007/978-3-030-38748-8_9
21. Méndez, J.I., Meza-Sánchez, A.V., Ponce, P., McDaniel, T., Peffer, T., Meier, A., Molina, A.: Smart homes as enablers for depression pre-diagnosis using PHQ-9 on HMI through fuzzy logic decision system. Sensors **21** (2021). https://doi.org/10.3390/s21237864
22. ANSI/ASHRAE: Standard 55-2017, Thermal environmental conditions for human occupancy (2017)
23. Peffer, T., Pritoni, M., Meier, A., Aragon, C., Perry, D.: How people use thermostats in homes: A review. Build. Environ. **46**, 2529–2541 (2011). https://doi.org/10.1016/j.buildenv.2011.06.002
24. Huchuk, B., O'Brien, W., Sanner, S.: A longitudinal study of thermostat behaviors based on climate, seasonal, and energy price considerations using connected thermostat data. Build. Environ. **139**, 199–210 (2018). https://doi.org/10.1016/j.buildenv.2018.05.003
25. Ponce, P., Meier, A., Mendez, J., Peffer, T., Molina, A., Mata, O.: Tailored gamification and serious game framework based on fuzzy logic for saving energy in smart thermostats. J. Clean. Prod. 121167 (2020). https://doi.org/10.1016/j.jclepro.2020.121167
26. Méndez, J.I., et al.: Designing a consumer framework for social products within a gamified smart home context. In: Antona, M., Stephanidis, C. (eds.) HCII 2021. LNCS, vol. 12768, pp. 429–443. Springer, Cham (2021). https://doi.org/10.1007/978-3-030-78092-0_29
27. Medina, A., Méndez, J.I., Ponce, P., Peffer, T., Meier, A., Molina, A.: Using deep learning in real-time for clothing classification with connected thermostats. Energies **15**, 1811 (2022). https://doi.org/10.3390/en15051811
28. Redmon, J., Farhadi, A.: YOLOv3: An Incremental Improvement. arXiv:1804.02767 [cs]. (2018)
29. Meier, A., Ueno, T., Pritoni, M.: Using data from connected thermostats to track large power outages in the United States. Appl. Ener. **256**, 113940 (2019). https://doi.org/10.1016/j.apenergy.2019.113940

Multimedia Environments
and Metaverse

Including Grip Strength Activities into Tabletop Training Environments

Christine Mégard[1], Sylvain Bouchigny[1(✉)], Samuel Pouplin[2,3], Céline Bonnyaud[3,4], Lucie Bertholier[5], Rafik Goulamhoussen[5], Pierre Foulon[5], Nicolas Roche[3,6], and Frédéric Barbot[7]

[1] CEA, LIST, Gif Sur Yvette, France
sylvain.bouchigny@cea.fr
[2] New Technologies Platform, Raymond-Poincaré Hospital, APHP, Garches, France
[3] Physiology and Functional Exploration Department, Raymond-Poincaré Hospital, APHP, Garches, France
[4] UVSQ, Research Unit ERPHAN, Paris-Saclay University, Versailles 78000, France
[5] Genious Healthcare, Groupe MindMaze, Montpellier, France
[6] Inserm Unit 1179, End: Icap Laboratory, Montigny-Le-Bretonneux, France
[7] CIC 1429 INSERM, Raymond-Poincaré Hospital, APHP, Garches, France

Abstract. We introduce a rehabilitation platform combining motor control of the upper-arm and force control exerted by the hand during the manipulation of a tangible object in a tabletop gaming environment. This interactive tangible object is equipped with force sensors distributed on its surface, allowing the game to include clamping and loosening forces of the hand as part of its gameplay. The paper presents the result of the evaluation with 14 chronic stroke patients. The experience of the patients is overall positive. Some patients mentioned the need to improve the hedonic aspect of the game and modify the tangibles to comply with their various handling capacities. An ergonomic analysis of the rehabilitation situation complements the results of the subjective evaluation from patients and introduces design principles for force management rehabilitation systems for stroke survivors.

Keywords: Force sensor · e-skin · Tangible interaction · Serious game · Rehabilitation

1 Introduction

One in six people worldwide will have a stroke in their lifetime. Although the age-standardized incidence of stroke has significantly decreased, the absolute number of people who have a stroke every year, stroke survivors, and the overall global burden due to the effects of stroke are great and increasing (respectively +68%, +84% and +12% in the past two decades) [1]. Fifty percent of stroke survivors are chronically disabled [2] with hemiparesis involving sensorimotor dysfunctions. Fifty-five percent of post-stroke patients experience disability [3] with various levels of severity in motor and strength control [4]. Rehabilitation of the paretic upper limb motor control is essential

S. Berretti and G.-M. Su (Eds.): ICSM 2022, LNCS 13497, pp. 261–271, 2022.
https://doi.org/10.1007/978-3-031-22061-6_19

for stroke patients to regain some form of autonomy. As the movement of objects is the major function of the upper limb, addressing the rehabilitation of object grasping is challenging as patients present paresis and spasticity. Otherwise, the training volume (number of repetition) and the intensity appear to be the most effective principles of rehabilitation [5]. Therefore, rehabilitation requires a lot of efforts and motivation from the patients. A fundamental issue concerning the need to carry out rehabilitation is the sustained repetition of gestures, a task often perceived as boring and tiring by patients [6].

Technological innovations are methods of improving motivation through pleasant and intriguing gaming environments [7]. However, previous developments using games associated with commercially available product (Nintendo Wii, Xbox Kinect or Sony Playstation) have shown mixed results with little impact on paretic upper limb function [8]. A way to improve the efficiency of these systems is however to increase the specificity of these games [9]. In addition, among existing devices, few integrate the rehabilitation of the upper limb and hand force management [10].

We intend to address this issue by developing a system specifically targeted towards upper limb movements and motor control grasping rehabilitation activities with a serious game deployed on an interactive tabletop. The patient interacts with the game through an instrumented tangible object with a flexible substrate force sensor system integrated on its entire surface. The patient must move, squeeze and release the tangible object according to the gameplay. The present study focuses on the results of the evaluation of the prototype with a sample of 14 patients. We introduce design principles for upper-arm stroke rehabilitation systems enabled with force components in the discussion.

2   Related Works

Interactive tabletops have interesting features for rehabilitation professionals. Their workspace is suitable for upper limb movements and combines the possibility to manipulate tangible objects into the digital space. Different studies involving interactive tabletops are observed in literature. Therapeutic exercises targeting hand movements were introduced into a game deployed on a tabletop with visual feedbacks [11]. An interactive table based application integrated rehabilitation movements with tangible objects such as cubes into a game for children with cerebral palsy [12]. A game developed on a pressure sensitive touchscreen used a real knife and a fork; and graphical and vibrotactile feedback regarding posture and compensations were provided to the patient through a vest [13]. The Contrast game provides a task-oriented training with the manipulation of every day physical objects into a gaming environment [14]. Classical rehabilitation of upper-arm activities such as the push-pull hand task, the squeeze hand task, key turning hand task, and the knob turn hand task were integrated into a gaming environment [15]. None of these works integrates hand force management activities [16], except [4] for home rehabilitation. One prototype proposing exercises of the upper arm (forearm, wrist and finger) into one single device integrated finger forces by pressing a button on a handle [17]. Some commercial devices propose to measure the maximum grip strength [18] or to incorporate various grip and pinch patterns into a gaming environment through a dedicated manipulandum without considering forces [19].

3 The Ergotact Platform

The Ergotact project introduces a post-stroke rehabilitation platform combining motor control of the upper limb with force control exerted by the hand during the manipulation of a tangible object in a tabletop gaming environment [16]. This platform focuses on grasping and displacement, two features that are not commonly available on current commercial systems that are mainly targeted toward shoulder and elbow movements [10].

The Ergotact prototype includes a portable capacitive tactile screen (that can be used as a tablet), a gaming environment and a tangible object. To keep the system cost effective and practical, the design avoids any additional sensors or devices. The table and the objet perform all required measurements and detections needed by the game design or requested by the therapist. The overall setup is presented Fig. 1.

Fig. 1. The Ergotact prototype. Left, overall setup with the object and the capacitive tactile screen. Right, in game rendering.

The gameplay is an action-adventure exploration game. The player performs different actions in the game through various manipulations of the tangible object. The player "walks" (slides the tangible object to a target), makes a "fan attack" (tightens the tangible object until vibration/graphic indicator then loosens the hand), "punches" (tightens and slides the tangible object), "jumps" (lifts and moves the tangible object over obstacles), "levitates" (lifts it above the table), "strikes a stick" and "rotates the stick" (rotates the tangible object clockwise/counter-clockwise). A virtual battle validates each level. A successful player will move to the next level in a different game environment such as the desert or an ice sea. The player must scroll the screen with the tangible object when the game reaches the edge of the table space. After each 30 min daily session, the rehabilitation stops. The score and percentage of successful actions in the game are displayed.

The parameters of the game are defined for each patient by a calibration and assessment phase performed at the beginning of each session. The parameters used for the calibration are the distance between two targets, the accuracy of the target's validation, and the percentage of the maximum capacity angle of rotation, the percentage of the maximum capacity anti-clockwise rotation angle and the percentage of the minimum/maximum clamping force capacity of the tangible object. After the calibration, the game records every minute the success rate for each type of target. The values of the parameters evolve during the game according to an adaptive law. The adaptive law adjusts the parameters according to the percentage of success over a temporal window of 1min. For a success rate between 0 and 30%, the difficulty of the parameters is reduced by a factor of 5%; for a success rate between 30 and 70%, the parameters keep unchanged; for a success rate between 70 and 100%, we increase the difficulty of 5% of all parameters.

4   Description of the Tangible Object

The tangible object has two purposes: interacting with the game and gathering information about the patient. To do so, the screen detects its position and orientation using capacitive coupling (the object is detected as "touches"). The object is covered by a layer of sensors measuring the grip strength of the patient as well as the distribution of this force on the surface. It communicates this information in real-time with the game software.

The main challenge from a technical point of view was to find a way to measure the grip strength with no constraints on the way the patient might handle the object, as this was a main request from the therapists. Two type of object were developed for the project. Object 1 is composed of two cylinders allowing two grasping sizes of 28 mm (top) and 68 mm (bottom) diameters (see Fig. 3). Object 2 is a cylinder with 50 mm diameter and 120 mm length (see Fig. 4). These shapes leave the grasping strategy free but the object must integrate the sensors on the entire surface so that the patient can hold it freely. To do so, we have developed a flexible tactile skin wrapped around the object. The skin consists of a network of parallel-plate capacitors implemented on two flex PCBs (Kapton foils) separated by a dielectric foam (Fig. 2). A picture of the skin is shown Figs. 3 and 4.

Fig. 2. Schematic view of skin sensor layers, presenting 3 parallel plate capacitors. (a) FlexPCB layer 1 with shielding (top) and copper plate (bottom). (b) Deformable dielectric layer (EPDM tape). (c) FlexPCB layer 2 with shielding (bottom) and copper plate (top). The copper plates on layer 1 are smaller than on layer 2 to avoid any shear force effect.

The force applied on the skin is measured by sensing the deformation of this dielectric layer. The foils themselves integrate the capacitive plates, wiring and shielding for an overall thickness of 0.8 mm. 8 units with 8 sensors (top part) and 16 units with 9 sensors (bottom part) are wrapped around the object. Each unit is connected to a PCB board inside the object where a Smartec UTI microchip measures the capacitance of each sensor using Smartec MucC01 multiplexers. The system provides 212 measurement points of the normal force in a range of 0.1 to 20 N for each point and at a rate of 10 Hz. Finally, an additional PCB and an Arduino microcontroller provides Bluetooth connection with the PC and interfaces a linear resonant actuator, providing tactile feedback.

The detection and localization of the object on the screen is done with three copper conductive plates at the bottom of the object and connected to the ground circuit of the

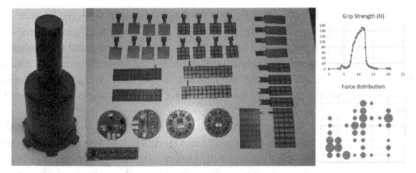

Fig. 3. Ergotact object 1 with protective layer and extended base to avoid tilting during manipulation. Center, smart skin electronics. Right, grip strength measurement and force distribution on the top cylinder (unfolded as a plane, the bottom left points show the pressure exerted by the thumb; the other points show the pressure from the other fingers).

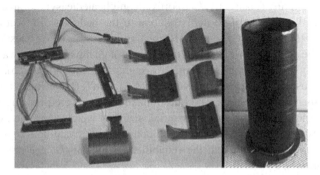

Fig. 4. Ergotact object 2. Left shows the electronics and the FlexPCB sensors.

sensors. With this specific technology, the system does not need any electrical contact with the user, as it is usually the case with tangibles or stylus on capacitive tactile screen. This approach lowers the cost of the system and any material can be used to cover the object, which is of primary importance in clinical environments.

5 Results

5.1 Evaluation with Stroke Patients

The objectives of the study and the planned procedure were presented to the patients during a preliminary visit at the hospital. They were informed that they had to come to the hospital five working days per week during two consecutive weeks for 30-min rehabilitation sessions. The Personal Protection Committees designated for our study approved the protocol of the evaluation. The participants had to sign the informed consent presenting the objectives and the conditions of the evaluation before they could participate. Fourteen chronic stroke survivors (8 female and 6 males) gave their consent. The mean age of the patients was 55 years (between 31 and 80). Mean time from stroke was 12 months (between 3.3 and 50 months). A therapist from the hospital in a quiet room handled the evaluation. At the end of the last session, the patients completed a user experience questionnaire. An ergonomic analysis complemented the evaluation during the rehabilitation sessions to identify possible difficulties encountered by patients and therapists regarding the gameplay and the use of tangibles. We gathered observations, comments, and all other situations of exchanges with the therapist in charge of the assessments.

A good user experience is an important aspect of any system or product. In rehabilitation, scrutiny on the evaluation of the user experience needs to be thorough as it may have serious impacts on the motivation of the patient. The user experience is assessed through self-evaluation questionnaires but the methodology should be carefully suited to the objectives and constraints of the study [21]. As the protocol was quite demanding, we decided to use a short version of the User Experience Questionnaire (UEQ) [22] complemented by ad-hoc questions intended to gather qualitative comments on the prototype. The questionnaire is based on 4 items relative to the pragmatic dimension regarding usefulness and usability (obstructive/supportive; complicated/easy; inefficient/efficient; confusing/clear. We selected two items relative to the hedonist dimension about the attractiveness of (boring/exciting and not interesting/interesting). The questionnaire was completed by the evaluation with the intent of use through three items ("I want to use the device during my rehabilitation" (1: strongly disagree to 9: strongly agree); 'How often do you wish to use the system during your rehabilitation?' (Never/once a month/once a week/more than once a week). The questionnaire also covered the subjective evaluation of tangibles (about the shape, material, size and weight) and the evaluation of the gameplay with a 9 levels Likert scale ("I consider the rehabilitation as a game"/"I am interested in the story"/"I find the rehabilitation amusing"/"The rehabilitation is sufficiently varied").

5.2   Results to the Questionnaires

Eleven patients from the initial corpus of 14 participated in the whole protocol. The object type 2 was used for this ergonomic evaluation. The results to the questionnaire were analyzed with the UEQ analyzing tool [22]. The theoretical extreme values to the questionnaire are between –3/3 for the worse/best result. Values between –0.8 and 0.8 represent a neutral evaluation of the corresponding scale; values over/under 0.8 represent a positive/negative evaluation. Results above 1.5 are positive estimates due to the general lack of extreme values. The raw data were rescaled between –3 and + 3 to fit the UEQ tool. The Alpha-Coefficient for the pragmatic and hedonic scales indicates a good consistency of the results (0.74 for the pragmatic scale and 0.95 for the hedonic scale). The overall user experience is positive (weighted mean = 1.2). This positive value is mainly due to a good estimation to the pragmatic scale (mean = 1.7).

The hedonic dimension is low (mean = 0.7). The application is not engaging enough (mean = 1.0, SD = 1.9), and the application does not interest the participants so much (mean = 0.5, Standard Deviation (SD) = 1.7). Despite this, all participants but one intent to use it for their rehabilitation (mean = 6.8, SD = 1.7).

The tangible object is considered moderately adapted to their abilities (mean = 5.54, SD = 2.2). Most of the criticisms stem from the size of the tangible object. The diameter is perceived as too large (mean = 4.09, SD = 2.07); the cylindrical shape is also criticized (mean = 4.81, SD = 2.22).

The polymeric material of the object (mean = 5.27, SD = 2.00) and its weight were considered suitable for rehabilitation sessions (mean = 7.27, SD = 1.27). The detection of the object on the table was not always effective due to the difficulty for the patient to maintain the object flat on the screen, a mandatory condition for the objet to be correctly detected. The tangible object is adapted to the objectives of the rehabilitation (mean = 6.36, SD = 1.28) and is well integrated into the game (mean = 7.09, SD = 1.37). There were no particular comments about the use of vibrations apart one patient who declared that he could not feel the vibrations but that it was not a problem due to the double coding of the instructions (vibrational and visual coding) in the game. The observations show that some very spastic patients can only grab the tangible with the phalanges. In that case, the object tends to slip as the game progresses and the patient must correct its position using the non-paretic hand. This becomes irritating over time and seems to be particularly detrimental for some actions in the game such as tightening and loosening the object.

5.3 Ergonomic Evaluation

One observation was initially planned for each patient. The observation of the first five patients was consistent and did not reveal additional elements so we decided to stop the observation after the 6th patient.

General set up: The setup is easy to install and start (plug and play). This aspect is very positive for therapists who have to dedicate time both to the rehabilitation itself, but also to the necessary interactions with the patients.

The game: The Patients understand the aim of the game, the way to progress into the game and the instructions after a first introduction by the therapists. The audio and graphic instructions include a statement in accordance with the script of the game and the required actions. One patient systematically rotated the tangible on each target in a compulsive way at the beginning of a session but finally understood the audio instructions. The double coding of the instructions allows the game to be adapted to different types of disabilities induced by stroke. The patients appreciate the diversity of movements although they are repetitive. The possibility of earning a bonus by "catching" a moving diamond integrated in the game with the tangible object was particularly appreciated by some patients "I love the Diamond… this application makes me hungry!" says one patient with a big smile. The application provides positive feedback, which is very important to some patients. When the app says "Great !!!" one patient was so happy and proud that she raised her fist in jubilation.

The adaptive control law: The purpose of the adaptive control law is to tune the difficulty of the game to the capacities of the patients without having to change the game settings on the fly. Sustained effort management is particularly tiring for stroke patients. Almost all patients expressed some fatigue at mid-session (after 15 min). As in any rehabilitation, it is quite possible that the patients were trying to do their best at the beginning of the sessions, but that the accumulated fatigue led to a significant drop in performance that the control law could hardly take into account in the implementation used. This law is based on averaging performance over an interval of time (here 1 min), with a 5% adjustment of the performance requirement (either increasing or decreasing); this law favors an adjustment for values that fluctuate only slightly, but has difficulty in proposing a suitable adjustment for a sudden drop in performance. For example, the "virtual battle" used to change level in the game giving access to a new game environment were difficult to reach. For example, one session failed 4 times in a row, which was very frustrating for the participant.

The interaction between patients and therapists: The observations show that therapists play a fundamental role in encouraging the patients during the game. They support the patient with frequent messages such as "Go ahead! Great! I think it's good" "Go ahead, you're almost there" when the patients show that they are tired; or "Jumps … you can do it!" to support the patient when the movement was challenging.

Digital rehabilitation systems provide scores that aim to show the performance level. Most patients rely on the comments of the therapists on their results rather than looking at their score by themselves; only one patient checked the data in detail and took a picture of the results. The scores were certainly too complex for the patients to be checked rapidly.

6 Discussion and Design Principles for Upper-Arm Stroke Rehabilitation Systems with Force Component

The rehabilitation was demanding for the patients, as they had to come to the hospital every working day for two weeks, especially for those living far away. The requirements of the rehabilitation program have certainly introduced a bias by involving highly motivated patients. This might benefit the results to the user experience and the intent of use, but can give more weight to the criticisms and qualitative assessments. The analysis shows the extent to which rehabilitation devices must tackle various deficits of post-stroke patients for stroke instructions, biofeedback, and interactions with the game.

The analysis also complements existing design principles of digital environments for Upper Limb Motor Rehabilitation of Stroke [19]. The double coding of the instructions (graphic icons and audio instructions) is particularly useful to deal with cognitive impairments that may accompany stroke. The application must also comply with possible shifts of attention focus during the game. The patient must be able to replay an instruction if necessary. This recommendation is in line with [23] and can be extended to feedback for the same reasons. Each action in the game triggers visual and verbal feedback, which clearly indicate a success or failure.

Activities provided in different storyboards: Rehabilitation technologies provide the patients with a panel of various activities. Changes of game environments introduce a diversity that breaks the monotony of regular rehabilitation sessions. Rehabilitation applications should also benefit from game heuristics [24] such as the possibility to get a bonus; the presence of surprising effects and rewards should also help maintain a satisfactory level of motivation. The game elements must be adapted and validated with stroke patients to find a good balance between motivating aspects and distraction.

Adaptive control law: Adaptive laws in rehabilitation aim to provide achievable performance objectives and avoid failure for patients if well-tuned and validated with a large sample of patients. However, adaptive laws should not act like a black box.

Therapists must be aware of their rules to interpret performance and assess progress of the patient in a transparent manner.

Tangibles: Tangible objects link real manipulation with digital environments, but object handling is often critical in stroke patients. Stroke patients can be spastic and/or stiff leading thus to high strength levels requiring much more robust tangibles. We have measured instantaneous cumulative efforts as high as 90N. Others cannot easily open the hand to grasp the tangible. In that case, when the tangible is grabbed with the phalanges. In these cases, the contact surface between the hand and the object is reduced and high efforts can be applied punctually. A future version should make the most of patients' grasping capacities. It should include a hand rest to prevent the hand from slipping and to avoid additional hand fatigue. The diameter of the tangible should be reduced to be adapted to all and provide the possibility of adapting the orientation of the handle to the capacity of the patients. A limitation of the force sensors appeared during the evaluation. Distributed force sensors offer the possibility to investigate the grip distribution on the object but their limited individual range (20 N), that may seem sufficient for standard grip, but may reach saturation in clinical evaluation where some patients exert a strong force over a very small area. The range of the sensors may be adapted to fix this issue by selecting a different material for the dielectric foam. However, some other patients exert a very small grip on the object so that, for future investigations, different type of skin will have to be developed to fit the different clinical conditions. We will also investigate the use of vibrations as feedback for those who still perceive the vibrations.

Scores for patients and for therapists: The scores available at the end of each session were useful to the therapist as a tool to explain the performance to the patients; but they were too complex for the patients themselves. Future developments should provide two performance indicators, the progression of a global score over successive sessions for the patient and a detailed score for therapists.

7 Conclusion

The design process involved stakeholders since the beginning of the project. They influenced the design of the application and of the tangible objects. The evaluation of patients during a long and demanding assessment protocol brings additional inputs that will be used to inform the final design. The analysis presents the patient's point of view. In spite of criticisms about the variety of the game, and problems with the robustness of the tangible object prototype, the very concept of upper limb and hand rehabilitation including a digital environment and tangible interface is a path endorsed by patients.

References

1. Feigin, V.L., Forouzanfar, M.H., Krishnamurthi, R.: Global burden of diseases, injuries, and risk factors study 2010 (GBD 2010) and the GBD stroke experts group. In: Global and regional burden of stroke during 1990–2010: findings from the Global Burden of Disease Study 2010. Lancet. 2014 Jan 18, vol. 383, pp. 245–254 (2010)
2. Donkor, E.S.: Stroke in the century: A Snapshot of the burden, epidemiology, and quality of life. Stroke Res. Treat. **2018**, Article ID 3238165, 10 (2018). https://doi.org/10.1155/2018/3238165

3. Yoon, S, Gutierrez, J.: Behavior correlates of post-stroke disability using data mining and infographics. Br. J. Med. Res. **11**(5), BJMMR.21601 (2016)
4. Kytö, M., Maye, L., McGookin, D.: Using both hands: Tangibles for stroke rehabilitation in the home. In: Proceedings of the 2019 CHI Conference on Human Factors in Computing Systems (CHI '19), p. 14, ACM, New York, NY, USA, Paper 382 (2019)
5. Oujamaa, L., Relave, I., Froger, J., Mottet, D., Pelissier, J.Y.: Rehabilitation of arm function after stroke. Literature review. Ann. Phys. Rehabil. Med. **52**(3), 269–293 (2009)
6. Rijken, P.M., Dekker, J.: Clinical experience of rehabilitation therapists with chronic diseases : A quantitative approach. Clin. Rehabil. **12**(2), 143–150 (1998)
7. Delbressine, et al.: Motivating arm-hand use for stroke patients by serious games (2012)
8. Thomson, K., Pollock, A., Bugge, C., Brady, M.: Commercial gaming devices for stroke upper limb rehabilitation : A systematic review. Int. J. Stroke **9**(4), 479–488 (2014)
9. Thomson, K., Pollock, A., Bugge, C., Brady, M.C.: Commercial gaming devices for stroke upper limb rehabilitation : A survey of current practice. Disabil. Rehabil. Assis. Technol. (2015)
10. Merrill, D., Kalanithi, J., Maes, P.: Siftables: Towards sensor network user interfaces. In: Proceedings of the 1st international conference on Tangible and embedded interaction (TEI '07). ACM, New York, NY, USA, pp. 75–78 (2013)
11. Boulanger, C., Boulanger, A., de Greef, L., et al.: Stroke rehabilitation with a sensing surface. In: Proceedings of the SIGCHI Conference on Human Factors in Computing Systems, April 27–May 02, 2013, Paris, France (2013)
12. Li, Y., Fontijn, W., Markopoulos, P.: A tangible tabletop game supporting therapy of children with cerebral palsy. In: Proceedings of the 2nd International Conference on Fun and Games, Springer-Verlag, Berlin, Heidelberg (2008)
13. Delbressine, F., et al.: Motivating arm-hand use for stroke patients by serious games. In: Proceedings of the Conference of the IEEE Engineering in Medicine and Biology Society, San Diego, USA (2012)
14. Jacobs, A., Timmermans, A., Michielsen, M., Plaetse, M.V., Markopoulos, P.: CONTRAST: Gamification of arm-hand training for stroke survivors. In: CHI '13 Extended Abstracts on Human Factors in Computing Systems (CHI EA '13), pp. 415–420. ACM, New York, NY, USA (2013)
15. Vandermaesen, M., De Weyer, T., Feys, P., Luyten, K., Coninx, K.: Integrating serious games and tangible objects for functional handgrip training: A user study of handly in persons with multiple sclerosis. In: Proceedings of the 2016 ACM Conference on Designing Interactive Systems (DIS '16). ACM, New York, NY, USA, pp. 924–935 (2016)
16. Mégard, C. et al.: Ergotact: Including force-based activities into post-stroke rehabilitation. In: CHI 2019 Late-Breaking Work, May 4–9, 2019, Glasgow, Scotland, UK
17. Yang, Z., Jie, S., Shiqi, L., Ping, C., Shengjia, N.: Tangible interactive upper limb training device. In: Proceedings of the 2018 ACM Conference Companion Publication on Designing Interactive Systems (DIS '18 Companion). ACM, New York, NY, USA, pp. 1–5 (2018)
18. K-FORCE. https://www.k-invent.com/k-force/
19. Ramírez-Fernández, C., Morán, A.L., García-Canseco, E., Orihuela-Espina, F.: Design factors of virtual environments for upper limb motor rehabilitation of stroke patients. In: Proceedings of the 5th Mexican Conference on Human-Computer Interaction (MexIHC '14). Association for Computing Machinery, New York, NY, USA (2014)
20. Yousef, H., Boukallel, M., Althoefer, K.: Tactile sensing for dexterous in-hand manipulation in robotics – A review. Sensors Actuators A: Phys. **167**(2), 171–187 (2011)
21. Obrist, M., Roto, V., Väänänen-Vainio-Mattila, M.: User experience evaluation: Do you know which method to use?. In: CHI '09 Extended Abstracts on Human Factors in Computing Systems (CHI EA '09). ACM, New York, NY, USA (2009)

22. UEQ: User Experience Questionnaire. https://www.ueq-online.org/
23. Willems, L. L., Tetteroo, D., Markopoulos, P.: Towards guidelines for the design of patient feedback in stroke rehabilitation technology. In: Proceedings of the International Conference on Health Informatics, vol. 1, blz., pp. 60–68, January 12–15, 2015, Lisbon, Portugal (2015)
24. Desurvire, H., Caplan, M., Toth, J.A.: Using heuristics to evaluate the playability of games. In: CHI '04 Extended Abstracts on Human Factors in Computing Systems (CHI EA '04), ACM, New York, NY, USA (2004)

Matrix World - A Programmable 3D Multichain Metaverse

Xinyao Sun[1,2(✉)], Xiao Wu[2,3], and Shuyi Zhang[1]

[1] Matrix Labs Inc., Vancouver, Canada
{asun,tim}@matrixlabs.org
[2] University of Alberta, Multimedia Research Centre, Edmonton, Canada
xinyao1@ualberta.ca
[3] WhiteMatrix Tech Ltd., Vancouver, Canada
wuxiao@whitematrix.io

Abstract. Matrix World is a decentralized open virtual world that lets users interact with immersive 3D applications simultaneously running on different blockchains. The world consists of Lands, issued as Non-Fungible Tokens (NFTs), which permanently persist on blockchain networks such as Ethereum and Flow. Thus, Matrix Lands are tradable and transferable via blockchain networks and their owners retain complete control over the creations on their Lands. These Lands can be bound to a sandbox in the Matrix Network which has its own compute and storage resources. These resources allow landowners and creators to govern the visual appearance, inner properties, and lifecycle logic of the Land's creations, ultimately resulting in the production of an immersive 3D application that can operate in perpetuity in the Matrix World cyberspace. In Matrix World, users can take advantage of standard 3D open-world features such as building 3D architectures, hosting virtual meetings, exhibiting digital assets (e.g., NFTs), etc., along with more advanced functionality such as creating and hosting 3D decentralized applications (DApps) using Matrix's built-in creator services and computational resources. These DApps include 3D games and 3D marketplaces, among others. Our ultimate goal is to create unlimited possibilities and opportunities in the metaverse by eliminating the boundary between blockchains and building a next generation 3D open-world DApp platform on top of Matrix World.

Keywords: Blockchain · Metaverse · Query · 3D · Cyberspace

1 Introduction

The metaverse can be seen as an iterative network of three-dimensional virtual worlds, which are persistent online computer-generated spaces in which several individuals from disparate physical places can interact in real-time for work or play, as stated in [4]. [5] discusses in general how technologies have been enhanced and should be further developed to completely facilitate the functional metaverse. Each field has advanced in its own right over the last few decades, gradually altering people's lifestyles. The metaverse has recently gained popularity for two primary reasons: 1) as a result of the COIVD-19 pandemic, and 2) as a result of the adoption of blockchain technology [9].

Speaking of open-world games, it is always inevitable to mention Minecraft [3]. Minecraft offers the freedom for players to build a customized 3D world. Additionally, it exposes powerful and robust backend interfaces that enable users to implement their own game or application in that universe. More significantly, it introduces the Bring Your Own Device (BYOD) concept to the game [2]. As a creator, you can purchase your own compute resources, namely a Minecraft server, to run your Minecraft-based minigame. The majority of blockchain-based 3D open-world platforms are inspired in some way by Minecraft, and we have already seen several well-implemented products [10]. Some of these products have concentrated on gaming and game creation, others have focused on 3D social networking and showcasing assets (NFTs), and some have offered a thin layer of 3D visualization while allowing the community to add additional extensions to enrich their ecosystems [6].

The availability of metaverse-related studies and surveys has increased over the past few years. However, none of the existing blockchain-based open worlds have demonstrated the same customization and user experience level as Minecraft. Despite its obvious potential, blockchain-based research is restricted, and its widespread adoption for many applications is not yet available [1]. In other words, none of the current blockchain-based 3D open worlds have provided a Turing-complete world capable of creating infinite possibilities. A 3D world built entirely on assets is not a real living Metaverse, but a new revolution is underway to enrich the community. The transition from a collection of separate virtual worlds to an interconnected network of 3D virtual worlds, or the Metaverse, depends on advancements in four areas: immersive realism, ubiquity of access and identification, interoperability, and scalability [4].

2   Proposed Programmable Multichain Metaverse

2.1   Programmable Objects

We breathe life into Matrix World's objects by introducing the concept of programmable objects, known as Matrix Objects, that run on a canonical virtual machine called the Space Virtual Machine (SVM). The SVM keeps track of a group of Matrix Objects in one or more Lands and manages each object's attributes, appearance, and lifecycle. Meanwhile, users can define custom functionalities to customize the behavior of these Matrix Objects. Using the Matrix Object and Scene Editor, authorized creators can easily program the actions of their Matrix Objects and create complicated 3D applications by combining multiple programmed/scripted Matrix Objects which can then interact with one another. As illustrated in Fig. 1, the ideal SVM enables hierarchical objects to operate jointly by following predetermined scripted operations. The requested SVM CPU and memory capacity dictates the maximum number of objects and the complexity of application logic. At the same time, the Matrix Lands determine the virtual physical space in the Matrix World.

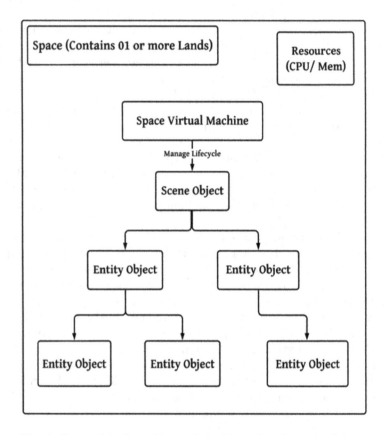

Fig. 1. Space virtual machine and the hierarchy of matrix objects

Additionally, multiple SVMs can connect to form a Matrix Network, as shown in Fig. 2. Each SVM is responsible for operating a Scene on a single Land or Space (a collection of merged Lands) that provides various features similar to a game server. Not only may creators construct 3D applications on a single SVM, but also cross-SVM for multiscene applications.

As defined earlier, Matrix World is a Turing-complete metaverse, which means that each object can be programmed as a living entity. Additionally, we provide developers and creators with tools and services to aid in the production of Scenes and DApps. Thousands of 3D applications will operate simultaneously in the form of fully programmable objects and will be decentralized. That is how Matrix World integrates apps into the actual metaverse.

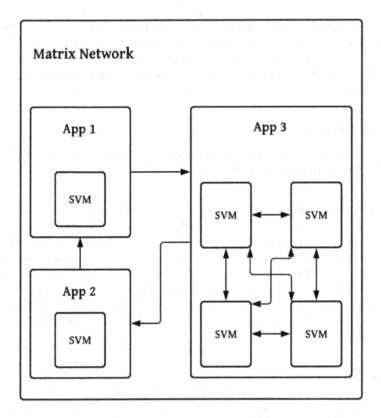

Fig. 2. Applications in matrix network

2.2 Multichain Connectivity

The Matrix World technology facilitates the maintenance of identities across several blockchains. The system currently supports Ethereum and Flow, but it can easily be expanded to incorporate new blockchains as its persistent and consensus layers. Moreover, creators can also connect Matrix Objects to external smart contracts running on multiple blockchain networks. For instance, game developers and content creators can build a 3D Uniswap ATM by crafting its appearance and defining the action that invokes Uniswap's swap function in Ethereum. Futhermore, they can place such an ATM in a virtual NBA Top Shot museum that features a 3D video gallery powered by the Flow network. In addition to blockchain networks, Matrix World can also connect to external services. In the first instance, we propose integrating NFT marketplaces on multiple blockchains such as OpenSea, Rarible, MyNFT, as well as storage protocols like IPFS and Arweave, and streaming services such as Twitch and YouTube. To this end, creators can directly connect to the aforementioned services using both programmable 3D objects and API endpoints. Notably, the entire system is designed to be extensible, allowing for the continuous addition of new services to enrich the world. Matrix World's multichain connectivity enables the creation of unique immersive social networks. Users can experience a world with unlimited possibilities by connecting to multichain DApps and other resources. Matrix World will bolster interoperability among multichain DApps and streamline communication across several metaverses.

2.3   Architecture

The Matrix World system is illustrated in Fig. 3, which consists of 3 layers: Frontend, Backend, and External Services.

Frontend: The frontend of Matrix World is a public-facing web or mobile client. It contains: 1) The Identity Client, a multichain identity client that users will use to log in and get authorized to access Matrix World's services via their blockchain credentials (cold wallet or hot-wallet services). 2) The Content Rendering Client, a WebGL-based client that renders the visual appearances of 3D Matrix Objects in the browser. 3) The Interaction Client, the client responsible for interacting with Matrix Object models managed by the Matrix Network. The Matrix Network routes the signal to a specific SVM to perform the update in the backend.

Backend: The backend is responsible for two major data workflows, as shown in Fig. 3. The first is synchronizing the data from external services like blockchain networks (Ethereum [11] and Flow [12]) and other centralized services (e.g., OpenSea [8]). These data will be collected in the Matrix World state and aggregated events will flow to the Event Bus and be subscribed to by the frontend clients. The second workflow consists of the instructions input from the frontend. These data will stream into the Matrix Network, update the world state, and trigger external service calls.

A Matrix Object is the minimal unit of Matrix World as illustrated in Fig. 4. It has default properties, such as its physics, textures, 3D transformations, and attached NFT media. Creators can also define custom attributes for Matrix Objects via scripts or the Editor UI. Matrix Objects are programmable so that the user can define their lifecycles and action functions. Action functions will be triggered when an interaction signal is sent from the client and lifecycle functions will be triggered when the Matrix Object's state changes. These functions can have multiple purposes like updating the 3D visuals, transforming and making payments, invoking another Matrix Object's action, and performing external calls. Matrix Objects can also subscribe to events from the Event Bus.

The Space Virtual Machine (SVM) is a sandbox server that manages all of the Matrix Objects' states and lifecycles for a specific Matrix Space (Fig. 1). Its computation and storage resources are equal to the sum of the resources of its Lands. Matrix is currently hosting all SVMs ourselves, but we plan to release self-hosted SVMs for landowners for decentralized land management and DApp hosting.

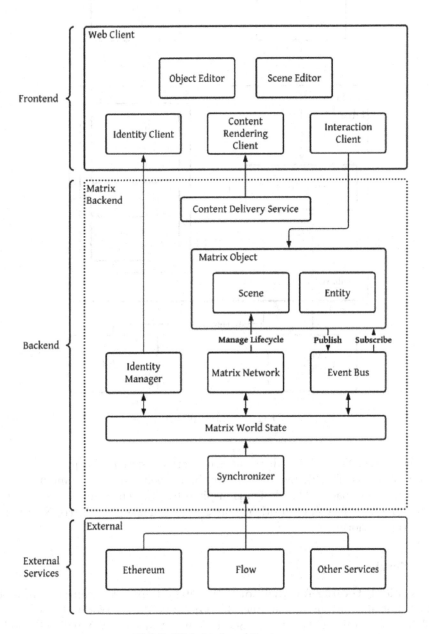

Fig. 3. High-level architecture

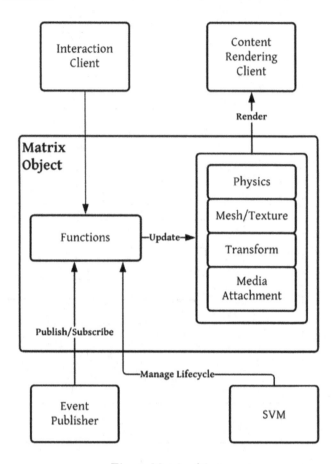

Fig. 4. Matrix object

The Matrix Network is a distributed network consisting of applications built on top of SVMs (Fig. 2). Landowners can build their DApps by running SVMs bound to their Spaces. This design is inspired by Ethereum and allows all SVM public functions to call each other. Our goal is to make the Matrix Network a 3D DApp platform

2.4   Services

Convenient Tooling for Creators: We intend to release a collection of tools that will enable anyone to construct 3D objects in Matrix World. The first tool will be a cloud-based What You See Is What You Get (WYSIWYG) model editor that allows users to alter a 3D model's geometry, material, lighting, and shaders [7]. Along with the model editor, we also plan to include a Scene editor, which will allow creators to import the numerous 3D objects they have built into a Scene or directly use other external resources to ramp up and augment a Scene. Apart from handcrafting models from scratch, the we are also working on an automated tool that will enable users to generate Scenes using pre-defined blueprint templates. Additionally, creators can offer their blueprints to Matrix World's marketplace to accelerate the building process for other players while also monetizing their works.

NFT Ecosystem: Matrix World will offer users convenient functions to import NFTs into the metaverse. These functions will allow users to directly import 2D images, 3D models, videos, audios, and other multimedia NFTs into the 3D world. All imported NFTs will be transformed into Matrix Objects with original appearances (e.g., a 2D image will be transformed into a 3D picture with a frame).

Since each of the NFTs imported into Matrix World becomes a Matrix Object, creators can program the loaded NFTs to make them more functional. For instance, if the creator is a game maker, he or she can construct playable characters for a mini-game using Loot NFTs and create a mini-game using these characters. On the other hand, if the creator is an artist that knows about programming, he or she can convert a static image into animation by applying a dynamic animating filter.

Apart from importing existing NFTs, Matrix World is capable of creating NFTs from Matrix Objects. As a result, Matrix World will have a first-party 3D marketplace that will connect to Ethereum and Flow marketplaces. With a few clicks, users will be able to easily publish NFTs from Matrix Objects.

3   Case Study

3.1   Immersive 3D DApps

Matrix World's development tooling and programming language will allow users to develop immersive 3D applications. Creators can define 3D objects' shapes, attach materials and textures, create lighting, add transformation functions, attach scripted and AI-based behaviors, make payments, invoke external services, etc. The players can go into the 3D NFT buildings to check details such as textures. They can finish shopping in the immersive marketplace without external links, bring the NFT home, and use an auto construction code pack to see the NFT building auto-build on their Land.

3.2   Programmable NFTs

Matrix World allows users to directly import 2D images, 3D models, videos, audios, and other multimedia NFTs into the 3D world with convenient functions. Each imported NFT will be transformed into a Matrix Object with an original appearance (e.g., a 2D image will be transformed into a 3D picture with a frame). Since all of the NFTs imported into Matrix World become Matrix Objects, creators can program the loaded NFTs to make them more functional.

For instance, if the creator is a game maker, he or she can construct playable characters for a mini-game using Loot NFTs and create a mini-game using these characters. On the other hand, if the creator is an artist that knows about programming, he or she can convert a static image into an animation by applying a dynamic animating filter.

3.3 Cross-Chain Socializing and Trading

Matrix World is a metaverse space that connects multiple blockchain networks and services. It provides a place for different players from various blockchain networks to gather and meet.

1. Cross-Chain Conferences and Meetings: Matrix World's public social facilities and buildings can authorize people with identities from different blockchain networks to attend meetings and conferences. We believe cross-chain communications will be more productive than single-chain ones.
2. Cross-Chain Advertising: Brands that want cross-chain influence will find Matrix World to be the best fit for them. They can attract potential customers from multiple blockchain networks by posting advertisements in Matrix World's multichain visible public area.
3. Cross-Chain Trading: The Matrix World will develop a mechanism to allow cross-chain trading and payment to increase the liquidity of the assets (e.g., NFTs and in-game objects) in the Matrix World.

4 Conclusions

Matrix World is an open virtual world that empowers users to build 3D immersive applications on top of different blockchains. The world is made up of Lands, which are issued as Non-Fungible Tokens (NFTs). In Matrix World, users can build 3D architectures, host virtual meetings, exhibit NFTs, as well as create their own 3D decentralized applications using Matrix Wolrd's built-in computation resources. Matrix World is designed as a 3D DApp platform. The Matrix Network consists of a number of SVMs, each of which is responsible for running a Scene on a specific Land that offers various functionalities. Matrix World will provide creators with development and automation tools to facilitate Scene and DApp development. We hope that in the future, thousands of 3D applications will operate concurrently on Matrix World's DApp platform.

References

1. Bolger, R.K.: Finding wholes in the metaverse: posthuman mystics as agents of evolutionary contextualization. Religions **12**(9), 768 (2021)
2. Callaghan, N.: Investigating the role of minecraft in educational learning environments. Educ. Media Int. **53**(4), 244–260 (2016)
3. Cipollone, M., Schifter, C.C., Moffat, R.A.: Minecraft as a creative tool: a case study. Int. J. Game-Based Learn. (IJGBL) **4**(2), 1–14 (2014)

4. Dionisio, J.D.N., Burns III, W.G., Gilbert, R.: 3D virtual worlds and the metaverse: current status and future possibilities. ACM Comput. Surv. (CSUR) **45**(3), 1–38 (2013)
5. Duan, H., Li, J., Fan, S., Lin, Z., Wu, X., Cai, W.: Metaverse for social good: a university campus prototype. In: Proceedings of the 29th ACM International Conference on Multimedia, pp. 153–161 (2021)
6. Gadekallu, T.R., et al.: Blockchain for the metaverse: a review. arXiv preprint arXiv:2203.09738 (2022)
7. Guo, H., Mao, N., Yuan, X.: WYSIWYG (what you see is what you get) volume visualization. IEEE Trans. Vis. Comput. Grap. **17**(12), 2106–2114 (2011)
8. Kiong, L.V.: DeFi, NFT and GameFi made easy: a beginner's guide to understanding and investing in DeFi. NFT and GameFi Projects, Liew Voon Kiong (2021)
9. Koo, H.: Training in lung cancer surgery through the metaverse, including extended reality, in the smart operating room of Seoul National University Bundang Hospital, Korea. J. Educ. Eval. Health Prof. **18**, 33 (2021)
10. Ryskeldiev, B., Ochiai, Y., Cohen, M., Herder, J.: Distributed metaverse: creating decentralized blockchain-based model for peer-to-peer sharing of virtual spaces for mixed reality applications. In: Proceedings of the 9th Augmented Human International Conference, pp. 1–3 (2018)
11. Wood, G., et al.: Ethereum: a secure decentralised generalised transaction ledger. Ethereum Proj. Yellow Paper **151**(2014), 1–32 (2014)
12. Zaucha, T., Agur, C.: Newly minted: non-fungible tokens and the commodification of fandom. New Media Soc. 1–22 (2022)

Matrix Syncer - A Multi-chain Data Aggregator for Supporting Blockchain-Based Metaverses

Xinyao Sun[1]([✉]), Yi Lu[2], Jinghan Sun[3], Bohao Tang[2], Kyle D. Rehak[1], and Shuyi Zhang[1]

[1] Matrix Labs Inc., Vancouver, Canada
{asun,kyle,tim}@matrixlabs.org
[2] Dapper Labs, Vancouver, Canada
{amberluyi,bohao.tang}@dapperlabs.team
[3] Chinese University of HongKong, Shenzhen, China
jinghansun@link.cuhk.edu.cn

Abstract. Due to the rising complexity of the metaverse's business logic and the low-latency nature of the metaverse, developers typically encounter the challenge of effectively reading, writing, and retrieving historical on-chain data in order to facilitate their functional implementations at scale. While it is true that accessing blockchain states is simple, more advanced real-world operations such as search, aggregation, and conditional filtering are not available when interacting directly with blockchain networks, particularly when dealing with requirements for on-chain event reflection. We offer Matrix Syncer, the ultimate middleware that bridges the data access gap between blockchains and end-user applications. Matrix Syncer is designed to facilitate the consolidation of on-chain information into a distributed data warehouse while also enabling customized on-chain state transformation for a scalable storage, access, and retrieval. It offers a unified layer for both on- and off-chain state, as well as a fast and flexible atomic query. Matrix Syncer is easily incorporated into any infrastructure to aggregate data from various blockchains concurrently, such as Ethereum and Flow. The system has been deployed to support several metaverse projects with a total value of more than $15 million USD.

Keywords: Blockchain · Metaverse · Query · Indexing · Event · Reflection

1 Introduction

Metaverse is a portmanteau of the prefix "meta" (which means "beyond") and the suffix "verse" (shorthand for "universe"). It is derived from Neal Stephenson's science fiction novel Snow Crash [12] and literally refers to a cosmos beyond our physical world. As stated in [2], the metaverse is predicated on advancements in four key areas: immersive realism, ubiquitous access and identification, interoperability, and scalability. To enable the construction of a completely functional metaverse, each of those areas must be adequately developed. The development can be grouped into several sub-domains, including numerous multimedia technologies - network transmission and prototyping, computer graphics, image processing, virtual reality, and augmented reality [2].

Globalization has increased the volume of international communication and cooperation on a global scale, however geographic distance is an objective hindrance that increases costs. Additionally, as a result of the COVID-19 pandemic, many events have been suspended to comply with pandemic preventive standards [5]. These stringent requirements have created a major opportunity for initiatives including teleconferences and virtual gatherings, in which the metaverse could provide significant accessibility to meet those social requirements [5]. Moreover, decentralization has been described as a critical component of initiating the fifth phase of metaverse development [2]. Decentralized development has resulted in the decoupling of the client and server sides of a virtual world system, which the blockchain protocol has facilitated in recent years. The metaverse is expected to connect everyone on the entire globe. Study [3] asserted that blockchain technology is critical for ensuring the sustainability of metaverse ecosystems by ensuring decentralization and fairness.

From a macro perspective, a three-layer metaverse architecture is outlined in [5] as 1) ecosystem, 2) interaction, and 3) infrastructure. Each layer is composed of distinct modules that work together to make a comprehensive, interactive, and functional metaverse. As a result, having an interoperable, robust, and low-latency middleware between each module becomes critical for ensuring that diverse virtual worlds are able to seamlessly connect and overlap. As we progress toward a more decentralized metaverse, we must examine the solutions we wish to emphasize in the metaverse and other advancements.

2   Related Works

2.1   Traditional Web 2.0 Metaverse

Numerous companies, game creators (*Roblox*[1] and *Epic Games*[2]), software giants (Microsoft, Amazon), social media conglomerates (Facebook - now Meta, Twitter), and graphics processor manufacturers (Nvidia) are involved in the metaverse, beginning with online events. Extensive research and development has been conducted to optimize virtual gathering in the metaverse; NVIDIA's Omnivers[3] is a scalable, multi-GPU real-time reference development platform for 3D simulation and design collaboration; and Alibaba's *Cloud Metaverse*[4] has been released for the purpose of utilizing cloud computing to construct the entire virtual world as a service. Meta recently launched *Horizon Worlds*[5], a

---

[1] https://www.roblox.com/.
[2] https://www.epicgames.com/.
[3] https://developer.nvidia.com/nvidia-omniverse-platform.
[4] https://www.alibabacloud.com/solutions/metaverse.
[5] https://www.oculus.com/facebook-horizon/.

virtual-reality social networking platform that allows up to 20 avatars to explore, hang out, and build in the virtual realm, as well as a number of revolutionary gadgets, controllers, and supporting hardware, including VR gloves with haptic feedback. However, all of these advancements in multimedia technology are web 2.0 based. The service providers are centralized identity providers, and users' digital identities are produced and held centrally. These advancements are not sufficient to create a transparent, stable, and sustainable digital economy, where digital properties belong to the users, not the operators [10].

2.2   Blockchain-Based Web 3.0 Metaverse

The success of Bitcoin [8], as a decentralized transaction system has garnered significant attention. Later, in 2013, Vitalik Buterin proposed Ethereum [1], a decentralized computation platform that introduced the use of smart contracts to execute programs autonomously and transparently on the blockchain. Since then, a variety of different public blockchain networks have been established, including *Flow*[6], *EOS*[7], etc. each of which supports the development of decentralized applications (DApps) and has its own design philosophy aimed at improving the user experience and system performance. Numerous DApp-based metaverses, such as *Sandbox*[8] and *Decentraland*[9], have attracted increasing attention and consumers in recent years, resulting in significant revenues [5]. This demonstrates how decentralization's power could ensure that digital properties are unique, permanent, and transferable, which benefits the metaverse's development and enables the construction of a fair, free, and sustainable society [6]. Web 2.0's inadequacies, along with the existence of public blockchain technology, have gradually increased public awareness of privacy, data rights, censorship, and identity difficulties. These factors have facilitated the transition of users to a more decentralized Web 3.0 metaverse.

2.3   Indexing Blockchain Data

We are still in the early stages of the development of decentralized technologies. When developing a decentralized application for the metaverse, the developer frequently encounters constraints on on-chain compute power and storage, along with a wide variety of public blockchain interfaces [7]. Due to the increasing complexity of the metaverse's business logic and the requirements of low-latency user experiences, properly reading, writing, and retrieving historical on-chain data in order to support their functional implementation at scale has always been an inevitable challenge [4]. The *Graph*[10] is a popular decentralized protocol for indexing and querying data on Ethereum-based blockchains. It enables

---

[6] https://www.onflow.org/.
[7] https://eos.io/.
[8] https://www.sandbox.game.
[9] https://decentraland.org/.
[10] https://thegraph.com/.

the querying of data that was inaccessible directly before. However, its support provides only a restricted set of interfaces for transforming data from an on-chain structure to a GraphQL-compatible schema. It offers fast and efficient querying of historical blockchain data, but does not address reflective requirements such as making an external service request or initiating transactions in response to on-chain states and received events.

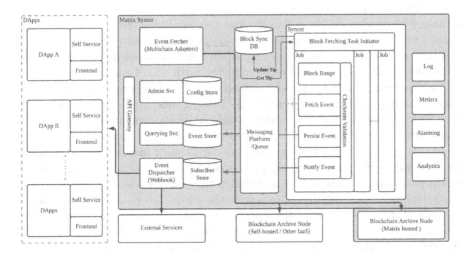

Fig. 1. An overview of the Matrix Syncer architecture

2.4   Web3 Infrastructure as a Service

Typically, developers can establish their own web 3.0 infrastructure and communicate directly with a self-hosted blockchain node in order to fetch the on-chain state change (event) and perform reflective processing on their own services. Unfortunately, it is widely known that setting up a self-hosted node is costly and time-consuming [11]. There is a demand for utilities that decrease the entrance barrier and make blockchain data more accessible. Infrastructure-as-a-Service (IaaS) offerings are among the most critical. *Infura*[11] is leading the charge, providing developers, decentralized application teams, and corporations from a variety of industries with a suite of tools for connecting their apps to blockchain networks. *Alchemy*[12] further extends Infura by adding support for Ethereum layer-2 and other public blockchains (e.g., Flow). The most significant ultility they have introduced is the Alchemy Notify API, which uses webhooks to trigger external actions, allowing for on-chain reflective implementations. However, neither of these projects offer a proper way to retrieve historical on-chain events through traditional query requests.

---

[11] https://infura.io/.
[12] https://www.alchemy.com/.

2.5   Gap of Research and Development

Top traditional IT businesses are attempting to improve the virtual social gathering user experience in order to make it a more effective and secure new way of living. To meet the demands of a decentralized metaverse, developers will need to overcome some key issues caused by current web 3.0 limitations. Numerous products and services have been developed to facilitate the development and deployment of smart contracts in the modern era, such as ChainIDE [13]. One of the key constraints on the scalability and accessibility of the majority of decentralized projects is the inability to index and react effectively to on-chain events that trigger self-business logic after the contracts have been deployed. Furthermore, blockchain is currently a niche industry, with one of the key causes being that blockchains lack interoperable infrastructure, which prevents applications from being deployed on a large scale [7]. Thus, a solid infrastructure that connects web 2.0 and web 3.0 development stacks and supports multiple blockchains is essential to accelerate the decentralized metaverse revolution.

In this work, we present Matrix Syncer, the ultimate middleware that bridges the data access and event trigger gaps between blockchains and end-user services and applications. Matrix Syncer is intended to make it easier to consolidate on-chain data into a distributed data warehouse while also enabling customizable on-chain state transformation for scalable storage, access, and retrieval. It provides a uniform layer for both on-chain and off-chain states, as well as quick and flexible atomic querying and event triggering. Furthermore, the Matrix Syncer can easily be integrated into any infrastructure to aggregate data from multiple blockchains simultaneously.

3   Proposed Framework

3.1   System Design

The metaverse, we believe, is an event-driven environment in which each user interaction should trigger a chain of reflections altering the state of related properties. Such events are typically required by an application in the metaverse to execute reactive activities in response to the current world state and system logic. Matrix Syncer is a cloud-based IaaS platform that allows developers to effortlessly develop the aforementioned workflows without having to worry about blockchain-related setups or maintenance. The overview of the system architecture is shown in Fig. 1. Matrix Syncer can help developers to create web 3.0 compatible applications while maintaining their familiar web 2.0 development style.

3.2 Event Registration

As seen in the figure, developers can configure their EOIs by registering them in the Block Sync DB, which registers the EOI's global unique identity derived from the chain type, contract address, and event signature. Each event has its own set of variables that control how the block syncer distributes jobs; the three most critical ones are *initBlockHeight*, *syncedStartBlockHeight*, and *syncedLatest-BlockHeight*. Developers define the *initBlockHeight* when registering a new event. It serves as the starting point for synchronizing a specific event, which is typically the block height of the contract deployment. This is necessary because the contract may have previously been deployed and performed transactions before enrolling on our platform. Here, *syncedStartBlockHeight* refers to the block height at the time this event was first synced, and *syncedLatestBlockHeight* refers to the block height that the block syncer scanned for this event the most recently.

3.3 Block Synchronization

The block syncer distributes **regular** synchronization jobs in parallel for each event with the batch size $K = min(mB, cL - \gamma - syncedLatestBlockHeight)$, where mB denotes the maximum batch size for a given chain and cL indicates the most recently minted block height from connected archive node. mB is an adaptive parameter that can be adjusted based on the transaction per section (TPS) rate as well as the rate of finalized blocks in order to ensure that the processing capacity of the entire system is sufficient for each chain's throughput. γ is a chain-specific parameter that prevents the most recent confirmed blocks from being reverted, as is the case with Ethereum when an archive node receives an uncle block and later reverts it to adhere to the longest chain protocol [9]. As a result, γ can be set to a number (e.g., 5 for Ethereum, it is the default number of blocks for confirmation of freshly minted blocks defined by the Go Ethereum client) to ensure that the synced block information is already persistent in the network. When *initBlockHeight* is less than *syncedStartBlockHeight*, Block Syncer will distribute a dedicated group of **backfilling** jobs to rapidly scan for the target event between [*initBlockHeight*, *syncedStartBlockHeight*] in parallel. Once the **backfilling** task has been completed, the *syncedStartBlock-Height* of this event will be set to equal to the *initBlockHeight*.

3.4 Event Fetching

The fetcher is implemented using corresponding interfaces according to different blockchains. It reads all EOIs for a certain block range and converts them to a meaningful data structure that can be stored in a database. The majority of public blockchains provide an interface or SDK for interacting with their nodes in order to perform simple chain state or event queries. For example, on Ethereum, we can use *ether.js*[13] to fetch events by block range from an archive node. It's a

---

[13] https://docs.ethers.io/.

little more complicated when dealing with Flow since Flow's blockchain data is segmented via Sporks. Each spork stores data for a specific block range. When we use the official Flow endpoint, we can only request blocks or events from the current Spork. To address these issues, we designed the ultimate event fetcher as a middleware module that automatically subdivides large block ranges into smaller sections with associated Spork endpoints and then merges all fetched events together to provide a level of usability similar to Ethereum's. The source code is available at [GitHub]. (reveal later to respect to the double-blind rule). In our instance, we just need to construct an adapter for each chain using a suitable library and connect to an archive node capable of retrieving historical data. The user can specify the type of nodes they wish to employ. They can either create their own self-hosted node or provide the endpoint for third-party IaaS platforms such as Infura or Alchemy. Additionally, Matrix Syncer offers its own archive node for multiple chains, providing developers more freedom for cost-effectiveness optimization.

3.5 Event Persistence and Indexing

When registering an event, developers have the option of specifying a database scheme for converting on-chain events into a database (event store). All event fetchers will feed EOIs into a persistent module as decoded data. The persistent module converts feed-in events into various data structures defined by the developers and saves them in the event store. All subsequent queries will be performed at the data model specified in the schema. We adopted $DynamoDB^{14}$ as a cloud database solution to achieve 1) horizontal scaling via managed partitioning and sharding, 2) high availability with assured SLAs, and 3) high-level consistency. After the backfilling process is completed, developers will be able to do advanced queries, such as filtering, pagination, sorting, grouping, and joining result sets. Matrix Syncer is designed to accommodate multiple chains. Hence, it brings the important advantage of allowing a unified data structure for DApps, which support similar business logic on multiple chains. Users can define the same schema for different chains in order to convert heterogeneous raw events into a standard format and facilitate interoperability in the metaverse.

3.6 Event Data Integrity Assurance

We assure end-to-end data completeness for each block fetching task by performing a checksum verification at the end of each step. During the event fetching and persistence step, we calculate the number of persisted events grouped by registered event type and the number of non-persisted events. The checksum equation is:

$$count(allEvents) \iff count(nonPersistedEvents)$$
$$\sum_{type} count(persistedEventsPerType) \quad (1)$$

---

[14] https://aws.amazon.com/dynamodb/.

If this equation holds, we persisted all registered events, and non-registered events are correctly skipped. In the Nofity Event phase, we will check if the $count(notificationSent)$ equals to $\sum_{type} count(persistedEventsPerType)$, which is the checksum we calculated in the previous step to make sure no registered events are missing during the notification step. These checksums are persisted in Block Sync DB as structured data and can be used for analytic and monitoring purposes. For instance, any checksum verification failure will result in an instant alarm. Developers can then use the persisted checksums of each step to expedite debugging and root cause analysis.

3.7   Reflective Hook and Event Queue

To build a fault-tolerant robust system, we employ *Apache Kafka*[15] on the side to manage inter-module communication and jobs. It permits asynchronous communication, which optimizes the data flow throughout the system. Queues make our intermediate event data persistent, which improves reliability and reduces errors when different pieces of our system are unavailable. Additionally, it can be quickly scaled to distribute workload among a fleet of users during peak periods. Once the event has been passed from the fetcher to the event store, if there are registered subscribers for a synchronized EOI, the event dispatcher will notify them of the structured event information via a webhook. Webhooks enable users to be notified when an EOI occurs on the blockchain. Rather than querying the server continuously to see whether the state has changed, webhooks deliver information as it becomes available, which is far more efficient and advantageous for developers. Webhooks operate by registering a URL endpoint to which notifications should be sent when specified ROI occurs. The developer maintains complete control over the endpoint, which could be a third-party service or one of their own.

Table 1. Market statistics of supported metaverse projects

Project	Blockchain	# Tokens	# Trans	$ Init Sales	$ VolumnTraded
RiverMen	Ethereum	10.0 K	3327	1.6 M	4.5 M
MatrixWorld	Ethereum	1.8 K	1416	5 M	4 M
	Flow	2.0 K	1518	5 M	1.5 M

4   Operation Cases and Statistics

Matrix Syncer has been deployed in the industry to serve a variety of decentralized metaverse projects, we'll look at two of them here: *RiverMen*[16] and *MatrixWorld*[17], which are both representative projects with good operation

---

[15] https://kafka.apache.org/.
[16] https://www.rivermen.io/.
[17] https://matrixworld.org/.

cases. Table 1 provides an overview of their market statistics. Rivermen is known as the world's first metaverse project dedicated to exporting traditional Chinese culture and first released in August. MatrixWorld is our first-party project, which was launched in October and is the first multi-chain support metaverse project. It currently supports the Ethereum and Flow networks and is among the top three projects by transaction volume (unofficial) on the Flow network.

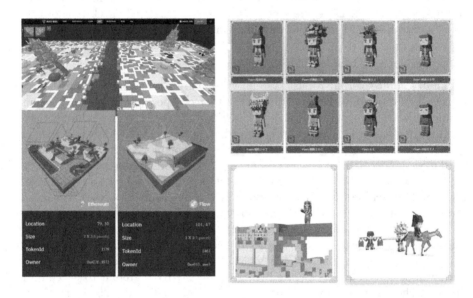

Fig. 2. Left: MatrixWorld project, where users can interact their lands which are minted on Ethereum and Flow blockchains, Right: RiverMen pawn tokens (above) can be used to mint new RiverSpace Token (below).

4.1 Event Data Indexing

It is commonly established that dApps must provide a front-end user interface as shown in Fig. 2 in order to provide an interactive user experience. River-Men's website has a feature to display users' own tokens and perform advanced queries among their tokens as well as the entire collection. The typical smart contract's interface is insufficient for those scenarios. However, by defining a mapping scheme in Matrix Syncer, all events emitted by minting and transferring are well tracked in the event store with a timestamp. To meet the aforementioned needs, RiverMen's front-end can easily call query APIs to our event query service endpoint. Additionally, because all events are timestamped when they are saved to the store, developers can perform advanced queries, such as retrieving the entire transaction history of a particular token or obtaining stats such as the number of transactions of a given token within a specified time window. Similarly, in MatrixWorld, users can locate and investigate all land information directly from the map interface Fig. 2 (left). We defined minting and transfer mapping schemas for the Ethereum and Flow blockchains and unified their data structures in the event store. Later on, our front-end can simply query the needed information without having to write two data parsers for the distinct chains. We demonstrated that Matrix Syncer platform can significantly reduce the effort required to onboard a new project with historical on-chain data querying requirements, regardless of the blockchain it is built on or the number of chains it supports concurrently.

4.2 Event Reflection

Apart from advanced on-chain state queries, event reflection processing is critical for many products with more complicated business logic. RiverMen introduced the *RiverSpace*[18] token, which is capable of minting new tokens through the fusion of a set of RiverMen tokens as shown in Fig. 2 (right). The 3D model of the newly created RiverSpace token is built and rendered dynamically based on the metadata of the RiverMen tokens. However, due to the restricted compute capacity on-chain, these procedures must be performed off-chain. Here, the RiverMen team registered an event triggered by the RiverSpace contract that contained information about the token IDs of each RiverMen token used to fuse the newly minted RiverSpace token. By registering the target event and the webhook on Matrix Syncer, they only need to develop rendering and metadata services to handle off-chain processing. Once their services receive the event, they will begin rendering the new 3D model based on the metadata of the component tokens, which can be fetched through their token IDs. Once the rendering process is complete, they can update their own metadata services to enable the front end to display RiverSpace 3D mode. Here, developers are able to concentrate entirely on their own business logic without having to implement or maintain any of the servers or infrastructure required to interact with blockchains.

4.3 Monitoring and Analytics

Following the best practices of web services development, Matrix Syncer uses *GPL*[19] (Grafana/Prometheus/Loki) stack for metrics, alarms, and log viewer for regular DevOps purposes. Aside from that, we also utilize our persisted checksums for real-time alarming and analytics. We set up instant alarms during the Block Fetching Job execution and stream the checksum data to Grafana dashboard to have double insurance. In MatrixWorld, the checksum failure alarm helped us discover that the data sync error was due to an unstable data node provider. In RiverMen, the checksum stats mismatch helped us find a bug in

---

[18] https://opensea.io/collection/river-space.

[19] https://www.opencue.io/docs/other-guides/monitoring-with-prometheus-loki-and-grafana/.

the event processing chain. Furthermore, we ran an analytics query on our persisted checksum stats to help our customers and our team understand each typed event's processing rate and distribution over time (Fig. 3).

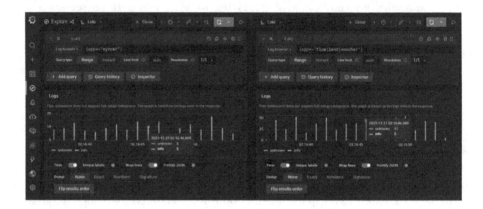

Fig. 3. Monitoring dashboard and data analytics with Grafana

5   Conclusions

With instant setup, multi-chain support, and cloud-based IaaS, our proposed Matrix Syncer is designed to be the ultimate solution for developing complex metaverse DApps that demand advanced on-chain data operation and event reflection. All of these advantageous infrastructures can help hasten the construction of a functional metaverse that is more user-friendly, interoperable, and accessible. By using the elasticity, high availability, and flexibility of cloud computing, Matrix Syncer bridges the barrier between web 2.0 and web 3.0 by leveraging the strengths of each, and enabling the rapid development of a functional, decentralized metaverse with enhanced user-friendliness, interoperability, and accessibility. The future direction will be to integrate additional decentralized protocols into the present system because, as with the advancement of the blockchain's fundamental technology, the current web 3.0 shortcomings will undoubtedly be resolved in the future.

References

1. Buterin, V., et al.: A next-generation smart contract and decentralized application platform. White Pap. **3**(37), 1–2 (2014)
2. Dionisio, J.D.N., Burns III, W.G., Gilbert, R.: 3D virtual worlds and the metaverse: Current status and future possibilities. ACM Comput. Surv. (CSUR) **45**(3), 1–38 (2013)

3. Duan, H., Li, J., Fan, S., Lin, Z., Wu, X., Cai, W.: Metaverse for social good: a university campus prototype. In: Proceedings of the 29th ACM International Conference on Multimedia, pp. 153–161 (2021)
4. Farmer, C., Pick, S., Hill, A.: Decentralized identifiers for peer-to-peer service discovery. In: 2021 IFIP Networking Conference (IFIP Networking), pp. 1–6. IEEE (2021)
5. Han, Y., Oh, S.: Investigation and research on the negotiation space of mental and mental illness based on metaverse. In: 2021 International Conference on Information and Communication Technology Convergence (ICTC), pp. 673–677. IEEE (2021)
6. Harwick, C., Caton, J.: What's holding back blockchain finance? on the possibility of decentralized autonomous finance. Quart. Rev. Econ. Finan. **84**, 420–429 (2020)
7. Liu, Z., et al.: Make web3. 0 connected. In: IEEE Transactions on Dependable and Secure Computing (2021)
8. Nakamoto, S., et al.: Bitcoin. A peer-to-peer electronic cash system (2008)
9. Ritz, F., Zugenmaier, A.: The impact of uncle rewards on selfish mining in Ethereum. In: 2018 IEEE European Symposium on Security and Privacy Workshops (EuroS&PW), pp. 50–57. IEEE (2018)
10. Ryskeldiev, B., Ochiai, Y., Cohen, M., Herder, J.: Distributed metaverse: creating decentralized blockchain-based model for peer-to-peer sharing of virtual spaces for mixed reality applications. In: Proceedings of the 9th Augmented Human International Conference, pp. 1–3 (2018)
11. Singh, J., Michels, J.D.: Blockchain as a service (baas): providers and trust. In: 2018 IEEE European Symposium on Security and Privacy Workshops (EuroS&PW), pp. 67–74. IEEE (2018)
12. Stephenson, N.: Snow crash. Metropolis Media (1992)
13. Wu, X., Qiu, H., Zhang, S., Memmi, G., Gai, K., Cai, W.: ChainIDE 2.0: facilitating smart contract development for consortium blockchain. In: IEEE INFOCOM 2020-IEEE Conference on Computer Communications Workshops (INFOCOM WKSHPS), pp. 388–393. IEEE (2020)

Construction and Design of Food Traceability Based on Blockchain Technology Applying in the Metaverse

Liping Bian[1(✉)], Rong Xiao[1], Ying Lu[2], and Zhibin Luo[2]

[1] Jiangsu Academy of Agricultural Sciences, No. 50 Zhongling Street, Xuanwu District, Nanjing, Jiangsu, China
blp_njau@163.com

[2] Nanjing Zhimai Information and Technology Co., Ltd., Room 1102, No. 8 Xingzhi Rd., Pukou District, Nanjing, Jiangsu, China

Abstract. The production and distribution methods in human society are gradually changed since the opening of the Metaverse Era. Meanwhile, the construction process of elements in the virtual world is being pushed on, and it activates the technology upgrade and application development of the agricultural food industry in the digital virtual field. This article proposes the construction, design, and development schema of food traceability based on blockchain technology applied in the Metaverse. (i) This paper discusses the application scene and significance of the derivative concept of digital virtual food and agricultural food traceability technology in the Metaverse. (ii) This paper discusses the technical principle of blockchain-based agricultural food traceability. (iii) State the functional processes of food traceability application and the design of relevant smart contracts in the Metaverse. (iv) This paper summarizes the application feasibility, advantages, and future sustainable development possibility of Food Traceability Technology and its application in the Metaverse world.

Keywords: Food traceability · Blockchain · The Metaverse · Smart contract · Virtual consumption · Social significance of food

1 Introduction

The word the Metaverse firstly came from the fiction "Snow Crash" (1992), which wrote by famous American fiction novelist Neal Stephenson. The book describes a networked world named "the Metaverse" which is like the real world, and each person, in reality, would own a mirrored identity in the network world. Now we define the Metaverse, with conceptions including VR/AR technology, Internet, game, and SNS, all these conceptions could be mixed up to create a new appearance of the Internet. "The Metaverse" may be the third revolution of productivity, quite a lot of behaviors in contemporary society will completely change in this case.

S. Berretti and G.-M. Su (Eds.): ICSM 2022, LNCS 13497, pp. 294–305, 2022.
https://doi.org/10.1007/978-3-031-22061-6_22

Due to the 5th STI (Science, Technology, and Innovation) Basic Plan from the Japanese Cabinet Office website, the Plan paper firstly proposed the conception of Society 5.0. And in the 6th STI Basic Plan, defined Society 5.0 as "A human-centered society that balances economic advancement with the resolution of social problems by a system that highly integrates cyberspace and physical space [1]." Fig. 1 shows the stage change in society. The definition of Society 5.0 includes the part that is more consistent with the concept of the Metaverse, that is, the highly virtualized immersive Internet pattern. It can be seen that the fields of information technology and digital multimedia technology have taken the lead in making more consistent predictions about the changes in social patterns.

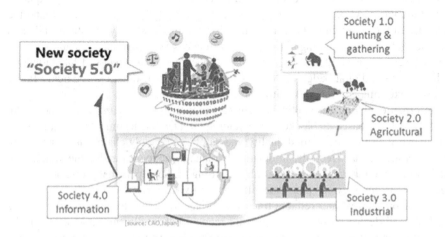

Fig. 1. Conception development from Society 1.0 to Society 5.0 [1]

When the productivity changes from physical strength to intelligence, the company replaces the landlord, the vote replaces violence, and the state, the people, and everything jump with the productivity. From today to the future, when human intelligence (the ability to operate machines) is no longer important and machines themselves have strong enough productivity, human society has finally ushered in our "dreamed" the Metaverse – all productivity is performed by machines, based on the distribution model of distribution and decentralization. It has virtual original governance institutions, major "political bodies" that monopolize computing power, the concept of virtual citizen, as well as education, "work" and life in the virtual world. All these are the truth of the Metaverse.

Clothing, food, shelter, and transportation are still important fields of life elements and interest communities that citizens cannot escape in the Metaverse. Among them, the force of food in social activities is undoubtedly friendly and powerful. In addition to the popular simulation restaurant operation game for more than 20 years, there are still a large number of feverish players. In the game world of "Cyberpunk 2077", there are street snacks and Thai Restaurant LED boards. Takemura's (a role in the game) collection of "night city food guide" and chicken balls may also impress you. Retailing, gradually changes its form from online to virtual, providing more service experience possibilities for the Metaverse consumer market [2].

Vegetarians, organicists, and Michelin restaurant lovers can establish community circles, theme activities, and collection exhibitions respectively in the Metaverse. They can continue to advocate the concept and practice of a healthy diet and high-quality diet in the virtual world, find user groups interested in the same topic and expand the influence of the community. Of course, it's also possible to show your life attitude and food aesthetics by consuming "junk food" with excessive actual calories and sugar in the Metaverse. Spiritual food can't fill your stomach, but it can meet self-label identity. Believe that the digital badge of "Coca Cola lovers" can also help you quickly find a group of like-minded Metaverse citizens.

Informational food traceability has a history of more than 20 years all over the world. Food traceability based on blockchain technology is a heating technical topic at present. Whether it is used to trace food quality and safety and supply chain links, or to store consumer market sales records, automatically manage the food stored in the digital twin refrigerator, and purchase food materials according to users' personal preferences, food traceability technology can penetrate all aspects of virtual life.

Based on the real and reliable traceability data collecting terminals and channels (such as IoT device terminals, AI graphic recognition intelligent cameras, etc.), the traceability and information integrity of food can be highly guaranteed. We can trace back to the source ingredients, logistics and transportation, fresh-keeping storage, processing, and cooking methods of each salad and plate of advanced cuisine, even the hygiene qualification and Michelin star rating of cooking restaurants can be permanently recorded in a traceability file, and the blockchain will keep distributed accounts for the file.

All these can be seamlessly connected to the virtual world of the Metaverse to form an organic recipe or a Michelin restaurant experience record. Users can pay to browse these unique online file data and conduct a virtual experience of a Michelin restaurant (even an immersive experience of dishes with 3D virtual and VR media files). In case of traceability records that do not meet the food quality and safety standards, warnings or penalties will be triggered through the automatic verification of smart contracts. For another example, this recipe makes you eager to try, and you can't easily reach the restaurant tens of kilometers away. You can place an order with one click and send it home directly according to the source information of food materials recorded by traceability, to restore the healthy and good taste of the recipe as much as possible. You can even participate in the virtual planting project launched by the farm to immerse yourself in the growth and processing of food ingredients and remotely fertilize and water them.

Since 2017, the Jiangsu Academy of Agricultural Sciences · Institute of agricultural product quality and safety has been committed to the research of agricultural product traceability informatization and has made outstanding research achievements in the intelligent agricultural IoT, the complete process quality control model of agricultural products, and the multi-level in-depth traceability of agricultural products.

Nanjing Zhimai Information and Technology Co., Ltd. has been deeply engaged in the field of agriculture, forestry, animal husbandry, and fishery. With the intelligent information software development technology and the research results of Jiangsu Academy of Agricultural Sciences, it has realized a set of agricultural products in-depth traceability software system with the functions of safety early warning, source traceability, flow traceability, information inquiry, and product recall.

The "deep traceability system of agricultural food quality and safety based on blockchain" was jointly developed by Jiangsu Academy of Agricultural Sciences and Nanjing Zhimai Information and Technology Co., Ltd., it can realize the above traceability and intelligent evaluation functions. The application has won the digital governance award of 2020Y digital Jiangsu construction's excellent practical achievements (China) [3, 4]. In addition to traceability, the application platform can also manage and control the complete process of agricultural productivity growth and processing based on the agricultural product growth model, digital twin technology, and whole process control technology system.

The system has the following features.

I. A whole process quality and safety control model covering the before, middle, and after production of agricultural products is constructed, which can automatically generate SOP standard operation process control middleware according to the production and operation standards of different agricultural products, use the middleware to evaluate the production data, guide users to standardize the operation, and store the pre-processing and post-processing data through the blockchain. Ensuring that the real content of each standardized operation can be viewed in the final traceability results (Fig. 2).

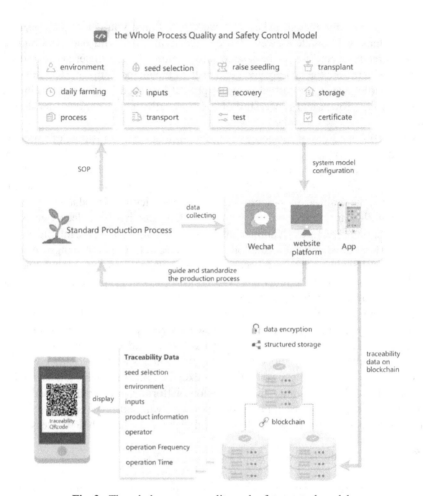

Fig. 2. The whole process quality and safety control model

II. It provides blockchain penetrating supervision technology in the scenarios of agricultural product traceability, supervision, and certification. Through our agricultural blockchain BaaS platform, we can quickly deploy smart contracts, realize horizontal cross-category and vertical cross-level penetrating smart supervision, prompt and accurate feedback on possible risks to regulators or producers, improve the production management level, reduce the quality and safety risks and possible losses of agricultural products.

2   Principle of Agricultural Food Traceability Technology Based on Blockchain

The essence of traceability is information transmission, and the blockchain itself is also information transmission. The data is made into blocks, and then the private key is generated according to the relevant algorithm to prevent tampering, and then the chain is formed by timestamp, which exactly conforms to the process production mode of the commodity market.

The unique decentralized storage of blockchain technology does not depend on an organization or individual. All information is publicly recorded in the "public ledger" by using trusted technical means. The data on the chain is time-stamped and cannot be tampered with. Once the immutable information is set up, it is equivalent to finding the unique identity of real-world goods in the Internet world, the corresponding information will also be permanently recorded on the chain, and all tracking and recording based on this identity flow are realized. Most of the current landing application projects of blockchain anti-counterfeiting and traceability are based on public blockchain or consortium blockchain.

All data of production, supply, and marketing of agricultural products are encrypted and stored on the blockchain (including variety, place of origin, inputs, growth environment, collecting, processing, packaging, storage, detection, qualification label, and operator of each link). Each block of stored data will be associated with the front and rear blocks through hash value, to realize the whole process traceability of quality and safety data. Because the information cannot be tampered with at will, the information of each link from production to transportation to final sales must be recorded on the blockchain to ensure the uniqueness of goods. Therefore, the information about fake goods and defective products cannot enter the blockchain system. The commodity information on the chain, whether it is standardizing the production process or safe sales logistics, can be determined through the automatic verification of the smart contract. The nonconforming operations will be exposed in real-time, warned, and permanently recorded on the blockchain.

3 Functional Processes of Food Traceability Application in the Metaverse: Take a Traceable Dish as an Example

A complete food traceability application system generally has at least three program terminals: back-stage, terminal for enterprise, and terminal for the consumer. Enterprises punch in the whole-link-supply chain traceability information of a finished dish through the application, and consumer users upload the dish consumption experience or query all the traceability information data through the terminal for consumers, as well as claim virtual planting projects. The backstage needs to monitor and manage the users of the consortium blockchain, the performance of the underlying blockchain, smart contracts, the real-time operation status of function modules, and traceability information compliance. All the above terminals work together to build up the application service framework as Fig. 3 shown below.

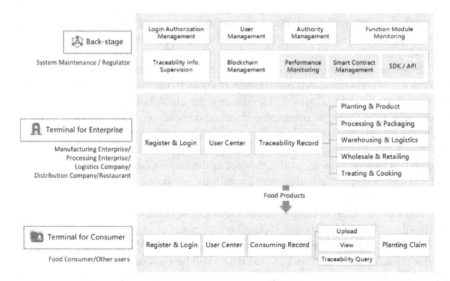

Fig. 3. Application service framework of food traceability application system in the Metaverse

As illustrated in the above figure, the backstage is mainly operated by the system maintenance and regulators who are in charge of food quality and safety. Its login authorization management, user management, and authority management function modules synchronously control the registration and login of the other two terminals and the functions of the user center in real-time. The traceability information supervision module can give a real-time alarm to the data on blockchain that does not meet the standard so that the personnel of the regulatory authority can understand the risk to food quality and safety. The function module operation monitoring module mainly serves the application system operation and maintenance personnel to understand the real-time performance of each application terminal and function module. The most critical and special module is the blockchain management module, which includes a series of core functions such as monitoring the performance of the blockchain, managing smart contracts, and managing the interfaces to external systems such as the Metaverse.

The planting and production companies, distribution and logistics companies of food materials, and the restaurant that finally processes dishes can use the terminal for enterprise to record the traceability information of each link of the supply chain. Enterprises that are diligent in recording and committed to ensuring the correctness and integrity of food traceability information, can be rewarded with memorable NFT or blockchain certification.

Consumers and other users can upload food consumption experiences or query all the traceability information data through the terminal for consumers, as well as claim virtual planting projects. Based on all the traceability information of food and ingredients and digital twin technology, consumers can even experience the whole traceability report through VR, and intuitively watch the production process and processing process of food.

4   Design of Food Traceability Smart Contract

In the entire system, the key work processes of the blockchain with key confirmation, traceability records storing, and verification functions are shown in Fig. 4 below.

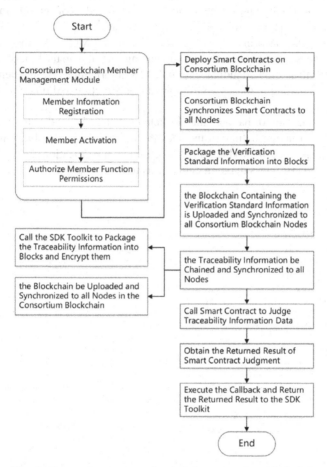

Fig. 4. Key work processes of underlying consortium blockchain

The deployment of a smart contract on the consortium blockchain is to link the created smart contract file to the consortium blockchain in the form of a deployment transaction. Then send the binary code and bytecode of the smart contract file into the input data of the blockchain transaction, edit the receiving address starting with 0x0, and package it into the transaction block. According to the address header of 0x0, the blockchain miner determines that the transaction type in this transaction block is "send the contract, deploy on-chain". After verifying the correctness of the information, execute the deployment of the block onto the chain, which is called deploying the smart contract. Currently, the smart contract has the contract address and exists in the chain.

The smart contract deployment transaction will be synchronized to all nodes by the consortium blockchain, that is, all nodes can read and call the smart contract through EVM (Ethereum Virtual Machine). When the traceability information of agricultural food is chained, the operation mechanism of smart contract verification and early warning is mainly divided into information chaining and synchronized to all nodes. The nodes call the smart contract to judge the traceability information data of agricultural products, obtain the judgment return results and execute a callback.

Smart contracts can primarily serve two roles in the Metaverse food traceability applications, one is to standardize the data uploading format onto the blockchain and the other is to perform automated judgment of uploaded information. The elements of smart contracts corresponding to these two roles are a data structure and function design.

I. Standardize the data format onto the blockchain, that is, define the structure of traceability data. Structure definition includes the definition of traceability information structure of agricultural products and the definition of verification standard structure. The possible UML class diagram could be as follows (Fig. 5):

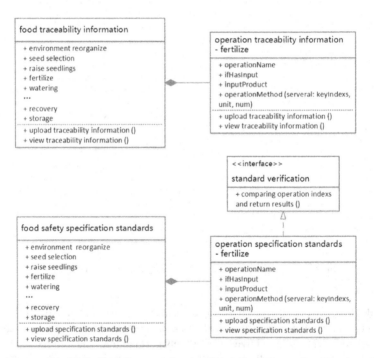

Fig. 5. The possible UML class diagram of traceability information and verification standard

Taking the definition of agricultural product traceability information structure as the above example, it is the standard information entry structure created for the data to be uploaded at the key nodes of agricultural food quality and safety.

II. Function design involves combing and splitting the syllable objects that need to be judged and splitting the output. Below flow chart shows the functional design that judges whether the uploaded traceability information meets food safety specification standards (Fig. 6):

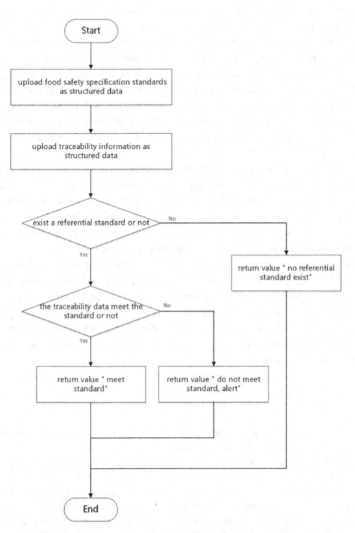

Fig. 6. The flow chat of traceability information judgment

Through this smart contract, we can achieve (i) return "no referential standard" if the operation name in the traceability information structure is not found in the verification standard structure; (ii) return "met standard" if the structure of single traceability information and the verification standard agree perfectly; (iii) If a single traceability information structure and a verification standard structure exist with at least one syllable value that does not agree, return "do not meet the standard, alert ".

5 Conclusions and Future Research

The exploration of stepping into the Metaverse has just begun. The clothing, food, housing, transportation, leisure, and entertainment projects in the virtual world rely on different innovative technologies to gradually become plump and complete. The new decentralized society environment in which "silicon-based" life is located will inevitably give birth to the evolution of the essence of things. The demand for traceability of food safety will not decrease with the change in social patterns, on the contrary, due to technologies such as blockchain, traceability information will become more complete and more reliable. The application of food traceability in the Metaverse connects reality and virtual, and increases the experience of food, which is no less than the feeling of simply enjoying finished food in real society.

We should pay more attention to the users' experience and trust in the application when designing the food traceability application of the Metaverse. We should make full use of the technical characteristics of blockchain, the reliable distributed record storing method is used to record the traceability data information of the food supply chain, and the smart contract is used to standardize the data structure on blockchain and automatically figure out whether it meets the food safety standards.

On the other hand, in the future, we should continue to explore the performance optimization and compatibility of applications in the Metaverse, constantly break through the upper limit of throughput on the blockchain infrastructure, and combine with advanced multimedia technology for virtual experience, to coordinate the functional experience of application software with the virtual life experience of the Metaverse. Only by adapting to the attributes of the Metaverse can food traceability applications run and grow in the virtual world for a long time.

References

1. Cabinet Office, Government of Japan. Society 5.0. (2021). https://www8.cao.go.jp/cstp/english/society5_0/index.html. Accessed 26 Mar 2021
2. Bourlakis, M., Papagiannidis, S., Li, F.: Retail spatial evolution: paving the way from traditional to the Metaverse retailing. Electron. Commer. Res. **9**, 135–148 (2009)
3. Li, Y., Wang, P.: The blockchain technology of Jiangsu Academy of Agricultural Sciences has won the digital governance award of 2020Y digital Jiangsu construction's excellent practical achievements (China). Jiangsu Agricultural Science and Technology Newspaper (CN 32-0001), vol. 26 (2021)
4. Zhang, W., Li, Y., Wang, P.: Use blockchain technology of Jiangsu Academy of Agricultural Sciences to trace the origin of agricultural products (2021). https://www.xyshjj.cn/detail-45826.html. Accessed 2 Apr 2021
5. Choi, H.S., Kim, S.H.: A content service deployment plan for Metaverse museum exhibitions – Centering on the combination of beacons and HMDs. Int. J. Inf. Manage. 37(1), 1519–1527 (2017)

6. Hassouneh, D., Brengman, M.: Retailing in social virtual worlds: Developing a typology of virtual store atmospherics. J. Electr. Comm. Res. **16**(3) (2015)

7. Volker, R.S., Lakshmi, G., Mattia, C.: The Metaverse as mediator between technology, trends, and the digital transformation of society and business. J. Virtual Worlds Res. **8**(2) (2015)

8. Kim, J., Nevelsteen, L.: 'Virtual World', defined from a technological perspective, and applied to video games, mixed reality and the Metaverse. CoRR, abs/1511.08464 (2015)

9. Gadalla, E., Keeling, K., Abosag, I.: Metaverse-retail service quality: A future framework for retail service quality in the 3D internet. J. Market. Manage. **29**(13–14), 1493–1517 (2013)

10. Ludlow, P., Wallace, M.: The Second Life Herald: The virtual tabloid that witnessed the dawn of the Metaverse. MLT Press, 0262122944 (2007)

11. Dionisio, J.D.N. III, Burns, W.G., Gilbert, R.: 3D virtual worlds and the Metaverse: Current status and future possibilities. In: ACM Computing Surveys (CSUR), vol. 45, pp. 1–38 (2013)

12. Computer Companies; Patent Issued for System and Method for Attending a Recorded Event in the Metaverse Application (USPTO 9165426). Computer Weekly News (2015)

13. Lee, S.-G., Trimi, S., Byun, W.K., Kang, M.: Innovation and imitation effects in the Metaverse service adoption. Service Bus. **5**(2), 155–172 (2011)

14. Marion, G., Horst, T.: The influence of Blockchain-based food traceability on retailer choice: The mediating role of trust. Food Control **129** (2021)

15. Hernandez San Juan, I.: The Blockchain technology and the regulation of traceability: The digitization of food quality and safety. Eur. Food Feed Liter. Rev. **15**, 563 (2020)

Deep Learning on Video and Music

Motion Segmentation Based on Pixel Distribution Learning on Unseen Videos

Zhiyuan Lu⬤, Youwei Chen$^{(\boxtimes)}$⬤, and Chenqiu Zhao⬤

Department of Computing Science, University of Alberta, Edmonton AB T6G2R3,
Canada
{zlu6,youwei1,chenqiu1}@ualberta.ca
https://www.ualberta.ca/index.html

Abstract. Motion segmentation plays an important role in many applications including autonomous driving, computer vision and robotics. Previous works mainly focus on segmenting objects from seen videos. In this paper, we present a novel approach based on pixel distribution learning for motion segmentation in unseen videos. In particular, optical flow is extracted from consecutive frames to describe motion information. We then randomly permute these modified motion features, which are used as input of a convolution neural network. The random permutation process forces the network to learn the pixels' distributions rather than local pattern information. Consequently, the proposed approach has a favorable generalization capacity and can be applied for unseen videos. In contrast to previous approaches based on deep learning, the training videos and testing videos of our proposed approach are completely different. Experiments based on videos from the KITTI-MOD dataset demonstrates that the proposed approach achieves promising results and shows potential for better motion segmentation on unseen videos.

Keywords: Machine learning for multimedia · Multimedia vision · Video processing

1 Introduction

A movement of an object that carries temporal statistics and spatial information that can be tracked through a computer or camera even when they undergo several transformations. As humans, we can easily identify each moving object by observing its color, shape, motion trail, etc. Such a task, named Motion Segmentation, is a fundamental problem in the computer vision field.

The motion segmentation is useful in surveillance, action recognition, safe autonomous driving, scene understanding, and many other applications. Prior approaches to motion segmentation usually use optical flow and scene information to segment static background and moving objects, which requires a large

---

Z. Lu and Y. Chen—Equally contributed to this paper.

© The Author(s), under exclusive license to Springer Nature Switzerland AG 2022
S. Berretti and G.-M. Su (Eds.): ICSM 2022, LNCS 13497, pp. 309–321, 2022.
https://doi.org/10.1007/978-3-031-22061-6_23

number of labeled images as training input. Meanwhile, training models in a specific dataset to achieve satisfactory performance could potentially lead to lower accuracy in other completely unseen datasets. To handle it, we proposed a novel motion segmentation method in which a pixel-wise motion distribution descriptor named Random Permutation of Spatial Pixels Portion (*RPoSPor*) is used as the input of the network for learning the statistical distribution of spatial pixels. More specifically, spatial pixels from the obtained optical flow images perform random permutation with surrounding pixels forming patches.

Usually, well-labeled motion segmentation masks required many human efforts for annotation. Besides, labeling tasks are easily fraught with potential errors. Both of these factors limited the usage of deep learning networks in real applications of motion segmentation. Previous approach usually devised an end-to-end network with the input of optical flow and the output of binary masks. These methods require a lot of annotated frames for training, which is not suitable for real applications. Meanwhile, the performance of these methods various when the testing videos are compared from a scene which is not included in the training set. Therefore, in this paper, we proposed to learn the statistical distribution of spatial pixels from optical flow images to handle these limitations. Compared to previous methods, the proposed approach only assumed limited labelled masks for training. In addition, due to the well generalization ability of distribution information, the proposed approach is even effective under the condition that training videos and testing videos are captured from different scenes.

The proposed approach is straightforward. The distributions of spatial pixels from optical images are extracted to construct the Random Permutation of Spatial Pixels Portion features in which the pixels are randomly permuted to guarantee that only statistical information contains in the feature. Then a simple classification network is devised to classify if the distribution belongs to the foreground or background. In particular, we measure the distribution of spatial pixels from multiple scales instead of only one scale. Benefiting from the multi-scales strategy, the proposed approach achieves better results with even fewer training data. Compared to previous methods, our network not only works from the training dataset but also in more general circumstances. With more computational resources, our robust randomness processing method could be used in more complex models and environments, that potentially achieve a better result. Overall, our proposed method shows a well-performed generalization ability that can train on a specific dataset and maintained a satisfactory result in other completely unseen datasets. This expands the performance boundary of deep learning-based methods.

2   Related Work

Motion segmentation separates the independently moving objects(pixel-wise) from the static background and serves as an important procedure in many applications, such as physical rehabilitation exercises [9], robotics [8], and Intelligent

Visual Surveillance [6]. In this section, we present three essential topics for our work: Optical Flow Extraction, Motion Segmentation Method, and Pixel Distribution Learning Algorithm

2.1   Optical Flow Extraction

Optical flows describe the motion of individual pixels(objects) on the image plane. It serves as a good approximation of the true physical motion projected onto the image plane [21]. Thus, a precise optical flow algorithm plays an essential role in our proposed method.

Traditionally, Horn and Schuck derived optical flow based on the constraints of constant image brightness and flow smoothness, followed by minimizing these constrained equations which yield the flow field [3]. At the same time, Lucas-Kanade optimized the Horn-Schunck algorithm's computational complexity [10]. Though traditional methods of optical flows are generic, both Horn-Schunck and Lucas-Kanade algorithms assume constant brightness in the time frame, they would not work well in different illumination scenes.

An alternative method to reduce the constraints described above is by using neural network architectures. Dosovitskiy et al. proposed the FlowNetS by stacking a pair of images together, spatially compressing the images down, and then extracting the feature maps to a higher resolution [2]. Another neural network structure that Dosovitskiy et al. introduced, dubbed FlowNetC, learned two images' features separately at first and then combined them into a "correlation layer" to compare the feature maps [2]. Ranjan et al. combined the traditional pyramid methods with deep learning methods. The neural network structure, a spatial pyramid network, is 96% smaller and has a lower error on benchmarks than FlowNet [14]. Ilg et al. proposed FlowNet 2.0 by stacking multiple FlowNetS and FlowNetC architectures with warping layers. As a result, the FlowNet 2.0 has nearly 2x more parameters than that of the FlowNet [5]. Sun et al. extract and wrap feature representations of two images from L-level pyramids. Construct a cost volume layer that stores the matching costs of a pixel with its matching pixel at the next frame, as the input to the network [16]. Hui et al. proposed the LiteFlowNet, constructing the architecture with smaller model size and faster inference speed than FlowNet 2.0 by encoding the two pyramids of high dimensional image features, and sending the same resolution features at each pyramid level to the decoder to estimate the flow field [4].

Recent work on optical flow focuses more on large displacements between two images, as large motions would change patch appearance, and the search space of the same pixel in two images increases. The cost volume, which is a four-dimensional match between all pixels in two images, is considered to be inefficient to compute. Z. Teed and J. Deng proposed RAFT architecture using cost volume, too [17]. Raft architecture uses a GRU-based update operator that retrieves the motion priors from 4D multi-scale correlation volumes and iterative updates the flow field so that the motion in the update operator is not handcrafted but learned. In addition, like the LiteFlowNet, Raft architecture does not draw down the input's resolution, which could prevent the network from missing the small and fast-moving objects [17].

2.2   Motion Segmentation Method

For motion segmentation, previous approaches include traditional geometry analysis, tracking , and learning-based methods. The geometry-based method involves clustering image features together that belongs to moving objects. For example, Bideau et al. performed seed pixel selection based on optical flow cues, then use feature matching and geometry methods to correct the seed pixels. [1] Tracking-based method mainly focuses on calculating the similarity of image pixels using historical data [7,12,19]. In Wong's work, he designed two statistical tests to determine the motion segmentation and tracking algorithm that follows the shift and align paradigm [19]. Margret Keuper et al. promote group points trajectories with multiple objects tracking by clustering of bounding boxes [7].

In the last decade, the deep learning method has shown substantially better performance and is computationally efficient during the testing phase. At the semantic motion segmentation level, Johan Vertens et al. proposed SMSnet that learns from motion features in optical flows and semantic features, and they also extended their work to an efficient multi-task model [18]. SMSnet achieved state-of-the-art performance on the KITTI dataset. At the instance level, Eslam Mohamed et al. adapted YOLACT to a motion-based instance segmentation model to separate connected segments of moving objects [11]. But deep learning methods in the motion segmentation field generally required sufficient dataset to train the model. Due to the lack of large public datasets, Eslam Mohamed and Mahmoud Ewaisha created a new InstanceMotSeg dataset [11]. Vertens et al. also introduce a Cityscapes-Motion dataset in their paper [18] Moreover, previous neuron network-based approaches are mainly focused on training and testing among the same dataset, with less performance on the other datasets. This brings us to the task of defining a general approach to improve the robustness and generalization ability of the learning-based method. Pixel distribution learning describes the hidden information in both temporal and spatial views, and is able to cluster both local and global information by stacking distributions. This pre-processing method pushes deep learning models' to a higher generalization ability potential.

2.3   Pixel Distribution Learning

The Pixel Distribution Learning model, which analyzes the random permutation of pixels, is previously proposed to solve problems related to the background subtraction [23]. Inspired by this model, the multi-scale deep pixel distribution learning method (MS-DPDL) is proposed by Wu and Zhao for concrete crack detection [20]. To improve performance, they also introduce a multi-scale patch generation method that provides the networks with better information to learn the distribution. Zhao et al. further applied this method to vessel segmentation [22]. Different from previous work, for this method, Random Permutation of Spatial Portion ($RPoSPor$)is proposed by shuffling both images and pixels group randomly, creating groups of patches under multiple scales by N times, and dividing the intensity of the center pixel. Then, the patches are fed into a convolutional neuron network. More specifically, patches with different scales go into our shallow and wide convolutional neuron network and achieved high accuracy given few training images. With testing on entire unseen and unrelated other dataset sequences, our model still generates satisfactory results. Also, our *RPoSPor* model analyzed optical flow scenes brings more temporal information, and it only requires a few training images and a small number of epochs. However, due to the nature of random permutation, the network may generate noises when the *RPoSPor* features in the testing phase are never used in the training procedure. We found this method can also be potentially combined with more complex deep learning models, by designing Random Permutation of Spatial Portion layers that involve sufficient global image information and scene information.

3   Motion Segmentation Based on Pixel Distribution Learning

In this section, the details of the proposed approach are discussed. The flow chart of the proposed approach is shown in Fig. 1, in which multi-scale randomized spatial pixels are extracted from the flow image, as the input of the network. The motion segmentation network would classify the distribution of spatial pixels in the flow image into "has motion" or "no motion" labels.

Table 1. Details of the Network Architecture

Type	Filters	Layer size	Datasize
Input Data			$6 \times 16 \times 16$
Convolution	512	$6 \times 16 \times 16$	$1 \times 1 \times 512$
ReLU			
Linear	2	1×512	1×2
Log-Softmax			

Since the *RPoSPor* features in the proposed approach are captured from optical flow images, we used the "RAFT" [17] to generate the flow images. To segment the moving object from the static background, the Random Permutation of Spatial Pixels Portion (*RPoSPor*) method is proposed. In particular, as shown in Fig. 1, the distributions of pixels in the moving objects and the static background are significantly different, so it is necessary to compare each pixel on the flow image with its neighboring pixels. Hence, instead of subtracting the center pixel with its surrounding pixels introduced in [23], we divide the center pixel which achieves a better recall score. In addition, since a single scaled patch of pixel distribution in a particular region may not cover enough spatial information, we use a multi-scaled patch to extract pixel distribution features

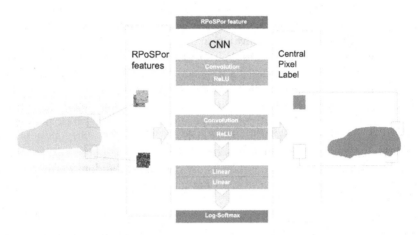

Fig. 1. The flow chart of the proposed method. The optical flow image is generated by "RAFT" network [17].

from different sized patches within the same pixel. The stacked patches with the randomized pixel distribution feature for a particular pixel are feed-forward to the network. The output of the network would classify the input multi-scaled patch for such pixel into "moving" or "static".

The procedure to capture $RPoSPor$ is shown in Fig. 2. Starting from a flow image $I(x,y)$, where x and y are the locations of an arbitrary pixel on the image, a small square patch with size P_1 needs to be created with its neighboring pixels, which its center locates at (x, y). Then, all pixels in such a patch are divided by the center pixel value. After the patches are created for all pixels in image I, we randomly permute each patch. To generate other patches with different sizes P_o, we could repeat previous steps, and randomly select same number of pixels from patch P_1 and patch P_o. Finally, the patches with multi-scaled features could be linked and fed into the network. Mathematically, the procedure can be summarized as follows:

$$RPoSPor_{x,y}(R_i, R_o) = \frac{I(x + r(m), y + r(n))}{I(x,y) + \epsilon} \tag{1}$$
$$m, n \in [\,1\,R_i\,], r(m), r(n) \in [\,1\,R_o\,]$$

where $RPoSPor_{x,y}(R_i, R_o)$ denotes the patch with multi-scale pixel distribution feature extracted from the image located at (x, y). ϵ is a very small number to prevent $I(x,y) + \epsilon$ to be zero. m, n are the indices of an entry in a patch and r() is the random permutation. R_i and R_o are the parameters to control the size of the extracted patch under multiple scales. In particular, several patches with different radius R_o with the center located at (x, y) are extracted first. To stack these patches with different radius, we select first $R_i \times R_i$ pixels from each flattened multi-scale patch. Thus, the multi-scale patches could be reshaped back to the size of $R_i \times R_i$ and linked together.

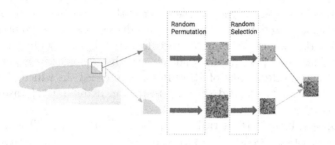

Fig. 2. The procedure for capturing Random Permutation of Spatial Pixels Portion features.

Another problem that we need to deal with is fake randomization, as machine-generated random values still follow a certain pattern. Indeed, it is hard to generate truly random values. To prevent the network overfits such patterns, we would not pre-generate all input patches before training. In contrast, we generate new random indices for all input patches in the training epoch so that the network would focus on learning the statistical pixel distribution information, rather than learning the randomization patterns. After pre-processing the flow image, the stacked multi-scale randomized pixel distribution feature is fed to the network. The output of the network is the label's pixel corresponding to the central pixel of the input patch.

The network architecture in our work is detailed in Table 1. It is a classification network to determine whether the pixel in the flow image belongs to "moving" or "static". According to [23], wider and shallower networks tend to learn pixel distribution features better. Our crafted network consists of a learning block and a decision block. The learning block has two convolutional layers and two rectified linear layers, and the decision block has two fully connected layers followed by log-softmax as the output layer. To summarize, the advantages of the proposed method are:

1. Since we input the small patch into the network instead of the whole image, one can easily train or test the model on different video sequences without changing the network structure.
2. With only a small number of training data, the network achieves a high Fm score, as negative training samples are not needed.

3. The method shows a strong cross-dataset generalization, as one can train the model on one dataset and directly test it on another dataset.

4   Experiments

In this section, we present the datasets used, the experimental setup for the proposed method, and detailed experiment results on unseen datasets. In section I, the datasets that we use to train and test the network are introduced. In section II, we show the hyperparameters that we use throughout the experiment. In section III, the transferability of RPoSPor is presented, in which the network is trained on one dataset, and directly tested on a different dataset without changing any hyperparameters.

The proposed framework is trained on the challenging KITTI MOD dataset [15], which consists of 1950 frames in total. Since the proposed method extracts pixel-level features, many flow images in the same scenario are similar, plus negative training samples are excluded, we randomly choose less than 200 images to train. The DAVIS dataset which our network is tested comprises 50 video sequences with different kinds of motions [13]. For some frames in DAVIS, the camera moves in nearly the same direction as the moving object, thus the color in the foreground and background are quite similar, which brings another challenge to our proposed method.

4.1   Experimental Setup

Throughout the experiment, the Adam optimizer is used with the learning rate of $1e - 4$. The negative log-likelihood loss (NLLLoss) is used as the loss function, with the weight of 0.3 and 0.7 for the class "moving" and class "static", respectively, as there are significant amount of scenes relate to the static camera setting. We set the patch size(R_o) to be 16×16 for each pixel during $RPoSPor$ feature extraction step. Our network is trained for 50 epochs on an Nvidia GeForce GTX 1050Ti GPU. Finally, we compute the precision, recall, and F2-score for the evaluation metric in our experiment.

4.2   Transferability of RPoSPor

In this section, the transferability of the proposed $RPoSPor$ feature extraction method is presented. Various motion segmentation networks only perform well on the dataset which has a similar motion to the training set. In real-life scenarios, however, there are numerous kinds of motions between the camera and the object, so it is impossible to generalize all motions in one dataset. In contrast, the proposed $RPoSPor$ continues to work well with the dataset that the network is not trained on, as the proposed method segments the motion by the difference in pixel distribution, which is not directly related to how the object moves.

To demonstrate the transferability, we would test the proposed method on datasets that are different from the training dataset. The evaluation process is split into two parts. First, we perform a simple experiment to illustrate why the proposed method generalizes well with a small amount of training data. In this scenario, the network is trained on the KITTI MOD dataset with a few selected video frames and is evaluated on the KITTI Motion dataset which has similar motions to the training sequences. The training video sequence from the KITTI MOD dataset and the network inference results on the KITTI Motion are shown in Fig. 3. Second, we randomly pick some frames in the KITTI MOD dataset to train the network and directly test the network on the DAVIS dataset. The DAVIS dataset contains more motions and more diverse scenarios that do not appear in the training dataset.

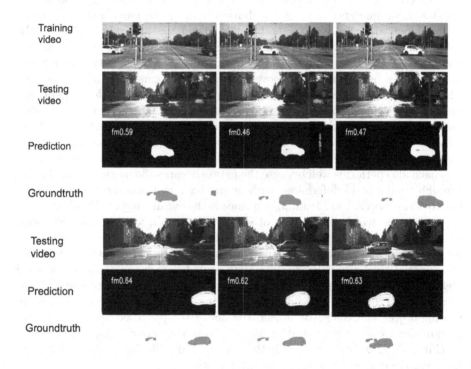

Fig. 3. The evaluation result of the proposed method in seen and unseen video sequences from KITTI Motion dataset. In particular, the network has already seen the right moving motion, but have not seen the left moving motion.

As depicted in Fig. 3, our model achieves an average Fm metric of 0.629 in a selected video sequence in KITTI Motion, in which we select some frames where the car moves to the right as the training set, and directly test the sequence where the car moves to the left. The proposed method could successfully segment the left-moving car, even if we do not tell the network how the "left-moving" should look in the flow image, and achieves similar Fm scores as the seen video sequence (right-moving car). In addition, the camera is static in the training dataset but it is moving in the test dataset, which would generate some noise in the background of the flow image. The proposed method could eliminate a majority of the noise and segment the background as static. With such a simple experiment, the transferability of the proposed method can be demonstrated, as it is redundant to teach the network similar motions, the network could generalize the unseen motions based on pixel distributions.

As shown in Fig. 4, the transferability of the proposed method is further discussed under the assumption of the differing training and test sets. Figure 4 shows the test result on the DAVIS dataset. With distinct scenes and various motions in the dataset, the proposed method performs surprisingly well. In particular, though some motions indeed do not appear in the training dataset, with the statistical difference in pixel distribution between the moving object and the static object, the proposed method could extract the moving object from the entire frame. For example, in the first scene "drift-chicane", though we still segment the car's motion, the motion type is different, where the network is trained on a sequence in which the car moves away from the camera in a straight path while tested on the sequence in which the car moves toward the camera in a curved path. Due to the clear pixel distribution difference between the moving car and the background, the proposed method achieves a satisfactory result. In scene 2 "Tennis" and scene 3 "Kitesurf", we can see that the scenarios in the training set and test set are completely different but with similar motions, the approach also performs well because the network learns the motion segmentation through spatial pixel differences, with no matter how the scene changes. In the fourth scene "soccer", the training sequence is the car moving to the left far from the camera, in which the moving object is quite small, whereas the test frame is a soccer ball moving in the same direction but closer to the camera. Since we select multiple patches with different scales in $RPoSPor$ feature extraction, more spatial pixel distribution information could be extracted, thus the proposed method could get a promising average Fm score in the whole sequence with the number of 0.7767. As a consequence, the larger the color difference between the moving object and the static object, the better the proposed method performs despite the scene changes. Rather than subtraction, the division of the center pixel in $RPoSPor$ better segments the smaller color difference between the moving foreground and the static background motions.

A few scenarios appear to be challenging to the proposed method, thus future works would apply. In some cases such as scene 5 in Fig. 4, where the moving motorcycle is much larger than our selected patch size, there is not enough spatial color distribution information to determine whether the pixels in the moving object are moving or static, resulting in a hollow in the moving motorcycle. One method we think might solve such a problem is to globally select $R_o \times R_o$ pixels from the entire frame and do the division again after $RPoSPor$ feature extraction, but it would take more time to pre-process the data.

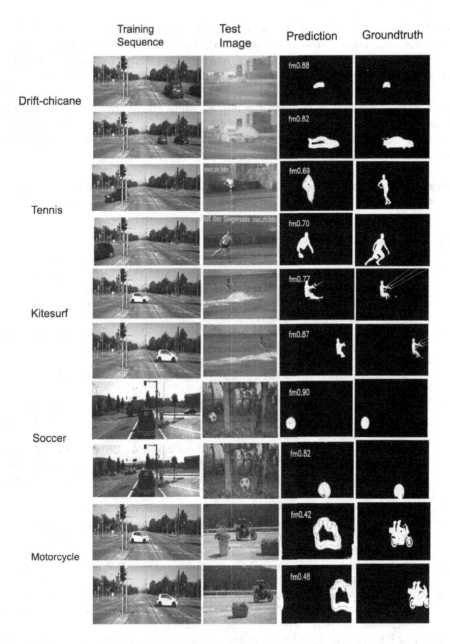

Fig. 4. The evaluation of the transferability of the proposed method on some frames from DAVIS dataset. In particular, the training frames are randomly selected from KITTI MOD dataset and testing frames are obtained from DAVIS dataset.

5 Conclusion

In this paper, we proposed a new method of motion segmentation by a multi-scale deep pixel distribution learning model. The model is trained on the KITTI Mod dataset with a small number of images and reaches competitive results. In addition, our motion segmentation neuron network is simple, with a limited number of parameters. The model also generalizes quite well on the DAVIS 2016 dataset without re-training the model, which shows strong transferability. Our pre-processing multi-scale deep pixel distribution learning is intended to be used for many other deep learning-based methods, to provide a satisfying generalization ability.

Acknowledgements. The authors would like to thank the reviewers and the following colleagues for helpful comments and discussions: Anup Basu and Chuqing Fu.

References

1. Bideau, P., Learned-Miller, E.: It's moving! a probabilistic model for causal motion segmentation in moving camera videos. In: Leibe, B., Matas, J., Sebe, N., Welling, M. (eds.) ECCV 2016. LNCS, vol. 9912, pp. 433–449. Springer, Cham (2016). https://doi.org/10.1007/978-3-319-46484-8_26

2. Dosovitskiy, A., et al.: FlowNet: learning optical flow with convolutional networks. In: Proceedings of the IEEE International Conference on Computer Vision, pp. 2758–2766 (2015)

3. Horn, B.K., Schunck, B.G.: Determining optical flow. Artif. Intell. **17**(1–3), 185–203 (1981)

4. Hui, T.W., Tang, X., Loy, C.C.: LiteFlowNet: a lightweight convolutional neural network for optical flow estimation. In: Proceedings of the IEEE Conference on Computer Vision and Pattern Recognition, pp. 8981–8989 (2018)

5. Ilg, E., Mayer, N., Saikia, T., Keuper, M., Dosovitskiy, A., Brox, T.: FlowNet 2.0: evolution of optical flow estimation with deep networks. In: Proceedings of the IEEE Conference on Computer Vision and Pattern Recognition, pp. 2462–2470 (2017)

6. Jiang, M., Crookes, D.: Video object motion segmentation for intelligent visual surveillance. In: International Machine Vision and Image Processing Conference (IMVIP 2007), pp. 202–202 (2007). https://doi.org/10.1109/IMVIP.2007.7

7. Keuper, M., Tang, S., Andres, B., Brox, T., Schiele, B.: Motion segmentation & multiple object tracking by correlation co-clustering. IEEE Trans. Pattern Anal. Mach. Intell. **42**(1), 140–153 (2018)

8. Kottinger, J., Almagor, S., Lahijanian, M.: Maps-x: explainable multi-robot motion planning via segmentation. In: 2021 IEEE International Conference on Robotics and Automation (ICRA), pp. 7994–8000. IEEE (2021)

9. Lin, J.F.S., Joukov, V., Kulić, D.: Classification-based segmentation for rehabilitation exercise monitoring. J. Rehabil. Assistive Technol. Eng. **5**, 2055668318761523 (2018)

10. Lucas, B.D., Kanade, T., et al.: An iterative image registration technique with an application to stereo vision. Vancouver (1981)

11. Mohamed, E., et al.: Monocular instance motion segmentation for autonomous driving: KITTI instancemotseg dataset and multi-task baseline. In: 2021 IEEE Intelligent Vehicles Symposium (IV), pp. 114–121. IEEE (2021)
12. Ochs, P., Brox, T.: Object segmentation in video: a hierarchical variational approach for turning point trajectories into dense regions. In: 2011 International Conference on Computer Vision, pp. 1583–1590. IEEE (2011)
13. Perazzi, F., Pont-Tuset, J., McWilliams, B., Gool, L.V., Gross, M., Sorkine-Hornung, A.: A benchmark dataset and evaluation methodology for video object segmentation. In: The IEEE Conference on Computer Vision and Pattern Recognition (CVPR) (2016)
14. Ranjan, A., Black, M.J.: Optical flow estimation using a spatial pyramid network. In: Proceedings of the IEEE Conference on Computer Vision and Pattern Recognition, pp. 4161–4170 (2017)
15. Siam, M., Mahgoub, H., Zahran, M., Yogamani, S., Jagersand, M., El-Sallab, A.: ModNet: moving object detection network with motion and appearance for autonomous driving. arXiv preprint arXiv:1709.04821 (2017)
16. Sun, D., Yang, X., Liu, M.Y., Kautz, J.: PWC-Net: CNNs for optical flow using pyramid, warping, and cost volume. In: Proceedings of the IEEE Conference on Computer Vision and Pattern Recognition, pp. 8934–8943 (2018)
17. Teed, Z., Deng, J.: RAFT: recurrent all-pairs field transforms for optical flow. In: Vedaldi, A., Bischof, H., Brox, T., Frahm, J.-M. (eds.) ECCV 2020. LNCS, vol. 12347, pp. 402–419. Springer, Cham (2020). https://doi.org/10.1007/978-3-030-58536-5_24
18. Vertens, J., Valada, A., Burgard, W.: SMSnet: semantic motion segmentation using deep convolutional neural networks. In: 2017 IEEE/RSJ International Conference on Intelligent Robots and Systems (IROS), pp. 582–589. IEEE (2017)
19. Wong, K.Y., Spetsakis, M.E.: Tracking based motion segmentation under relaxed statistical assumptions. Comput. Vis. Image Underst. 101(1), 45–64 (2006)
20. Wu, X., Ma, J., Sun, Y., Zhao, C., Basu, A.: Multi-scale deep pixel distribution learning for concrete crack detection. In: 2020 25th International Conference on Pattern Recognition (ICPR), pp. 6577–6583. IEEE (2021)
21. Zelkowitz, M.: Advances in Computers. Elsevier (2002)
22. Zhao, C., Basu, A.: Pixel distribution learning for vessel segmentation under multiple scales. In: 2021 43rd Annual International Conference of the IEEE Engineering in Medicine & Biology Society (EMBC), pp. 2717–2721. IEEE (2021)
23. Zhao, C., Cham, T.L., Ren, X., Cai, J., Zhu, H.: Background subtraction based on deep pixel distribution learning. In: 2018 IEEE International Conference on Multimedia and Expo (ICME), pp. 1–6. IEEE (2018)

Estimation of Music Recording Quality to Predict Automatic Music Transcription Performance

Markus Schwabe[1(✉)], Thorsten Hoffmann[1], Sebastian Murgul[2], and Michael Heizmann[1]

[1] Karlsruhe Institute of Technology, Institute of Industrial Information Technology, Hertzstraße 16, 76187 Karlsruhe, Germany
{markus.schwabe,michael.heizmann}@kit.edu
[2] Klangio GmbH, Alter Schlachthof 39, Karlsruhe, Germany
sebastian.murgul@klangio.com

Abstract. Music signals can nowadays be recorded and further processed by lots of different devices in order to extract additional information like instruments and genre or use parts of those signals in various applications. Thereby, music recording quality has a big impact on all kinds of Music Information Retrieval (MIR) signal processing and their results. In this work, the recording quality of piano music is estimated by three separate neural network approaches for background noise, sound disturbances, and reverberation. The approaches for background noise and sound disturbances estimate the resulting Signal to Noise Ratio (SNR) of the music piece, the first for constant SNR and the latter for the time-dependent case. Reverberation is estimated by means of the two room parameters reverberation time and early decay time. Exemplarily, the SNR estimation results are validated in the field of piano music transcription, where the impact of the estimated recording quality on the automatic transcription results is analysed. According to those results, the piano music transcription performance can be predicted by means of the recording quality parameters.

Keywords: Recording quality · Piano music · Music transcription · Neural networks

1 Introduction

Automatic music transcription (AMT), which is one part of the Music Information Retrieval (MIR) task, tries to create a human readable sheet of music from an input audio signal. Commercial products like 'Piano2Notes'[1] tend to output transcription results of varying quality, depending on the musical complexity and the quality of the recording. These products are available as mobile device applications and are used in different scenarios by both professional and amateur users. In most cases, the user can influence the recording quality for example by

---

[1] https://piano2notes.com.

S. Berretti and G.-M. Su (Eds.): ICSM 2022, LNCS 13497, pp. 322–337, 2022.
https://doi.org/10.1007/978-3-031-22061-6_24

the distance to microphone, the reduction of environmental sounds, or the choice of the recording room. Therefore, it is useful to estimate the recording quality for direct user feedback in order to give hints for possible improvements. Moreover, further processing algorithms like noise suppression could be used in advance of the MIR task in case of a low estimated recording quality to improve the recorded signal or reduce problematic interferences. Since the recording quality generally affects AMT results as well as other MIR tasks like music source separation or beat tracking, the approach based on the estimated recording quality for the AMT task in this work can be transferred to other MIR tasks as well.

Degraded music signal quality and its impact on MIR task performance has been investigated by Mauch and Ewert [14] by a toolbox with 14 controlled degradation units. Their experiments showed that no general relationship between music degradation and all MIR task performances can be found, but that performance strongly depends on the methods and degradations used. They analysed audio ID, score-to-audio alignment, beat-tracking, and chord detection as MIR tasks and suggested the development of more robust algorithms by means of their audio degradation toolbox [14]. Especially for data-driven approaches, robustness is achieved by the incorporation of diverse training examples, which was highlighted by Serizel et al. [20] for the case of sound event detection with noise and signal degradation. Additionally, robustness against adversarial attacks can be improved by simple methods like compression or addition of white noise [21]. Beside degradation, audio compression is a second impact on MIR results that was investigated by Hamawaki et al. [8] for content-based MIR and by Uemura et al. [23] for chord recognition. While chord recognition is not strongly affected by compressed input signals, the effects of different bit rates could be reduced by normalizing the MFCC feature in case of content-based MIR results.

Quality evaluation of audio signals is often achieved by human perception and judgement in literature, e.g. for compressed music [4] as well as for telephone speech signals [16]. Even if they aim to develop an objective framework for the quality evaluation, the human perception is not important for further signal processing algorithms. Therefore, objective criteria like Signal to Noise Ratio (SNR) suit better for this aim. In case of music signals, no SNR estimation approach is known by the authors, but for speech signals, the NIST SNR measurement [2] and the WADA-SNR algorithm [11] are used to estimate the SNR by exploiting the statistical characteristics of speech like the amplitude density and gamma distributions. As there are significant characteristical differences between speech and music, proven approaches for speech SNR estimation unfortunately lead to big errors in music SNR estimation, even for white noise.

Besides SNR estimation, Kendrick et al. [10] tried to rate the room influence by means of important room acoustic parameters that are calculated under the premise of a known speech or music signal. For unknown signals, blind estimation algorithms of the reverberation time have been presented only for the speech case. Eaton et al. [6] achieved a noise-robust estimation and Diether et al. [5] developed a real-time algorithm suitable to mobile applications. According to the different characteristics of speech and music, those algorithms are not suited for reverberation time estimation in music signals.

In this work, the recording quality of music signals and its impact on an MIR task is estimated by means of relevant objective quality parameters. Consequently, subjective human perceptions are not included in that quality definition. As the estimation should identify possible opportunities for improving recording quality, several quality parameters are estimated for the relevant signal degradation sources. Empirically, three main sources for a reduced AMT task performance caused by the recording quality have been identified: room reverberation (incl. echos), noise, and short interferers. These sources lead for example to inaccuracies in active notes' time estimation and increase the chance of false positives in case of AMT. Other audio degradations and audio compression only have very small impact, so they are neglected in this work.

We present three neural network approaches to estimate the influence of the identified degradation sources noise, short interferers, and reverberation in order to rate the recording quality of unknown piano music. Finally, we exemplarily analyse the impact of the recording quality on AMT algorithms using an implementation of 'Onsets and Frames' [9] in Sect. 6.

2   Music Data Processing

The pure recording process of music data can be described using three basic components: sound source x_S, sound transmission path $g(\cdot)$, and recorded sound y. It is assumed that the recording environment does not change. Therefore, the transmission can be modeled by a room impulse response (RIR) [13]. Mathematically, the discrete recording process can be described by

$$y[n] = x_S[n] * g[n] + r_{\text{disturb}}[n] + r_{\text{noise}}[n] \tag{1}$$

with the convolution operator $*$, the RIR $g[n]$ for the music source transmission path, the background noise $r_{\text{noise}}[n]$, and $r_{\text{disturb}}[n]$ for all disturbing short interferers which are transmitted to the recorder. As the transmission of noise or disturbing sound is not of interest, only their overlapping signal portions at the recorder are considered. Consequently, $r_{\text{noise}}[n]$ and $r_{\text{disturb}}[n]$ include the effects of RIR between background or disturbing sound sources and the recorder.

For preprocessing, the constant-Q transform (CQT) [18] is widely used in music signal tasks, because it defines a time-frequency representation with logarithmic frequency scale of the discrete signal $x[n]$. It is calculated by

$$X_{\text{CQT}}(m, k) = \sum_{n=m-\lfloor N_k/2 \rfloor}^{m+\lfloor N_k/2 \rfloor} x[n]\, a_k^* \left[n - m + \frac{N_k}{2}\right] \tag{2}$$

with time index m, frequency index k, frequency-dependent normalization factor N_k and the floor operator $\lfloor \cdot \rfloor$. The basis function $a_k[n]$ is defined by

$$a_k[n] = \frac{1}{N_k} w\left[\frac{n}{N_k}\right] e^{-j2\pi \frac{f_k}{f_A}} \tag{3}$$

with sampling rate f_A and the time-dependent window $w[n]$ at time step n.

3   Quality Metrics

In order to evaluate the quality of a music signal, several metrics can be used. The most common metric is the SNR which describes the ratio of the signal power P_{signal} to the sum of all noise or disturbance powers P_{noise} in a logarithmic scale:

$$\text{SNR}_{\text{dB}} = 10 \cdot \log_{10}\left(\frac{P_{\text{signal}}}{P_{\text{noise}}}\right). \tag{4}$$

Another metric for acoustic signals is the reverberation time t_{RT} of the recording room. It is calculated by means of the backwards integration

$$s_{\text{back}}[n] = \sum_{i=n}^{N_{\text{RIR}}} g^2[i] \,, \qquad 0 \le n < N_{\text{RIR}} \tag{5}$$

of the squared discrete RIR $g[n]$ [19]. Instead of infinity in the continuous case, the upper bound N_{RIR} of the sum represents the number of samples of the discrete RIR describing the sound transmission as in (1). Similar to the SNR calculation in (4), a logarithmic ratio

$$s_{\text{dB}}[n] = 10 \cdot \log_{10}\left(\frac{s_{\text{back}}[n]}{s_{\text{back}}[0]}\right) \tag{6}$$

is calculated which describes the steady decay rate of the signal. Then, t_{RT} is defined as the time span for the decay from $s_{\text{dB}}[n] = -5\,\text{dB}$ to $s_{\text{dB}}[n] = -25\,\text{dB}$. An alternative for the reverberation time is the early decay time t_{EDT} which describes the time span for the decay from $s_{\text{dB}}[n] = 0\,\text{dB}$ to $s_{\text{dB}}[n] = -10\,\text{dB}$. This can be useful for a detailed analysis of the early behaviour of the signal. Both times are extrapolated to a decay of 60 dB for comparison, like in [10].

The mean absolute error (MAE) is used as the main evaluation metric. For two discrete signals $z_1[n]$ and $z_2[n]$ of length N, it is defined by

$$\text{MAE} = \frac{1}{N} \cdot \sum_{i=1}^{N} |(z_1[i] - z_2[i])| \,. \tag{7}$$

4   Datasets

Several datasets have been used due to the different sound sources for a reduced recording quality. They can be split into the three parts piano music, noise sounds, and RIR.

Piano music is the basis for all quality analysis of this work. It is taken from the MAPS dataset [7], from which the first 30 s of all 270 music pieces (no solo notes) are extracted to get an equal length for all recordings. Those 270 music pieces consist of 210 synthesized piano songs that are used for training and 60 real recorded piano songs which are used for testing.

The noise dataset consists of generated white noise, once and double low-pass filtered white noise (often called pink and brown noise) and an additional recording of high frequency radio noise from [3]. These sounds have a fairly steady characteristic. Additionally, canteen and factory noise from [3] and several sound classes of the 'UrbanSound Dataset' [17] are used as disturbance noises with higher variances and more distinct separate events.

In order to simulate different recording conditions, a dataset with recorded RIRs of nine different rooms [22] is used. Within this dataset, two rooms (R112 and CR2) will be used exclusively for testing while the other seven rooms are used for training. Time intervals of $t_{RT} \in [0.4\,s, 2.2\,s]$ and $t_{EDT} \in [0.2\,s, 3.0\,s]$ for the training rooms and of $t_{RT} \in [0.4\,s, 2.0\,s]$ and $t_{EDT} \in [0.3\,s, 1.5\,s]$ for the test rooms have been calculated as ground truth room parameters.

5 Recording Quality Estimation

As room reverberation and background noise influence the whole music recording by different effects and short interferers are only present during a defined time interval, the estimation of the quality metrics is split up into three separate regression algorithms based on neural networks. Its schematic overview with the respective outputs is illustrated in Fig. 1. The neural network architectures were determined experimentally with focus on small but powerful networks. Therefore they are composed of several fully connected (FC) layers and some additional convolutional layers at the beginning if a dimension reduction is necessary.

All algorithms use the CQT of the music signal with 84 frequency bands, a minimum frequency of 32.70 Hz and a hop size of 512 as input for their preprocessing. The sampling frequency 22 050 Hz is common in audio processing. Furthermore, all networks are trained using Adam optimizer [12] and mean squared error loss with a batch size of 1024. The training is executed for 50 epochs. ReLU is used as activation function in the hidden layers and all fully connected layers are followed by a 40 % dropout to minimize overfitting. Each output layer is a single neuron with linear activation.

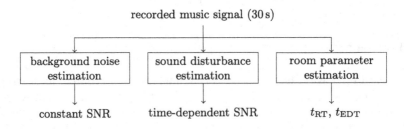

Fig. 1. Schematic overview of the recording quality estimation.

5.1   Background Noise Estimation

The first network estimates the SNR of a 30 s input song superposed by different background noise types. Therefore, white, pink, brown, or high frequency noise is scaled and overlapped with the original music to reach an SNR level in the interval $[-5\,\text{dB}, 20\,\text{dB}]$ with steps of 2.5 dB.

For the training dataset, every combination of MAPS piano song, noise type, and SNR level is created. In order to increase the amount of training samples, every recording is also resynthesized from its respective MIDI file with sound profiles of nine different instruments from the GM 1 sound set [1], followed by a similar data generation with all noise types and SNR levels. Acoustic grand piano (PC_1), church organ (PC_{20}), acoustic guitar (PC_{25}), acoustic bass (PC_{33}), viola (PC_{42}), trumpet (PC_{57}), tenor sax (PC_{67}), flute (PC_{74}), and banjo (PC_{106}) are chosen as synthesized instruments, for which the indices represent their respective MIDI program change (PC) numbers. For the test dataset, the 60 real recorded MAPS songs and the 9×60 resynthesized variants of them are considered in two separate cases with the same data generation as described above. In total, this yields 83 150 samples for training, 2640 samples for testing with real recordings, and 23 760 samples for testing with resynthesized songs.

On the basis of the CQT of each 30 s dataset sample, mean and variance are calculated for each of the 84 frequency bands during preprocessing. Thus, the input of the neural network is reduced to only 168 values which enables a very fast inference. The network consists of three hidden FC layers with 128, 64, and 32 neurons respectively which yields a network with 32 001 parameters. Its architecture is illustrated in Fig. 2.

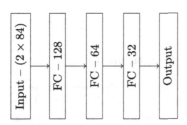

Fig. 2. Network structure for background noise estimation.

Table 1 shows the MAE results of the SNR estimation for the real piano recordings and the resynthesized test dataset for different noise types. The best results are obtained for brown noise and the worst ones for high frequency noise while all errors are very close within one test dataset. Between the two datasets, there is a distinct difference for all considered cases. For the real recorded songs, MAE values of 0.96 dB can be achieved, but for their resynthesized variants, the MAE almost doubles to 1.69 dB. One reason for this difference is the difficult noise estimation in case of specific instruments like church organ (PC_{20}) or acoustic bass (PC_{33}). In Table 2, the results of all synthesized instruments are compared to the real piano test recordings by means of the MAE and the mean standard deviation (STD). Since the STD lies within a single SNR step of 2.5 dB in most cases, the SNR estimation for background noise performs reliably. The best results can be achieved for the real recorded piano. Consequently, resynthesizing will not be considered for the following networks as it did not show better results and might also not represent a realistic scenario, because piano music synthesized by various instruments was used.

Table 1. MAE (in dB) of the background noise estimation for different noise types and test datasets (recorded and resynthesized).

	White	Pink	Brown	High freq.	Average
Real piano recordings	0.95	0.95	0.85	1.09	0.96
Resynthesized songs	1.80	1.67	1.54	1.77	1.69

Table 2. MAE and mean STD (in dB) of the background noise estimation for different instrument types of [1].

	Real	PC_1	PC_{20}	PC_{25}	PC_{33}	PC_{42}	PC_{57}	PC_{67}	PC_{74}	PC_{106}
MAE	0.96	1.19	3.06	1.15	2.14	1.41	1.65	1.75	1.12	1.78
STD	0.99	1.17	2.64	1.16	2.59	1.34	1.72	1.80	1.23	1.57

5.2 Sound Disturbance Estimation

The second network estimates the presence and the SNR values of overlapped impulsive noise sounds. As it is assumed that the disturbances are time-variant, short parts of 2 s length are analysed. For the dataset construction ten parts of each MAPS piano song are extracted and combined with a randomly chosen disturbance sound and SNR level in the range $[-5\,dB, 20\,dB]$ with steps of 2.5 dB. This leads to a total of 4.6×10^5 training and 13 200 test samples.

In order to consider time-dependency, the input of the neural network is the CQT of each 2 s dataset sample with 87 time bins. Figure 3 shows the network structure in which 'FC-Stack' consists of three fully connected layers with 256, 64, and 32 neurons respectively. The network has 71 473 parameters.

All results for the SNR estimation with different time-dependent noise types are listed in Table 3. The worst MAE value of 2.8 dB is detected in case of the air conditioner sound, the best MAE result of 1.75 dB is achieved for factory noise

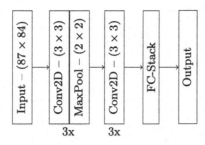

Fig. 3. Network structure for sound disturbance estimation.

disturbance. With an average MAE of 2.3 dB, most estimation errors remain within one SNR step of 2.5 dB. As the average STD of about 2.46 dB is also lower than one step, the estimation performs reliable.

Table 3. MAE and mean STD (in dB) of the sound disturbance estimation for air conditioner (a), car horns (b), playing children (c), dog bark (d), canteen (e), and factory (f) noise types.

	(a)	(b)	(c)	(d)	(e)	(f)
MAE	2.80	2.75	2.40	2.23	1.87	1.75
STD	3.05	2.51	2.68	2.48	2.04	1.97

In Fig. 4, the time-variant estimation is illustrated by the time-dependent average SNR values of several 30 s recordings with three dog barking disturbances and an overlap of 1 s between consecutive 2 s samples. The estimation shows a distinct break-in in fully disturbed parts and a slightly increase in partly disturbed samples (50 % is disturbed). Undisturbed samples generally show higher SNR values, which is expected, but the maximum SNR value of 20 dB cannot be reached in most cases. This effect could be explained by the real test recordings which maybe have included some additional noise caused by the recording conditions. Furthermore, higher SNR levels than 15 dB can hardly be discriminated which is illustrated exemplarily for dog barking in Fig. 5. Most SNR estimates are within one SNR step of 2.5 dB, but at higher SNR values this variance is enlarged. In those cases of high SNR, the overlapped sounds are too low for the neural network to detect the exact SNR level, so it estimates a value below the trained maximum of 20 dB. But although high SNR values are slightly underestimated, time periods of reduced SNR can be detected properly.

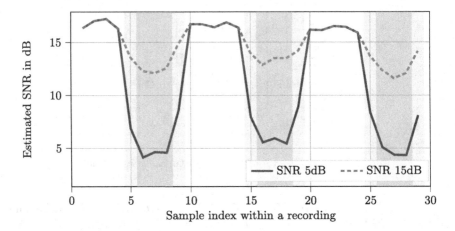

Fig. 4. Average estimated SNR values for fully (red), partly (50 % disturbed, light red), and undisturbed samples for overlapped dog barking at two SNR levels. The SNR levels specify those during the red area.

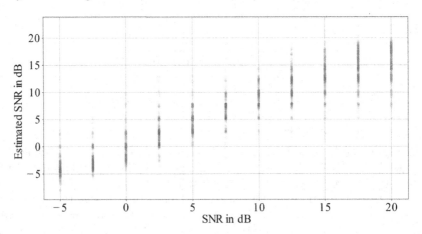

Fig. 5. Estimated vs. real SNR values for overlapped dog barking.

5.3   Room Parameter Estimation

Both room parameters reverberation time t_{RT} and early decay time t_{EDT} are estimated by the third neural network. To simulate recordings in different conditions, the MAPS piano songs are convolved with various RIRs of the RIR dataset. It is assumed that the room conditions may only slightly vary within a 30 s recording. Therefore, only a single estimation for each music piece is sufficient. As the data generation process is executed 500 times for each training song, about 9×10^4 unique samples are generated. For evaluation, each test song is convolved with randomly chosen RIRs out of the training and the test rooms to generate 1500 samples for each of the training and the test room dataset.

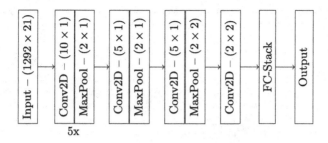

Fig. 6. Network structure for room parameter estimation.

Since the signal decay over time is influenced by the RIR, each 30 s dataset sample is preprocessed similarly to an onset detection [15] by a time differentiation. First, the CQT is transformed to logarithmic amplitude scale and then the time differentiation is performed which results in a decay value per time step. Moreover, the 84 frequency bands are reduced to 21 by summarizing blocks of 4 frequency bands respectively to get a more compact representation. The network has 35 745 parameters. Its structure is illustrated in Fig. 6 in which the same FC-Stack as in Sect. 5.2 is used.

Table 4 shows the results $t_{RT,60}$ and $t_{EDT,60}$ which represent the extrapolated estimation of t_{RT} and t_{EDT} to a 60 dB decay. The test room errors are smaller than the training room errors for both parameters because the ranges of their ground truth values are smaller. Furthermore, the estimation result errors are generally lower for $t_{EDT,60}$ because t_{EDT} is assumed to be both smaller in absolute numbers and easier to estimate. Figure 7 illustrates the distribution of all estimated $t_{RT,60}$ in relation to the real $t_{RT,60}$ values. The network generally underestimates higher values while lower values can be estimated decently. One reason could be the unbalanced training dataset which incorporates more rooms with moderate reverberation and therefore smaller time values.

Table 4. MAE of the estimated $t_{RT,60}$ and $t_{EDT,60}$ (in s) for training and test rooms extrapolated to a decay of 60 dB.

	Training dataset	Test dataset
$t_{RT,60}$	0.316	0.288
$t_{EDT,60}$	0.224	0.201

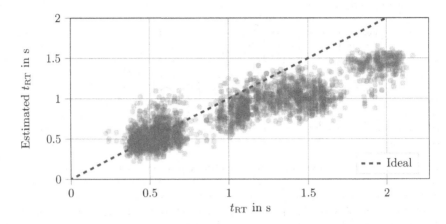

Fig. 7. Estimated vs. real tRT;60 for RIRs of all nine rooms of [22].

6   Experimental Results for AMT

In order to validate the quality estimation in a realistic MIR application, the relation between estimated SNR values and piano AMT results with the algorithm 'Onsets and Frames' [9] is investigated. As in the previous sections, real piano songs of the MAPS test dataset were superposed and convolved by different levels of noise or RIRs. Other MIR applications could benefit from the quality estimation as well, but are not considered in this work. The AMT result is given by the relative F1-Score

$$\text{F1}_{\text{rel}} = \frac{\text{F1}_{\text{d}}}{\text{F1}_{\text{p}}} = \frac{\text{TP}_{\text{d}} \cdot (\text{TP}_{\text{p}} + 0.5\,(\text{FP}_{\text{p}} + \text{FN}_{\text{p}}))}{(\text{TP}_{\text{d}} + 0.5\,(\text{FP}_{\text{d}} + \text{FN}_{\text{d}})) \cdot \text{TP}_{\text{p}}} \tag{8}$$

which is the resulting F1-Score for a disturbed recording F1_{d} in relation to its undisturbed 'pure' version F1_{p}. Both F1-Scores F1_{d} and F1_{p} are calculated by means of their respective correctly detected notes (true positives TP), falsely detected notes (false positives FP), and missed notes (false negatives FN). Consequently, a relative F1-Score of 100 % means that the analysed recording achieves the same transcription quality as the undisturbed recording.

The resulting relative F1-Scores for the background noise types are illustrated in Fig. 8 in relation to the estimated SNR values. As the SNR estimation has achieved appropriate results in Table 1 and 2, only the results for the estimated SNR values are given. Furthermore, the mean relative F1-Scores in Fig. 8 and those of the true SNR values showed similar characteristics in early experiments. AMT results are only marginally decreased in case of intense brown noise, whereas white, pink, and especially high frequency noise have a high impact on the investigated AMT performance. Due to the data distribution, outliers with an atypical relative SNR are possible. Those outliers can be explained by the different music pieces and their level of difficulty for AMT.

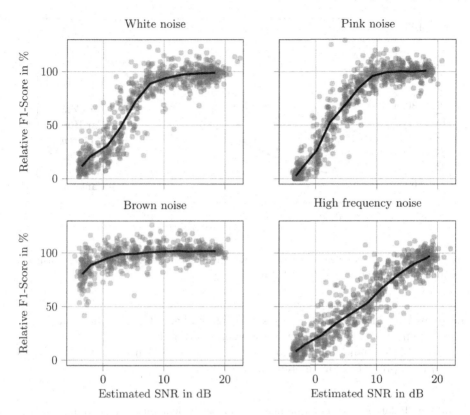

Fig. 8. Relative F1-Scores for piano AMT over estimated SNR values for dierent background noise types. The solid line represents the corresponding relative F1-Scores over the mean of all estimated SNR values for each ground truth SNR step.

In case of time-dependent sound disturbances, the relative F1-Scores are illustrated in Fig. 9 for various sound classes. The data distributions are comparable to those of high frequency noise in Fig. 8, so only the mean values are given here. During each music piece of 30 s, three disturbances of 4 s have been analysed, which results in 40 % disturbance per recording. All investigated sound disturbance classes have a comparable and nearly proportional effect on the AMT results. Consequently, the SNR estimations of background noise and sound disturbances can be used for predictions on AMT result declines and therefore AMT performance reduction due to lower recording quality.

The AMT results with room parameter estimation are illustrated in Fig. 10 for the early decay time because it has got slightly better results than for $t_{RT,60}$ in Sect. 5.3. Although a clear correlation between relative F1-Score and $t_{EDT,60}$ can be stated, the data variance is very high and no reliable AMT performance prediction is possible. One reason for that is the piano characteristic that includes controlled reverberation in its sound production, so the influence of low reverberation for piano AMT performance is insignificant. Therefore, only a classification

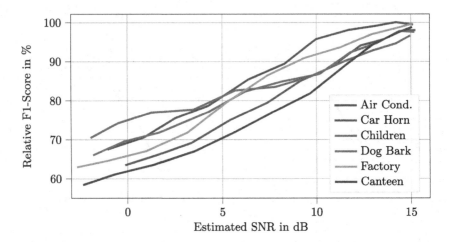

Fig. 9. Relative F1-Score for piano AMT over mean estimated SNR in case of disturbed samples (40% of the recording is disturbed).

of rooms with high or low $t_{EDT,60}$ is investigated here. The classification threshold is defined as 1.2 s according to the results of Fig. 10.

In Table 5, piano music transcription results classified by the predicted $t_{EDT,60}$ are presented for different SNR levels.

Table 5. Relative F1-Scores for piano AMT dependent on $t_{EDT,60}$ estimation and different SNR levels. Sound disturbance is present for 40 % of the recording, background noise and reverberation for the whole duration.

	SNR w.r.t. background noise	SNR w.r.t. sound disturbance		
		15 dB	7.5 dB	0 dB
$t_{EDT,60} < 1.2\,s$	15 dB	65 %	58 %	49 %
	7.5 dB	46 %	43 %	40 %
	0 dB	9 %	9 %	12 %
$t_{EDT,60} \geq 1.2\,s$	15 dB	34 %	31 %	25 %
	7.5 dB	19 %	18 %	15 %
	0 dB	4 %	4 %	4 %

The correlation of the estimated SNR values on piano AMT results is confirmed by those results. As different sound degradations have been used ensemble in this analysis, the relative SNR values are smaller than with only one degradation type. Additionally, a relation of the piano AMT performance and a high estimated early decay time can be stated because of the lower relative F1-Score for $t_{EDT,60}$ values above 1.2 s. Consequently, the performance reduction of piano

AMT due to lower recording quality and possible reasons for it can be predicted by the estimated quality parameters for background noise, short sound disturbances, and reverberation.

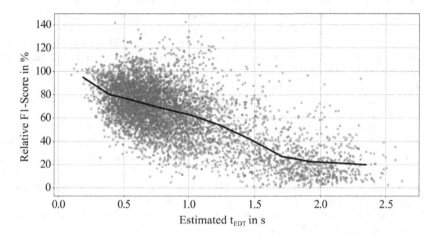

Fig. 10. Relative F1-Score for piano AMT over estimated tEDT;60. The solid line represents the corresponding sliding average.

7 Summary

Three neural network approaches for the estimation of piano music recording quality have been proposed. Each network concentrates on one of the recording quality degradation sources background noise, sound disturbances, or reverberation and estimates the respective SNR or room parameters. The results have been validated successfully in a realistic scenario of piano music transcription for which the quality estimation can be used to predict the performance reduction due to a lower recording quality.

In future works, quality estimation should be enlarged to other music genres than piano music. Furthermore, the presented quality estimation can be validated for other MIR tasks like music source separation or beat tracking.

References

1. GM 1 sound set. https://www.midi.org/specifications-old/item/gm-level-1-sound-set. Accessed 02 Sep 2021
2. NIST speech signal to noise ratio measurements. https://www.nist.gov/itl/iad/mig/nist-speech-signal-noise-ratio-measurements. Accessed 02 Sep 2021
3. Signal Processing Information Base (SPIB). https://spib.linse.ufsc.br/noise.html. Accessed 02 Sep 2021

4. Croghan, N.B.H., Arehart, K.H., Kates, J.M.: Quality and loudness judgments for music subjected to compression limiting. J. Acoust. Soc. America **132**(2), 1177–1188 (2012). https://doi.org/10.1121/1.4730881
5. Diether, S., Bruderer, L., Streich, A., Loeliger, H.A.: Efficient blind estimation of subband reverberation time from speech in non-diffuse environments. In: IEEE International Conference on Acoustics, Speech and Signal Processing (ICASSP), pp. 743–747. IEEE (2015). https://doi.org/10.1109/ICASSP.2015.7178068
6. Eaton, J., Gaubitch, N.D., Naylor, P.A.: Noise-robust reverberation time estimation using spectral decay distributions with reduced computational cost. In: IEEE International Conference on Acoustics, Speech and Signal Processing (ICASSP), pp. 161–165. IEEE (2013). https://doi.org/10.1109/ICASSP.2013.6637629
7. Emiya, V., Badeau, R., David, B.: Multipitch estimation of piano sounds using a new probabilistic spectral smoothness principle. IEEE Trans. Audio Speech Lang. Process. **18**(6), 1643–1654 (2009). https://doi.org/10.1109/TASL.2009.2038819
8. Hamawaki, S., Funasawa, S., Katto, J., Ishizaki, H., Hoashi, K., Takishima, Y.: Feature analysis and normalization approach for robust content-based music retrieval to encoded audio with different bit rates. In: Huet, B., Smeaton, A., Mayer-Patel, K., Avrithis, Y. (eds.) MMM 2009. LNCS, vol. 5371, pp. 298–309. Springer, Heidelberg (2009). https://doi.org/10.1007/978-3-540-92892-8_32
9. Hawthorne, C., et al.: Onsets and frames: dual-objective piano transcription. arXiv preprint arXiv:1710.11153 (2017)
10. Kendrick, P., Cox, T.J., Zhang, Y., Chambers, J.A., Li, F.F.: Room acoustic parameter extraction from music signals. In: IEEE International Conference on Acoustics Speech and Signal Processing Proceedings (ICASSP), vol. 5, pp. V801–V804 (2006). https://doi.org/10.1109/ICASSP.2006.1661397
11. Kim, C., Stern, R.: Robust signal-to-noise ratio estimation based on waveform amplitude distribution analysis. In: Ninth Annual Conference of the International Speech Communication Association. pp. 2598–2601 (2008)
12. Kingma, D.P., Ba, J.: Adam: a method for stochastic optimization. arXiv preprint arXiv:1412.6980 (2014)
13. Kuttruff, H.: Room acoustics. CRC Press (2016). https://doi.org/10.1201/9781315372150
14. Mauch, M., Ewert, S.: The audio degradation toolbox and its application to robustness evaluation. In: International Society for Music Information Retrieval Conference (ISMIR), pp. 83–88 (2013)
15. McFee, B., Raffel, C., Liang, D., Ellis, D.P., McVicar, M., Battenberg, E., Nieto, O.: librosa: audio and music signal analysis in Python. In: Proceedings of the 14th Python in Science Conference. vol. 8, pp. 18–25 (2015). https://doi.org/10.25080/MAJORA-7B98E3ED-003
16. Rix, A.W., Beerends, J.G., Hollier, M.P., Hekstra, A.P.: Perceptual evaluation of speech quality (PESQ) - a new method for speech quality assessment of telephone networks and codecs. In: IEEE International Conference on Acoustics, Speech, and Signal Processing (ICASSP). vol. 2, pp. 749–752. IEEE (2001). https://doi.org/10.1109/ICASSP.2001.941023
17. Salamon, J., Jacoby, C., Bello, J.P.: A dataset and taxonomy for urban sound research. In: Proceedings of the 22nd ACM international conference on Multimedia, pp. 1041–1044 (2014). https://doi.org/10.1145/2647868.2655045
18. Schörkhuber, C., Klapuri, A.: Constant-Q transform toolbox for music processing. In: 7th Sound and Music Computing Conference, Barcelona, Spain, pp. 3–64 (2010)
19. Schroeder, M.R.: New method of measuring reverberation time. J. Acoustical Soc. America **37**(6), 1187–1188 (1965). https://doi.org/10.1121/1.1939454

20. Serizel, R., Turpault, N., Shah, A., Salamon, J.: Sound event detection in synthetic domestic environments. In: IEEE International Conference on Acoustics, Speech and Signal Processing (ICASSP), pp. 86–90. IEEE (2020). https://doi.org/10.1109/ICASSP40776.2020.9054478
21. Subramanian, V., Benetos, E., Sandler, M.: Robustness of adversarial attacks in sound event classification. In: 4th Workshop on Detection and Classification of Acoustic Scenes and Events (DCASE), pp. 239–243 (2019)
22. Szöke, I., Skácel, M., Mošner, L., Paliesek, J., Černocký, J.H.: Building and evaluation of a real room impulse response dataset. IEEE J. Selected Top. in Signal Process. **13**(4), 863–876 (2019). https://doi.org/10.1109/JSTSP.2019.2917582
23. Uemura, A., Ishikura, K., Katto, J.: Effects of audio compression on chord recognition. In: Gurrin, C., Hopfgartner, F., Hurst, W., Johansen, H., Lee, H., O'Connor, N. (eds.) MMM 2014. LNCS, vol. 8326, pp. 345–352. Springer, Cham (2014). https://doi.org/10.1007/978-3-319-04117-9_34

Unleashing the Potential of Data Analytics Through Music

Jatin Dawar[1]([✉]), Prem Raheja[1], Utkarsh Vashisth[1], Nasim Hajari[2], and Irene Cheng[1]

[1] Department of Computing Science, University of Alberta, Edmonton, AB T6G 2R3, Canada
{jdawar,raheja,vashisth,locheng}@ualberta.ca
[2] Department of Computing Science, Concordia University of Edmonton, Edmonton, AB T5B 4E4, Canada
hajari@ualberta.ca

Abstract. One of the most significant challenges with sequential data is the identification of certain underlying patterns, which can be easily overlooked during human visualization. To overcome such problem, transforming data into sound (Sonification) has shown great potential to notify human about hidden patterns. Sonification is a process of mapping data into a non-speech audio format. In today's era, with data analysis being the backbone of most infrastructures, Sonification is gaining interests in various fields like data mining, human-computer interaction, exploratory data analysis, and musical interfaces. It presents a novel way of analyzing and interacting with data. In addition, it provides visually impaired people with an attainable alternative. A considerable amount of work has been done in the field of sonification and music generation. However, producing music from data using machine learning and deep learning techniques is inadequately explored. The conventional sonification methods require human involvement and knowledge of music to produce a tune that is appealing to the ears. This method is time-consuming and requires specialized expertise. In this paper, we aim towards developing a system that molds any time-dependent data into music while retaining the original characteristics of the data using deep learning techniques such as LSTM (Long-short Term Memory). The goals of this research are 1) generate music that is melodious and resembles the music composed by humans, and 2) help people not only auralize but also understand the associated data trend through the generated music. Quantitative and qualitative evaluations were used to validate our approach.

Keywords: Sonification · Machine learning · Data analysis · Human-computer interaction · Music · Data mining · LSTM (Long-Short Term Memory)

1 Introduction

"You can't see radiation, however If you'd like to measure it in some way that is handy and immediate, you can convert the radioactivity into sound." - John Neuhoff

© The Author(s), under exclusive license to Springer Nature Switzerland AG 2022
S. Berretti and G.-M. Su (Eds.): ICSM 2022, LNCS 13497, pp. 338–352, 2022.
https://doi.org/10.1007/978-3-031-22061-6_25

Data is growing exponentially with each passing day; hence the process of classifying, analyzing and conveying the embedded semantics has now become more challenging than ever. Science has proved that people without vision have better auditory senses [3,7]; and, sonification is a viable solution not only for visually impaired people to contribute effectively in today's data-driven world, but also for regular people to be notified when certain crucial events occur. Using sound to convey information is not a new idea. In many domains, auditory interfaces are well established. Some examples are tracking a patient's condition by using pulsoximeter during medical surgeries [2], acoustic alarms within airplane cockpits or the acoustic monitoring of single neuron cell activities, which is a standard method applied in neurobiology electro-physical laboratories. However, the number of Sonification applications today is insignificant compared to visualization applications if one considers how ubiquitous and profitable listening is in our everyday lives. Sound and Hearing assists us in drawing attention to various events, out of which some might be dangerous, and provides us with a better understanding of our environment. Leveraging various parameters of Auditory Perception like amplitude (height of the wave), spatial (the capacity of the auditory system to interpret or exploit different spatial paths by which sounds may reach the head), temporal (appreciation of the music), and frequency resolution (the ability to respond to a single frequency in an acoustic stimulus when other frequencies are present) can act as an alternative or complement to the existing visualization techniques. Rapidly growing computational power has made implementing neural network-based solutions easier and more feasible. With various algorithms like Convolutional Neural Network, Recurrent Neural Network, and Generative Adversarial Network, generating music and transforming one form of music into another has become conceivable. However, generating music leveraging time-series data remains highly challenging considering the variation in data.

Our goal is to develop a machine learning model that transforms data into music to assist in understanding the underlying trends and is pleasant to listen to. Recurrent neural networks (RNN) are a class of neural networks that are helpful in modeling sequence data and producing results that other algorithms can't with respect to sequential data. Our approach involves a RNN based model (LSTM) trained on piano melodies to generate musical notes. A survey of RNNs by Chung et al. [6] found GRUs and LSTMs to be the best recurrent networks for music modeling. The model is configured to learn temporal information such as start time, duration, volume and a pitch to generate music that is soothing to the ears and representing the data. Our objective is compute input values to influence the model output notes in order to generate certain human perceptions, e.g., awareness. In the following sections, we review existing techniques and different algorithms for sonification using Machine Learning, and describe the dataset used, proposed method and evaluation. We validate our approaches using a subjective survey designed to assess (1) whether the music generated is pleasing to hear, and 2) whether the music reflects the underlying data patterns or semantics.

2   Related Work

Sonification, the process of transforming data into sound, is not new. However, almost all of the current sonification systems require manual intervention or direct mapping of data into notes. There is an online software called TwoTone [20] available that allows users to upload data and generate music from it by directly mapping the data to the notes. These sonification techniques often lack natural melodies.

In recent years, Sonification has proved to be beneficial for visually impaired people. Mascetti et al. [12] used non-trivial computer vision techniques and conventional sonification techniques to recognize zebra crossing patterns and combined it with auditory guiding modes like speech messages, mono, and stereo sonification in order to help blind people in crossing the road. On observing the results, they found that 75% of the people prefer using sonification modes over the normal speech mode to cross the road. Similarly, Sonification has been useful for people inclined towards composing music but lack the artistry required to compose music. Hermann et al. [10] explored various traditional sonification techniques to perform exploratory data analysis. Hananoi et al. [19] developed a music composition system using Max/MSP software that enabled people to listen to a variety of time-series data. The Max/MSP software was used to map the values from the data onto the notes of the music. However, these methods do not generate delightful sound, and the data is simply directly mapped to the notes.

Recently, enormous efforts have been made towards generating music using machine learning approaches, especially using Generative models like encoders, LSTM, GRU, GANs. In these techniques, the desire has been to predict the next note of the music based on the previous notes. Many efforts have also been put to transfer musical style using LSTM, GRU, GANs [11]. For example, pop to classical and vice versa. In 1994, Mozer [13] generated music at every time step producing pitch, duration, chord using a recurrent neural network. The method aimed to encode music theory in the data representation. In 2002, Eck et al. [9] used LSTM to repeat a blues chord progression and play melodies over it. In these approaches, the music was generated using the relationship between the notes. In 2012, Boulanger-Lewandowski et al. [14] proposed a polyphonic music (independent notes) model which did not discriminate between chords and melodies and is not soothing to the ears. Recently, there are various approaches developed that take chord progression into the picture, e.g., Fidler et al. [5] generated soothing music using the hierarchical recurrent neural network and then generated chords and drums.

There has been impressive progress in the field of Sonification, and in parallel, there is music generated using machine learning. However, to the best of our knowledge, inadequate efforts have been made towards combining these two fields in order to advance the benefits of Sonification technology. This motivates us to introduce a novel model that would create application-centric music rhythms based on data input by the user using machine learning techniques.

3   Dataset

The music data used is in the MIDI format. MIDI is an abbreviation for Musical Instrument Digital Interface. It is a communication channel between computers and various musical instruments like piano, guitars, and other hardware. The dataset consists of 300 songs: classical piano pieces downloaded from http:// www.piano-midi.de/ [1]. Data preprocessing was done using the music21 [18], mido [15] and prettymidi [17] packages. The MIDI format incorporates features, such as the start of a note, pitch, tempo, rhythm and volume, that are required to decrypt a song successfully. In order to compare to the trajectory data used for testing, we simplified the MIDI files. We will explain the preprocessing step in more details.

We used various time-series datasets available over the internet to generate music using our trained model. Our use case datasets include population data of Canadian cities in 2019, Reliance share prices in the past one-year and Monthly minimum temperatures in Edmonton over the past 20 years.

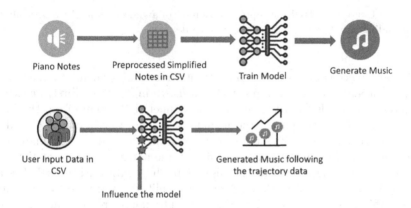

Fig. 1. Flow architecture

4   Proposed Method

An overview of our approach is shown in Fig. 1, First, we compose a midi file with the piano piece as our input data and extract notes from it with the help of a number of music libraries available in python. Next, we train an LSTM model to learn the relationship between the notes along with their musical properties. Post training, the model is capable of generating music based on the features it has learned. We then introduce a time-series data file to be analyzed with the help of music. Since the model only learned just to generate music, we need to develop a method to influence the model in such a way that the generated music reflects the underlying semantics of the user input time-series data. Based on this framework, we devised two approaches to generate music dictated by the user input data.

4.1 Pitch Based Approach

Preprocessing

Fig. 2. (a) Denotes a MIDI file representation of an audio file, and (b) represents the corresponding simplified version of the MIDI File.

A piano piece comprises of two different streams of sequential notes played by the right and left hands, respectively, as shown in Fig. 2(a). First, we extract the notes and chords from the right-hand channel of the piano from the MIDI files, as they generally contain the melody of a song, thus reducing the MIDI file into a single stream. Once the single stream is obtained, we eliminate some of the chords in the following fashion: if multiple notes are played at the same time in a chord, the note with the highest pitch is extracted as melodies are typically higher in sound pitch [4]. A highly simplified list of notes was obtained through this process, as shown in Fig. 2(b). The notes are then mapped onto their respective integer values using the notes to MIDI value conversion matrix, so that each note has a unique MIDI value and stored as a 1D NumPy array. The values are then normalized between 0 to 127 (Because the MIDI velocity range is from 0–127 with zero being silence and 127 being the loudest), with each value representing the different pitch values of the notes. We then split the NumPy array into multiple sequences of 50 consecutive notes each, and the 51st note is taken as an output. This is achieved by a convolutional approach where a 1D window of length 50 and stride 1 is slided across the NumPy array comprising all the notes, and a 50-note sequence is captured at every step.

During training, the data is split into 80:20 training/validation sets. The length of the sequence (50) is based on optimizing the available computing resources. While performing an initial exploratory analysis on the notes obtained from the training dataset, we noticed that pitches of the majority of notes lie between 50 and 90 on the pitch scale. Therefore, the testing time-series input data is also normalized so that the minimum and maximum values match the lowest and highest pitches of the training data, i.e. 50 and 90. Each note is represented by a timestep. Figure 3 shows an example of a normalized ground state trajectory.

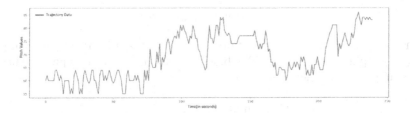

Fig. 3. Normalized reliance stock data for financial year 2018–2019

Model: LSTM- RNN

Long Short-Term Memory networks (LSTMs) [8] are a special kind of RNN, capable of learning long-term dependencies. The basic structure of LSTM is shown in Fig. 4. The core idea behind LSTM is that it maintains a memory cell ct and the ability to add or remove information to this cell is regulated by structures called gates (input, output and forget). Forget gate decides how much information to use from the previous cell state. The output from the last state h_{t-1} is concatenated with x_t and passed through a sigmoid unit. Sigmoid provides output between 0 and 1. 0 means 'Forget everything' and 1 means 'Retain everything'.

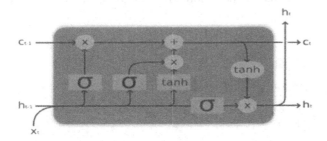

Fig. 4. LSTM architecture [16]

$$f_t = \sigma(w_f[h_{t-1}, x_t] + b_f) \tag{1}$$

where f_t is forget gate, σ is sigmoid function, w_f is weight for forget gate neurons, h_{t-1} is output of previous LSTM block at time $t-1$, x_t is input at current timestamp and b_f is bias for forget gate. The input gate decides which new information we are going to add to the cell state. The concatenated x_t and h_{t-1} are sent over a sigmoid unit which decides what values to update.

$$i_t = \sigma(w_i[h_{t-1}, x_t] + b_i) \tag{2}$$

where i_t is input gate, σ is sigmoid function, w_i is weight for input gate neurons, h_{t-1} is output of previous LSTM block at time $t-1$, x_t is input at current timestamp] b_i is bias for input gate

$$C_t = \tanh(w_C[h_{t-1}, x_t] + b_C) \tag{3}$$

where C_t is output of memory cell, tanh is hyperbolic tangent function, w_C is weight for input gate neurons, h_{t-1} is output of previous LSTM block at time $t-1$, x_t is input at current timestamp, b_C is bias for input gate. The output gate decides what information must be passed to the network in the next time step.

$$h_t = \sigma(w_C[h_{t-1}, x_t] + b_C) * \tanh(C_t) \tag{4}$$

where h_t is output of the model, σ is sigmoid function, C_t is output of memory cell, w_C is weight for input gate neurons, h_{t-1} is output of previous LSTM block at time t-1, x_t is input at current timestamp, b_C is bias for input gate, tanh is hyperbolic tangent function

We have developed a stacked LSTM model as shown in Fig. 5 with two LSTM layers followed by a fully connected softmax activation function layer. The first and second LSTM layer each comprises of 256 hidden units. The output generated from these hidden layers is then passed through the softmax activation function resulting in a MIDI value between 0–127, which is then converted into a musical note. Each input note is converted into a 1-D array and the model is then trained to predict the next note given an array of 50 consecutive notes.

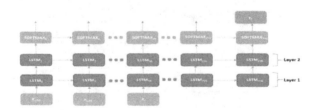

Fig. 5. Stacked LSTM model

Accuracy (ratio of predicted next note to the actual next note) is used as the evaluation matric to measure the performance of the model and the loss function used is cross-entropy. The model is trained for 700 epochs. The training accuracy for our stacked LSTM model is around 97% while the average validation accuracy is at 96%. We observed high training accuracy, which was essential for the model to be capable of generating satisfactory good quality music.

Music Generation

With the LSTM model trained to generate music, the next challenge was to influence the music generation towards matching the input data provided by the user. This process required influencing the prediction made by the model at the prediction time. During the training, the model learned to generate music by learning the relationship between the notes present in the MIDI files of our dataset. We used this relationship to force the predicted note to generate sound according to the user input data by taking a value from the input data and place it as the last element of the sequence. This way, the model would try to generate a note that has a pitch close to the input value but still contain musical patterns learned during the training. To start the music generation, we picked a random set of 49 notes from the validation set, converted them into MIDI values and appended the first value of the user input data as the last note of the sequence, thus making the size of the sequence to 50. This sequence was then fed into the trained model, and as a result, it generated a MIDI value, which replaced the 50th note in the sequence. We then concatenated the last 49 notes of the sequence with the next data-point of the user input data and followed the same process again, as illustrated in Fig. 6. Following the above pattern, the model is able to generate music with each note having a pitch similar to each value of the user input data, as illustrated in Fig. 7.

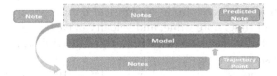

Fig. 6. Data influencing technique

Post Processing

Once the model had been influenced to generate music that was comprised of pitch values based on the user input data, we converted back the pitch values obtained from the model into notes, by leveraging notes to MIDI value conversion matrix again and produced a musical MIDI file representing the user data.

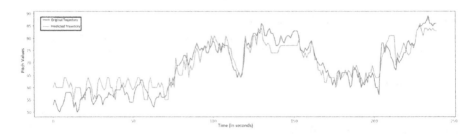

Fig. 7. Pitch based approach results

4.2   Temporal Based Approach

The results obtained from the pitch-based approach are promising, and the generated music reflects certain characteristics of the data provided by the user. However, we decided to incorporate more characteristics of the music. The generated music from the above algorithm has a fixed volume, and only a single note is played at a time. These factors motivated us to refine our approach so that the generated music has a better melody. We realized the need to consider other important characteristics of music, like duration, start time, volume, and pitch, so that the generated music is associated with smooth variations. Hence, we designed a Temporal-based approach, which is an enhanced version of the pitch-based approach.

Preprocessing

Initially, we read all the notes from our musical dataset comprising MIDI files. The start time, end time, note value, and velocity of each note are extracted and converted into a 98-dimensional vector from these MIDI files. An example of the feature vector is shown in Fig. 8.

where: 0th bit represents the pitch value, 1st to 16th bits are the duration, 17th to 96th bits are the start time bins, 97th bit is the velocity or the volume of the note.

The extracted notes from the MIDI files are converted into pitches using the MIDI note mapping and normalized by dividing it with the note's maximum value, just like in the pitch-based approach. We create 16 bins of 2.5 ms each for the duration. The intuition behind this is to accommodate the longest note, which takes 4 s. Similarly, we split the start time of each note into bins of 2.5 ms each, thus creating 80 such bins. 80 bins are selected for maximum utilization of the available computing resources. For the notes with a start time greater than 20 s, we calculate the bins by dividing the start time with 20 s and taking the remainder of it. The final bit, i.e. the 98th bit, represents the note's velocity/volume, which is normalized between 0 and 1. After creating a vector of 98 bits for each note, the input for the LSTM Model is created using 50 consecutive vectors and the 51th vector being the output for the model. In this approach, we

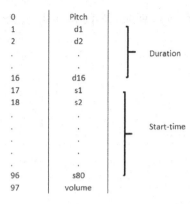

Fig. 8. Input vector for temporal approach

use a combination of classification and regression techniques. Classification for initial 97 bits and regression for the velocity of the note.

Model: LSTM-RNN

We leverage the same stacked LSTM model devised for the pitch-based approach. The training process is also identical to the initial approach. The training accuracy for our model is around 98.5%, while the average validation accuracy is around 97%. There is no change in the music generation process compared to the pitch-based approach followed.

The output obtained from the model is a 98-dimensional vector, just like the input. Once we get the output, the pitch value of the output is converted back into its corresponding notes. The value in the start time bins is further extracted and mapped to the start time of the notes. If the bin predicted by the model is less than the previously obtained output, we use the following equation to convert it into the start time:

$$start\_time[counter] = start\_time[counter - 1] + $$
$$((start\_time\_value) * 0.25) + (80 - \qquad (5)$$
$$start\_time\_index[counter - 1]) * 0.25$$

where start_time is an array of all the start time of the generated notes, start_time_value is the current bin value, which is predicted, and start_time_index is an array of all the previously predicted index of the start time notes. The total number of bins is 80.

The end-time is calculated using the start time and the duration of the notes. The velocity of the note is obtained from the 98th bin. Using this technique, all the obtained parameter values are merged, and the output MIDI file is generated, which comprises the music dictated by the user data. Figure 9 illustrates that the generated music not only represents the values of the user input data, but also has temporal information, which helps in making the music melodious and soothing to the ears.

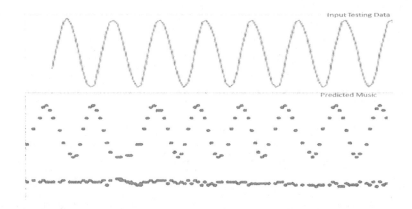

Fig. 9. Temporal-based approach results

5 Evaluation

We conducted both subjective and objective experiments to evaluate the quality of the generated music. We used two criteria for evaluation: how pleasant the generated music is to the user and how accurate the generated music follows the input data trend.

5.1 Quality Evaluation (Objective Evaluation)

Music Metrics
The output generated from the model was passed through several musical measurements, which evaluated the music quality.

- **Polyphony(P)** is defined as how often two tones are played at the same time.
- **Scale Consistency(SC)** is calculated by counting the fraction of tones in all standard scales and reporting the proportion for the best matching scale.
- **Tone Span(TS)** is the number of half-tones steps that are played between the highest and lowest notes played.

All these parameters were calculated for both the approaches and are compared with the music data, which was kept for the validation set.

As observed in Table 1, the music generated from the temporal-based approach is pretty close to real-life music.

Table 1. Musical quality evaluation by comparing with real music.

Approaches	P	SC	TS
Pitch-based approach	0.5374	0.8410	58.6
Temporal-based approach	0.3969	0.8265	45.5
Real music	0.3643	0.8268	57.1

5.2 User Study (Subjective Evaluation)

Quality of Music Generated

Different people have different preferences towards music. In addition to the objective evaluation, we also conducted a subjective study to validate the performance of our technique. To compare our temporal-based approach with an existing sonification technique, we passed the user input data to a sonification software called two-tone [20] and saved the MIDI file generated by the software. We created a survey inviting 34 participants to evaluate the results based on the following criteria:

- **Melody** - Is the generated music has proper melody?
- **Rhythmic** - Is the samples have proper rhythm?
- **Coherent** - Is the samples are coherent?
- **Soothing** - Is the samples are soothing to hear?

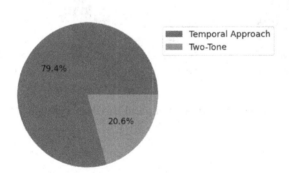

Fig. 10. Comparison between temporal-based approach and two-tone in terms of music quality.

We measured these parameters on a scale of 5. The participants were given different samples comprised of 2 sets of music input, out of which one was of music generated from our proposed model and the other one was created using two-tone software [20]. The music generated by our proposed model was preferred by 79.4% of the participants, as shown in Fig. 10. Based on the results, we concluded that the music generated with varying pitch, duration, start time, and volume is more soothing and melodious than the music generated by direct mapping using sonification software, i.e. two-tone.

Data Metrics

To evaluate whether the generated music represents the values of the input data, we provided the participants with the audio clip generated by our approach and two-tone software for pre-processed Bitcoin data from the year 2017–2020, as shown in Fig. 11, along with multiple trajectory graphs options as shown in Fig. 12. The participants were advised to listen to the given audio clips and match it to the trajectory from which it was generated. Figure 13 shows the average results obtained from the survey. Around 60% of the participants were able to match the generated audio from our temporal-based approach with the correct trajectory, whereas only 32% of participants were able to match the generated audio from two-tone with the correct trajectory. This demonstrates that our temporal-based approach outperforms two-tone [20] in reflecting the underlying data through music.

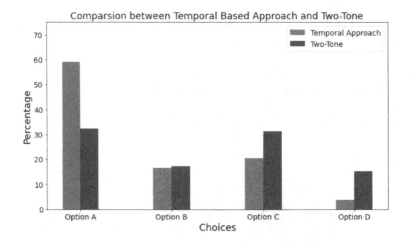

Fig. 11. Comparison between temporal-based approach and two-tone

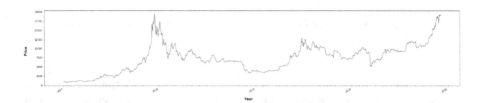

Fig. 12. Original image of bitcoin data provided

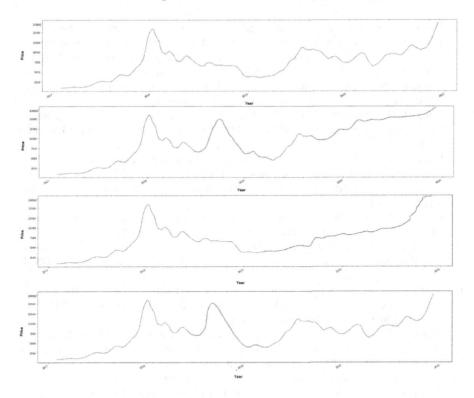

Fig. 13. Options provided to users for Fig. 11

6   Conclusion

In this paper, we introduce a novel learning-based approach to generate music from any arbitrary time-series numerical data. This can provide an alternative way to interact with and understand data that can be useful for visually impaired people. This also has immense potential to assist patients suffering from Parkinson's Disease to correct their unsteady gaits (movement pattern of the human limbs). Based on the experimental results, it is evident that our approach performs well compared to an existing sonification software. Our model can successfully generate pleasant music that represents the underlying time-series. In future work, we will expand the user study to evaluate the proposed model further including the evaluation of rhythm, tempo, and beats, to assess whether these factors will make the generated music more realistic and pleasing to hear.

References

1. Classical piano midi page website. http://www.piano-midi.de/

2. Device for auditory monitoring of pulse and blood oxygen level. https://www.medicalnewstoday.com/articles/318489#What-to-expect-from-pulse-oximetry
3. University of Washington. https://www.washington.edu/news/2019/04/22/brains-of-blind-people-adapt-to-sharpen-sense-of-hearing-study-shows/
4. Bestdigitalpianoguides.com: piano chords and melodies. https://bestdigitalpianoguides.com/piano-chords-and-melodies
5. Chu, H., Urtasun, R., Fidler, S.: Song from PI: a musically plausible network for pop music generation. CoRR abs/1611.03477 http://arxiv.org/abs/1611.03477 (2016)
6. Chung, J., Gulcehre, C., Cho, K., Bengio, Y.: Empirical evaluation of gated recurrent neural networks on sequence modeling. arXiv preprint arXiv:1412.3555 (2014)
7. Cohut, M.: Medical news today website. https://www.medicalnewstoday.com/articles/325032
8. Das, A.: RNN simplified- a beginner's guide. https://towardsdatascience.com/rnn-simplified-a-beginnersguide-cf3ae1a8895b. Accessed 07 June 2019
9. Eck, D., Schmidhuber, J.: A first look at music composition using LSTM recurrent neural networks. Istituto Dalle Molle Di Studi Sull Intelligenza Artificiale **103**, 48 (2002)
10. Sonification for exploratory data analysis. In: Hermann, T. (ed.) Sonification for Exploratory Data Analysis. Bielefeld University, Bielefeld, Germany (2002)
11. Magenta. https://magenta.tensorflow.org/
12. Mascetti, S., Picinali, L., Gerino, A., Ahmetovic, D., Bernareggi, C.: Sonification of guidance data during road crossing for people with visual impairments or blindness. CoRR abs/1506.07272 http://arxiv.org/abs/1506.07272 (2015)
13. Mozer, M.C.: Neural network music composition by prediction: exploring the benefits of psychoacoustic constraints and multi-scale processing. Connect. Sci. **6**(2–3), 247–280 (1994)
14. Boulanger-Lewandowski, N., Bengio, Y., Vincent, P.: Modeling temporal dependencies in high-dimensional sequences: application to polyphonic music generation and transcription. arXiv:1206.6392 (2012)
15. Mido. https://mido.readthedocs.io/en/latest/index.html
16. Phi, M.: Illustrated guide to LSTM's and GRU's: a step by step explanation. https://towardsdatascience.com/illustrated-guide-to-lstms-and-grus-a-step-by-step-explanation-44e9eb85bf21
17. Raffel, C., Ellis, D.P.W.: Intuitive analysis, creation and manipulation of midi data with pretty_midi. In: 15th International Conference on Music Information Retrieval Late Breaking and Demo Papers (2014)
18. Cuthbert, M.S., Ariza, C.: music21: a toolkit for computer-aided musicology and symbolic music data (2010)
19. Hananoi, S., Muraoka, K., Kiyoki, Y.: A music composition system with time-series data for sound design in next-generation sonification environment. In: International Electronics Symposium (IES), pp. 380–384 (2016)
20. TwoTone. https://twotone.io/

Haptic

Impact of PGM Training on Reaction Time and Sense of Agency

Enrique Calderon-Sastre⬛, Swagata Das⬛, and Yuichi Kurita$^{(\boxtimes)}$⬛

Graduate School of Advanced Science and Engineering, Hiroshima University,
Higashihiroshima, Japan
{enriquecs,swagatadas,ykurita}@hiroshima-u.ac.jp

Abstract. Sense of Agency (SoA) is the mechanism responsible for the feeling of control own actions. Given the nature of Sense of Agency, changes and disorders can occur to the detriment of individual health with disabling consequences. Pneumatic Gel Muscles (PGMs) are a type of soft actuator that, due to their design and construction, can operate with low pressures, removing the necessity of heavy compressors to actuate them, which makes PGMs ideal for wearable purposes. PGMs have proved to be helpful in imparting dexterity training tasks, and also, the soft properties of PGM allow a more natural and less restricted training. This paper presents the design, implementation, and testing of a training system based on Pneumatic Gel Muscles to speed up the user's Reaction Time in a Simple Reaction Time (RT) task, using a LED as a visual stimulus. Four different modes of PGM feedback training were tested for two participant groups that followed different training sequences. Fast PGM training was the training type that demonstrated the best improvement (defining improving as a speeding up in the RT measurement value), with an improvement of 20.6 ms for Group 1 and an overall improvement of 8.16 ms for all the data, which is very similar to improvement times achieved using other feedback type training.

Keywords: Sense of Agency · Haptics · Pneumatic Gel Muscle · Feedback · Simple Reaction Time

1 Introduction

Sense of Agency (SoA) is the neural and mental state which activates the feeling of being in control of the generation and the control of one's actions, and through them, the consequences and course of events in the external environment [6, 7,9]. Sense of Agency research is related to many fields in daily life, ranging from health problems to legal and philosophical issues. In the Human-Computer Interaction field, it is crucial to understand the Sense of Agency since a higher Sense of Agency is related to the design and creation of better interfaces and to how people interact with technology [15]. Also, sense of control is taken into account in the classic principles used for designing user interfaces [1].

S. Berretti and G.-M. Su (Eds.): ICSM 2022, LNCS 13497, pp. 355–368, 2022.
https://doi.org/10.1007/978-3-031-22061-6_26

There are two main paradigms for measuring the Sense of Agency: Implicit and explicit measures methods. Implicit measures assess a correlate of voluntary action and, based on that, infer something about the experience [16]. Intentional Binding is the most used implicit measure, and it is based on the compressed perception of time while being the agent of an action and the external sensory consequences [2]. On the other hand, explicit measures directly ask the participant about their agent experience using questionnaires.

Since Sense of Agency is neither a static nor immutable state [11], changes and disorders can occur to the detriment of individual health with disabling consequences. One of the most studied pathologies related to degradation in the Sense of Agency is schizophrenia, in which the individual shows difficulty in detecting modifications or his movements [20] because the patient does not feel he is the agent of his own's actions. In a similar way to schizophrenia, SoA impairments are related to psychosis [14]. Other disorders related to the degradation of Sense of Agency are the Obsessive-Compulsive Disorder [22], anosognosia for hemiplagia [16], or anarchic hand syndrome [2].

There exist various types of Reaction Times, such as Simple Reaction Time (RT), Recognition Reaction Time, Choice Reaction Time, and Serial Reaction Time. Simple Reaction Time is the "minimal time needed to respond to a stimulus" [23]. In Simple Reaction Time tasks, only one stimulus (visual, auditory, or other types) and one response are required. Simple Reaction Time can be divided into two big stages [23]: Stimulus detection time, which is the time required to perceive the stimulus, and the time required to push the response button, known as movement initiation time. RT acceleration or deceleration can have different implications, like the limitation of Go/Nogo reaction time in activities like baseball [13] or RT deficit linked to Parkinson's disease [4,10].

Pneumatic Artificial Muscles (PAMs) try to mimic human muscles in the sense that they only can contract, and it is required to use two opposing PAMs to achieve movement in two directions [5]. Pneumatic Gel Muscles (PGMs) are a low-pressure type of soft actuator that supposes less constrained movement than rigid structures. PGMs are composed in a similar way to McKibben actuators. PGMs are composed of two layers of materials. The first layer is an outer plastic mesh, while the second layer is an inner tube made from a customized styrene-based thermoplastic elastomer [17]. If compressed air is supplied, the inner tube will inflate until the limit of the outer mesh, and it will generate a linear force due to the actuator shrinking action. Unlike conventional PAMs, Pneumatic Gel Muscles do not require compressors to generate high forces; instead, the PGMs can function with CO_2 canisters, and due to their inner tube, PGMs can generate forces up to $40\,N$ with a $0.1\,MPa$ air pressure [17]. Also, the materials used make the actuator lightweight. These features make PGMs a great option to consider when developing wearable devices.

1.1   Related Works

Many research projects have successfully implemented haptic devices that actuate the user's body and are capable of imparting training sessions and helping skill acquisition. Goto et al., [5] developed a four-pneumatic muscle device to actuate the forearm of the user while practicing two-handed drumming. Richard et al., [18] had tested the impact of force feedback in embodiment during a drawing task in three conditions, obtaining the most favorable results with the force feedback approach. Sakoda et al., [19] presented a motion timing presentation method based on force stimulus and a VR environment.

Regarding the training for Reaction Time purposes, most of the previous work has been done using Electric Muscle Stimulation (EMS). Kasahara et al., [11] used preemptive force feedback to accelerate human reaction using EMS. Also they presented a model to see how the sense of agency is affected by a preemptive stimulus.

Kasahara et al., [12] presented a study using three different types of EMS stimulation and the performance obtained after removing the stimuli training. Tajima et al., [21] presented a framework to balance the haptic assistance-sense of agency trade-off while designing computer-touch interfaces.

Nobody had used PGMs before for this type of training purposes, as far as we know. PGMs leave the user with Agency as they neither contract nor activate the user's muscles and let the user use enough force to move against the PGM in case they wanted [5].

In this paper, we present the results of an experiment in which four different training sessions were provided to participants to accelerate their reaction times on a simple reaction task and, at the same time, preserve the sense of agency while receiving the training.

The remainder of this work is organized as follows: Sect. 2 details the Methodology used to design and develop the training device as well as the training schedule followed. Section 3 presents the results obtained during the experiment and in the exit questionnaire applied. Section 4 discusses the scope of the study and direction of future work, and Sect. 5 presents the conclusions of this work.

2 Materials and Methods

2.1 Task

The Simple Reaction Time task is a tap test consisting in touching a touch sensor using the middle finger as soon as a LED turns on. To start the trial, the user must touch the sensor once. After that, a random time between 4 and 10 s will pass, and then the LED will glow up, and it will remain on until the participant touches the sensor. The time elapsed since the LED glows up and the user touches the sensor will be registered as the Reaction Time for that trial. Most of the task was modeled following experiment protocols described in [11,12] but making the necessary adaptations to use PGMs.

2.2 Participants

A total of twelve participants (one female and eleven males, 11 right-handed, average age = 23.8 ± 2.29 years old) from our institution were selected. Participants were divided into two groups of six (labeled as Group 1 and Group 2).

Each group performed two training sessions per day for two days to complete the experiment. The experiment was performed according to the ethical standards of the Declaration of Helsinki.

2.3 Experimental Setup

A capacitive sensor module measured the user's touch to turn off the LED and register reaction time. This sensor is based on the TTP223[1] integrated circuit. Three PGMs were used to actuate the hand. The PGMs are manufactured by the DAIYA industry, Japan. A CO_2 canister made by mini Nippon Tansan Gas Co., Ltd. (NTG) was used as a compressed air source for the PGMs. One canister like the one used for the experiment can be used for multiple training sessions, as it is reported to actuate a 30 cm PGM 1280 times or 2240 times for a 20 cm PGM [3]. Two (one for each motion direction) SMC SYJ series solenoid valves were used to control the PGMs.

A 3D-printed case was designed to support the hand wearing the PGM for comfortable training sessions. This case also holds the LED required to make the reaction time task and has a slot for storing the touch sensor. Figure 1 shows some of the components used in this experiment.

(a) 3D-printed case.   (b) Capacitive sensor module.

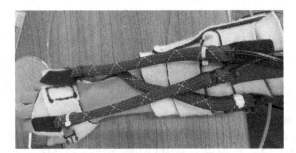

(c) Pneumatic Gel Muscle.

Fig. 1. Components of the training prototype.

The measurement device is controlled using an Arduino Uno. This board is used for making the following tasks: processing the start-trial signal (first user's touch), obtaining the random time, calculating the actuation-required times depending on the type of training selected, sending the actuation signal to the solenoid valve to activate the PGM system, recording reaction time, and sending it via serial port to a PC that uses a Python script to store, sort, clean and process the data.

Participants used a noise-canceling headset to avoid disturbances during the training sessions. Figure 2 shows the device used to record reaction times.

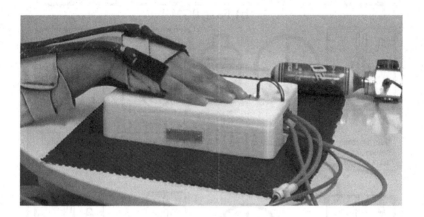

Fig. 2. Device used to record reaction times during the experiment.

2.4   Experiment Protocol

Training sessions included no PGM and three variations of PGM-based training:

   I. **No PGM** training: 2 complete training sessions with no PGM device used.

  II. **Fast PGM** training: 3 complete training sessions with PGM set up to give the stimulus 200 ms before the signal that turns the LED on appears.

III. **On time PGM** training: 3 complete training sessions with PGM set up to be synchronized with the LED signal.

IV. **Late PGM** training: 3 complete training sessions with PGM set up to be 200 ms late to the LED signal.

Group 1 performed the sequence I→ II → III → *IV* while Group 2 performed the reversed sequence IV→ III → II → I. Figure 3 shows the different types of training used during this experiment and how they work.

At the beginning of the first day of trials, participants performed a training session consisting of 10 trials to get familiar with the device before starting the session. After a 5-min rest, the base reaction time of the user was registered,

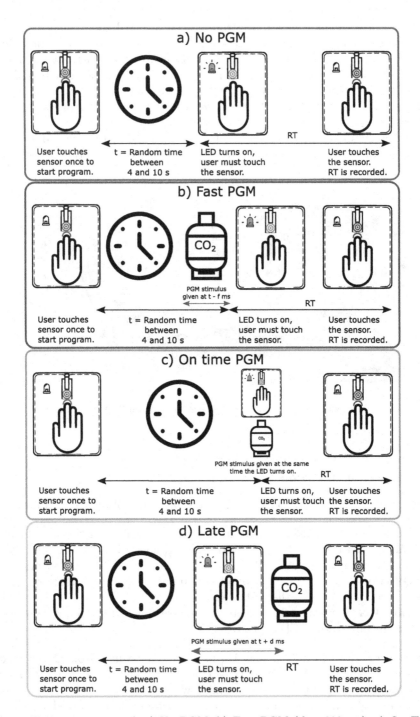

Fig. 3. Training types used: a) No PGM, b) Fast PGM ($f = 200$ ms), c) On Time PGM (stimulus synced with LED turn-on signal), d) Late PGM ($d = 200$ ms).

i.e., the reaction time before starting all training sessions. Only training sessions were recorded using PGM. Base, Pre, and Post recordings were made without feedback applied to measure the total effect of training on the participant.

After each training session of the experiment, the participant was requested to fill out an online questionnaire, including the NASA Task Load Index (NASA TLX) [8], to measure the mental workload that the Reaction Time task implied. All answers were given on a seven-point Likert scale. The online questionnaire also included two open questions ("1. Did you feel any change in your behavior between the start and end of this session?", and "2. do you have any other comment about the experiment?") for gathering open feedback at every stage of the experiments. Table 1 details the schedule followed for training sessions, where *Training_ type RT* corresponds to the training session in turn (No PGM, Fast PGM, On time PGM, or Late PGM). After completing all training sessions, 600 reaction time data points per participant were gathered in addition to the responses obtained for the exit questionnaire made at the end of every training session.

Table 1. Session order followed by experiment participants.

Frequency	Activity	Description	Number of trials
Once per participant	Familiarization training	Used to teach the participant how to use the device 5 min rest	10
Made at every training session (four times per participant)	Base RT	The reaction time recorded before starting current training session 2 min rest	20
	Pre RT	Reaction time recorded previous to start the training session	20
	Training_ type RT	Reaction time during the first training session	20
	Post RT	Reaction time recorded after the first training session 2 min rest	20
	Training_ type RT	Reaction time during the second training session	20
	Post RT	Reaction time recorded after the second training session 2 min RT	20
	Training_ type RT	Reaction time during the third training session	20
	Post RT	Reaction time recorded after the third training session Exit questionnaire about the training session	20

2.5 Statistical Analysis

Collected data were cleaned to delete outliers caused by distractions while doing the trial sessions. After that, data were sorted according to their respective stage of the experiment and belonging group.

A Shapiro-Wilk test was performed on all data to assess normality. After this test, we realized that not all the data followed a normal distribution, so we used a non-parametric test. Friedman test was used as a hypothesis testing method.

Finally, for making the plots in Fig. 6, the average value of all the trials performed during Pre and the last Post session per trial was plotted to summarize the individual effect of each participant through the experimental stages.

3   Results

3.1   Reaction Time Task

Group I. Table 2 shows the mean RT values obtained before and after training sessions with every type of feedback tested during this experiment. The mean

Table 2. Mean RT values obtained before and after training stages for Group 1.

Feedback type	RT Before (ms)	RT After (ms)
No PGM	300.29	321.32
Fast PGM	330.63	310.27
On-Time PGM	297.22	299.97
Late PGM	307.08	316.50

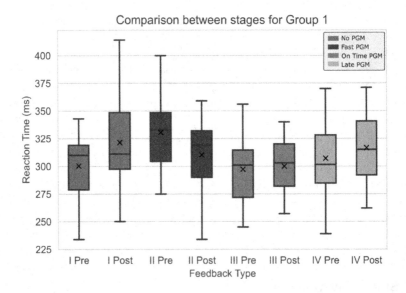

Fig. 4. Reaction Times plot for Group 1, including mean (marked with x) and median (marked with a line).

RT values obtained are within the typical values in different studies summarized in [23]. Fast PGM was the feedback type that got the best improvement between the value recorded at the beginning the training and the last recorded time after training. The improvement for these data group was 20.36 ms, and statistical significance in the analysis was achieved (p < 0.02). Figure 4 shows the comparison of previous to training and after training colorized according to the type of feedback provided during training.

Group II. Even when the mean RT values obtained are within the typical values in different studies summarized in [23], after analyzing data in this group, no statistically significant results were obtained. Therefore, the presented analysis will be limited to Group I and all the gathered data.

3.2 All Data

Table 3 shows the mean RT values obtained before and after training sessions with every type of feedback tested during this experiment. When merging the

Table 3. Mean RT values obtained before and after training stages for all data.

Feedback type	RT Before (ms)	RT After (ms)
No PGM	295.17	305.61
Fast PGM	309.16	301.00
On-Time PGM	292.19	298.63
Late PGM	301.91	303.77

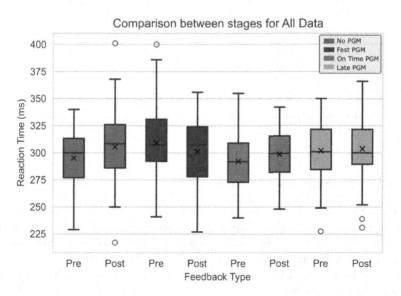

Fig. 5. Reaction Times plot for both groups, including mean (marked with x) and median (marked with a line).

data from the two groups, we get statistically significant results (p < 0.04), in which only the Fast PGM feedback type improved 8.16 ms in RT measurements. This value is very similar to the one presented by Kasahara et al., in [12]. Figure 5 shows the comparison of previous to training and after training colorized according to the type of feedback provided during training.

Figure 6 shows the comparison of the three types of training and the mean value per participant before starting training sessions (reaction times recorded without PGM actuation) and just after finishing the last training session with that PGM actuation type (also recorded without PGM actuation). From Fig. 6 we can see the evolution of each participant in the experiment across the three PGM feedback training sessions. It can be noticed that the most significant improvement occurred for P12 in the Fast PGM session, but also in P12, the most considerable deterioration occurred during the Late PGM session.

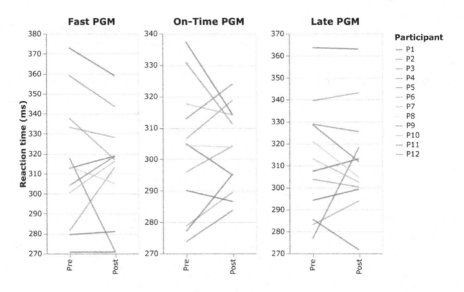

Fig. 6. Pre and Post mean reaction time values for each participant during the three types of PGM training given.

3.3 Exit Questionnaire

Figure 7 summarizes all the responses obtained through the exit questionnaire during the experiment. The graphs are designed to display the summary of responses categorized by feedback type and the corresponding percentage for seeing the general trend according to the feedback type used.

By analyzing each of the seven questions included in the NASA TLX questionnaire we can see that:

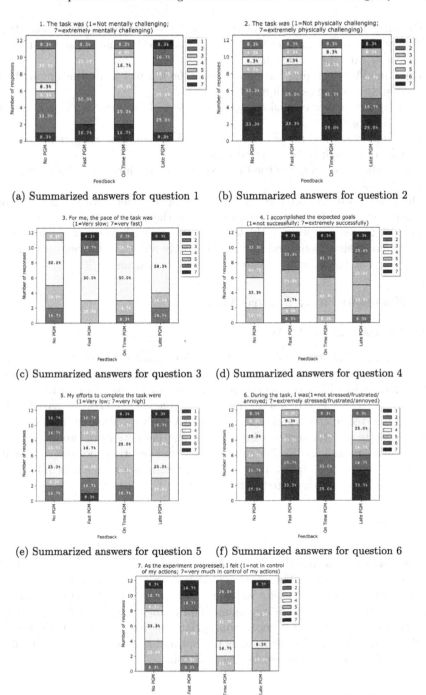

(a) Summarized answers for question 1   (b) Summarized answers for question 2

(c) Summarized answers for question 3   (d) Summarized answers for question 4

(e) Summarized answers for question 5   (f) Summarized answers for question 6

(g) Summarized answers for question 7.

Fig. 7. Answers for the exit questionnaire (NASA TLX) by all the participants during the 4 training stages of the experiment.

1. On Time PGM training was perceived as the less mentally challenging task. This matches with the fact that in On Time PGM training, the PGM signal is synced with the LED signal.
2. On Time PGM training was perceived as the less physically challenging task. It occurs the same as with question one.
3. Participants perceived a slower pace of the task during No PGM and Late PGM sessions.
4. On Time PGM was perceived as the training type that allowed the participants to accomplish the required task successfully.
5. Bigger efforts to complete the task were perceived during No PGM and Late PGM sessions.
6. On Time PGM was perceived as the training type that kept the participants the less stressed/frustrated/annoyed during the required task
7. Fast PGM was perceived as the training type with which the user felt more in control of their actions, and this can be closely related to the Sense of Agency perceived at the moment of executing the required task.

4 Discussion and Future Work

4.1 Limitations of the Work

There are some inherent limitations to the development of this work. However, once the positive impact of PGM in Simple Reaction Time has been assessed, these limitations will be considered in further research. One of the critical limitations of this work is how to obtain the Sense of Agency Preservation based on the method selected in the experiment. The NASA TLX questionnaire allows inferring indirectly how the user felt in control of his/her actions in two questions (4 and 7). We can also opt for designing a questionnaire that includes only Sense of Agency-related questions. The use of implicit measures such as Intentional Binding will also be considered in future work.

4.2 Future Work

In this experiment, we explored four different types of PGM training for accelerating participants' reaction times. As we have assessed that PGM helps to improve Simple Reaction Time, future work will be focused on doing more experiments with Fast PGM training variations that allow us to identify the optimal number of sessions to assess maximum improvement. Also, more experiments are required to identify the most appropriate intensity of the PGM-provided feedback and different factors such as the optimal number of sessions, intensity of the feedback signal, and rest between sessions.

5 Conclusions

After presenting the obtained results from this experiment, we can conclude that Pneumatic Gel Muscles in training aimed at accelerating the Simple Reaction

Time has a positive impact on the users. Also, based on the exit questionnaire results, we can conclude that PGM has no reduction in the Sense of Agency of the different participants.

References

1. Bergström, J., Knibbe, J., Pohl, H., Hornbæk, K.: Sense of agency and user experience: is there a link? ACM Trans. Comput.-Hum. Interact. **29**(4), 1–22 (2022). https://doi.org/10.1145/3490493. https://dl.acm.org/doi/10.1145/3490493
2. Braun, N., et al.: The senses of agency and ownership: a review. Front. Psychol. **9**(APR), 1–17 (2018). https://doi.org/10.3389/fpsyg.2018.00535
3. Das, S., Kurita, Y.: ForceArm: a wearable Pneumatic Gel Muscle (PGM)-based assistive suit for the upper limb. IEEE Trans. Med. Robot. Bionics **2**(2), 269–281 (2020). https://doi.org/10.1109/TMRB.2020.2990436. https://ieeexplore.ieee.org/document/9079448/
4. Gauntlett-Gilbert, J., Brown, V.J.: Reaction time deficits and Parkinson's disease. Neurosci. Biobehav. Rev. **22**(6), 865–881 (1998). https://doi.org/10.1016/S0149-7634(98)00014-1. https://linkinghub.elsevier.com/retrieve/pii/S0149763498000141
5. Goto, T., Das, S., Wolf, K., Lopes, P., Kurita, Y., Kunze, K.: Accelerating skill acquisition of two-handed drumming using pneumatic artificial muscles. In: Proceedings of the Augmented Humans International Conference, pp. 1–9. ACM, New York, March 2020. https://doi.org/10.1145/3384657.3384780. https://dl.acm.org/doi/10.1145/3384657.3384780
6. Grünbaum, T., Christensen, M.S.: Measures of agency. Neurosci. Conscious. **2020**(1), 406 (2020). https://doi.org/10.1093/nc/niaa019. https://academic.oup.com/nc/article/doi/10.1093/nc/niaa019/5890345
7. Haggard, P.: Sense of agency in the human brain. Nat. Rev. Neurosci. **18**(4), 196–207 (2017). https://doi.org/10.1038/nrn.2017.14. https://www.nature.com/articles/nrn.2017.14
8. Hart, S.G., Staveland, L.E.: Development of NASA-TLX (task load index): results of empirical and theoretical research. In: Hancock, P.A., Meshkati, N. (eds.) Human Mental Workload, Advances in Psychology, North-Holland, vol. 52, pp. 139–183 (1988). https://doi.org/10.1016/S0166-4115(08)62386-9. https://www.sciencedirect.com/science/article/pii/S0166411508623869
9. Jeannerod, M.: The mechanism of self-recognition in humans. Behav. Brain Res. **142**(1–2), 1–15 (2003). https://doi.org/10.1016/S0166-4328(02)00384-4. https://linkinghub.elsevier.com/retrieve/pii/S0166432802003844
10. Jordan, N., Sagar, H.J., Cooper, J.A.: Cognitive components of reaction time in Parkinson's disease. J. Neurol. Neurosurg. Psychiatry **55**(8), 658–664 (1992). https://doi.org/10.1136/jnnp.55.8.658. https://jnnp.bmj.com/lookup/doi/10.1136/jnnp.55.8.658
11. Kasahara, S., Nishida, J., Lopes, P.: Preemptive action: accelerating human reaction using electrical muscle stimulation without compromising agency. In: Proceedings of the 2019 CHI Conference on Human Factors in Computing Systems, CHI 2019, pp. 1–15. Association for Computing Machinery, New York (2019). https://doi.org/10.1145/3290605.3300873
12. Kasahara, S., Takada, K., Nishida, J., Shibata, K., Shimojo, S., Lopes, P.: Preserving agency during electrical muscle stimulation training speeds up reaction

time directly after removing EMS. In: Proceedings of the 2021 CHI Conference on Human Factors in Computing Systems, pp. 1–9. ACM, New York, May 2021. https://doi.org/10.1145/3411764.3445147. https://dl.acm.org/doi/10.1145/3411764.3445147

13. Kida, N., Oda, S., Matsumura, M.: Intensive baseball practice improves the Go/NoGo reaction time, but not the simple reaction time. Cogn. Brain Res. **22**(2), 257–264 (2005). https://doi.org/10.1016/j.cogbrainres.2004.09.003. https://linkinghub.elsevier.com/retrieve/pii/S0926641004002459

14. Krugwasser, A.R., Stern, Y., Faivre, N., Harel, E.V., Salomon, R.: Impaired sense of agency and associated confidence in psychosis. Schizophrenia **8**(1), 32 (2022). https://doi.org/10.1038/s41537-022-00212-4. https://www.nature.com/articles/s41537-022-00212-4

15. Limerick, H., Coyle, D., Moore, J.W.: The experience of agency in human-computer interactions: a review. Front. Hum. Neurosci. **8**(AUG), 1–10 (2014). https://doi.org/10.3389/fnhum.2014.00643. https://journal.frontiersin.org/article/10.3389/fnhum.2014.00643/abstract

16. Moore, J.W.: What is the sense of agency and why does it matter? Front. Psychol. **7**(AUG), 1–9 (2016). https://doi.org/10.3389/fpsyg.2016.01272. https://journal.frontiersin.org/Article/10.3389/fpsyg.2016.01272/abstract

17. Ogawa, K., Thakur, C., Ikeda, T., Tsuji, T., Kurita, Y.: Development of a pneumatic artificial muscle driven by low pressure and its application to the unplugged powered suit. Adv. Robot. **31**(21), 1135–1143 (2017). https://doi.org/10.1080/01691864.2017.1392345

18. Richard, G., Pietrzak, T., Argelaguet, F., Lécuyer, A., Casiez, G.: Studying the role of haptic feedback on virtual embodiment in a drawing task. Front. Virtual Real. **1**(January) (2021). https://doi.org/10.3389/frvir.2020.573167

19. Sakoda, W., Tsuji, T., Kurita, Y.: VR training system of step timing for baseball batter using force stimulus. In: Kajimoto, H., Lee, D., Kim, S.-Y., Konyo, M., Kyung, K.-U. (eds.) AsiaHaptics 2018. LNEE, vol. 535, pp. 321–326. Springer, Singapore (2019). https://doi.org/10.1007/978-981-13-3194-7_70

20. Synofzik, M., Thier, P., Leube, D.T., Schlotterbeck, P., Lindner, A.: Misattributions of agency in schizophrenia are based on imprecise predictions about the sensory consequences of one's actions. Brain **133**(1), 262–271 (2010). https://doi.org/10.1093/brain/awp291

21. Tajima, D., Nishida, J., Lopes, P., Kasahara, S.: Whose Touch is This?: Understanding the agency trade-off between user-driven touch vs. computer-driven touch. ACM Trans. Comput.-Hum. Interact. **29**(3), 1–27 (2022). https://doi.org/10.1145/3489608

22. Tapal, A., Oren, E., Dar, R., Eitam, B.: The sense of agency scale: a measure of consciously perceived control over one's mind, body, and the immediate environment. Front. Psychol. **8**(SEP), 1–11 (2017). https://doi.org/10.3389/fpsyg.2017.01552. https://journal.frontiersin.org/article/10.3389/fpsyg.2017.01552/full

23. Woods, D.L., Wyma, J.M., Yund, E.W., Herron, T.J., Reed, B.: Factors influencing the latency of simple reaction time. Front. Hum. Neurosci. **9**(MAR), 1–12 (2015). https://doi.org/10.3389/fnhum.2015.00131. https://www.frontiersin.org/Human_Neuroscience/10.3389/fnhum.2015.00131/abstract

Epidural Motor Skills Measurements for Haptic Training

Hugo Laplagne[1], Gwenola Touzot-Jourde[2], Delphine Holopherne-Doran[2], and Cédric Dumas[1,3]($\boxtimes$) (iD)

[1] IMT Atlantique, Nantes, France
cedric.dumas@imt-atlantique.fr
[2] ONIRIS, Nantes, France
[3] LS2N, Nantes, France

Abstract. Epidural procedure can be a challenging skill to learn. Veterinarians must feel how many ligaments they get through before performing the injection in the exact location, relying solely on haptic perception, on a wide range of animal sizes. Force, torque, and position sensors were mounted on a needle to measure an epidural injection procedure. We provide here the data analysis that led to the needle insertion force profile extraction, to be used in procedural haptic simulators.

Keywords: Epidural · Force feedback · Needle insertion · Haptic training

1 Introduction

There are many types of medical tools used in different procedures. Physicians conducting such procedures sense the interaction between the medical tool and the deformable specimen through haptic perception. One of the medical fields that requires high haptic sensitivity of the physicians is anaesthesiology. Epidural Analgesia is commonly performed in animals and taught in all veterinarian schools. This procedure is rarely image-guided and requires high haptic sensitivity to get through the tissue layers to perform the injection in the exact body space (the vertebral canal).

1.1 The Epidural Technique Data Collection

Epidural needle insertion was performed on an anaesthetized dog in a veterinarian school on two occasions. The needle was modified to contain position and forces sensors in order to get two sets of corresponding data. Each experiment was documented with video, and complemented with CAT-Scans of the dog from the side as well as from top showing the steps of the procedure during the second experiment. The time scale of the data extends on around two minutes and a half. However, during the second experiments breaks were

---

Supplementary Information The online version contains supplementary material available at https://doi.org/10.1007/978-3-031-22061-6_27.

taken to take the pictures via CAT-scans and therefore the force data cannot be used to reproduce the interventional skill with a high fidelity as the needle was left dangling. The purpose of this data collection was to build training material for the students. Haptic and spatial measurements were meant to model the epidural skill. This paper focus on the first dataset that can be used to characterize motor skills to perform adequately the epidural needle insertion. The data analysis provided here in terms of haptic measurements is meant to help building epidural skills training tools (virtual simulator, haptic device, or any other mean). This measurement consists of one directional force and position profile to characterize the sensitivity of layers perforations.

1.2   Anatomy

The epidural injection is a locoregional anesthetic technique that is commonly performed for pain management during veterinary surgical procedures on the hindlimbs as well as for humans mostly during labor. In small animal species like dogs and cats, animals are already under general anesthesia to ease the performance of this intrusive procedure on a still body. This operation consists in delivering a combination of local anesthetic and analgesic drugs in the epidural space in the low back at level of the intervertebral space between L7 and S1, which are respectively the last lumbar vertebra and the first sacral vertebra [1]. The very end of the spinal cord covered by the meninges sits at this level, making this procedure delicate and challenging. The needle has to cross 4 different tissues to get into the epidural space: the skin which provides for the first resistance; the supraspinous ligament and the interspinous ligament which offer the second resistance and then the ligamentum flavum that seems to provide the third and harder resistance. Three different Cat-Scans were taken during the second experiment (see Fig. 1) at each three key steps of the procedure: after the skin, before the ligamentum flavum; and after the ligamentum flavum in the epidural space. We can observe that the trajectory seems curved, due to the fact that the needle was left hanging before taking the pictures and the weight of the sensors made it tilt.

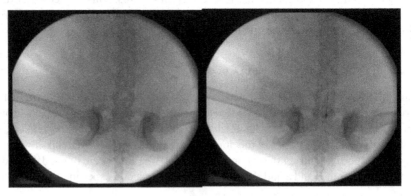

Fig. 1.   Top view before and after needle insertion

2 Related Works

Epidural injection techniques are a widely studied interventional skill (over 16,000 references in PubMed). In his survey on epidural simulators [2], Neil Vaughan concludes "the ideal epidural simulator [...] would contain a force feedback haptic device, with force data originating from measured Tuohy needle insertions from patient". Our study is meant to extract an accurate enough epidural insertion force profile to be trusted by experts when using in a simulator.

Despite recent advances in virtual and real (3D printed) simulators [3, 4], often for human size and anatomy, they do not focus on haptic accuracy: once the skin is passed with the needle, feeling the successive ligaments perforation is a real challenge for novices (or experts with infrequent practice). Notwithstanding ligament perforation can require little force [5], added frictions increase the epidural complexity: frictions between the needle and the animal tissue increase while being inserted forward. Our study focuses on procedural skills and training haptic sensitivity, as veterinarians only rely on haptic feeling to perform injection in the proper space, with the added constraints of being able to adapt to various animals size and anatomy.

3 Proposed Method

The measurements were setup in an operating room of the Nantes veterinarian school (see Fig. 2) with a C-arm, a motion tracking device, and a haptic sensor, described in the section below.

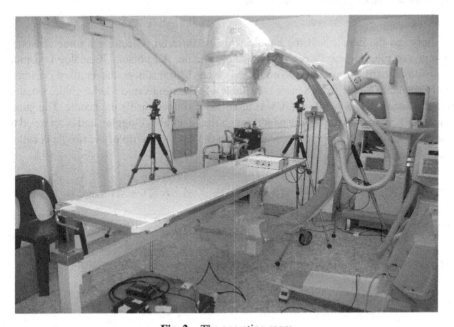

Fig. 2. The operating room

3.1 OptiTrack

The position data sensor is an OptiTrack system, which gets the position data with infrared reflectors mounted on the needle, and a set of three FlexV100 cameras around the operating table (Fig. 3). The infrared reflectors were placed on top of the needle. The four dots on the sensor constitute a tetrahedron of which centre is the point that is used to determine coordinates in the frame of reference. The tetrahedron indicates the rotational coordinates around each axis consecutively x, y and z. The frame of reference was placed on the table with axes x and z being horizontal and y being the vertical axis.

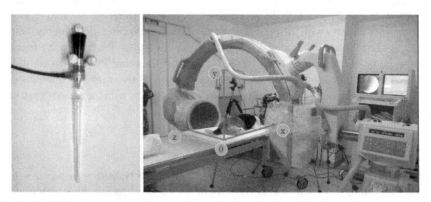

Fig. 3. (Left) The original epidural needle with mounted force sensor and infrared reflectors (right) motion tracking cameras frame of reference

Between the needle and the handle with the infrared reflectors, a force sensor is mounted to measure the forces applied to the needle. A Labjack U6 analog to digital acquisition board was used to record the ATI Nano17 transducer, the smallest commercially available 6-axis force sensor (see Table 1). The removable needle guide, situated into the lumen of the needle shaft, was modified and fixed inside the needle. The guide is usually here to help the needle pass through tissue without clogging and is removed to inject the drugs. For this sensor, the axis from the base to the tip of the needle is called the z axis.

Table 1. Force (F) and torque (T) characteristics of the ATI Nano17

Calibration reference	Sensing ranges				Resolution	
	Fx, Fy	Fz	Tx, Ty	Tz	Fx, Fy, Fz	Tx, Ty, Tz
SI-12-0.12	12 N	17 N	120 N mm	120 N mm	1/320 N	1/64 N mm

4   Results

4.1   Retrieving and Pre-treatment

The data is returned from the experiment in an xml file, using the open source HaCoE software [6] and the architecture seen in Fig. 4.

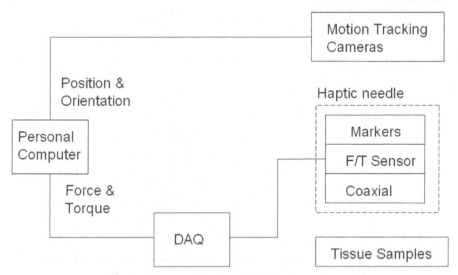

Fig. 4.   Block diagram of the apparatus

   With each timestamp in milliseconds, the position along each axis is stored in meters, then the orientation in degrees, then the forces in Newtons and the torques in Newton meter. A Python program was developed to retrieve the data from the xml file.

   After doing so, a quick plot of the data along the time shows that the position and orientation data. Position data are noised with a lot of its value being 9999.0 (no measurement) due to the occlusion of the needle from the cameras, that can happen during the procedure when the anaesthesiologist is moving the needle around. The force sensor has no such error with continuous measurements. A simple filter of the position data removing records with coordinate x above 100 was sufficient to remove the OptiTrack errors.

Synchronisation between videos, images and records was performed to identify which time window corresponds to the needle insertion. The recorded data file provides a hundred and fifty seconds long sequence, where the video of the procedure only lasts for thirty-eight seconds. To identify which part of the data coincides with the needle insertion, we looked at the force along the needle's axis and its fluctuations. By looking at this graph (Fig. 5) we can safely assume that the operation happens in the seventy-eight to one hundred seconds time frame.

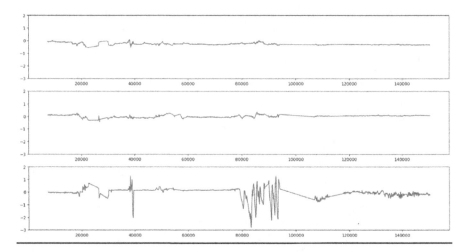

Fig. 5.   Graphs of the force insertion profile (Fx, Fy, Fz)

Once this time frame is identified, we just proceed to filter the torques and forces with a simple low-pass filter using the left rectangle numerical method:

$$s_i = s_{i-1} + \frac{Te}{\tau} \times (st_{i-1} - s_{i-1})$$

$$\tau = \frac{1}{2} \times \frac{1}{\pi \times f_c}$$

where s is the filtered signal, st is the original signal, f_c is the filtering frequency and Te is the sampling period. By trial and error, and looking at the graph, we use this filter to suppress high frequency noise without losing too much of the original information (Fig. 6).

We chose 0.75 Hz filtering frequency. Better filters and better frequency determination methods could be used to lose even less of the data and to provide a better filtering method.

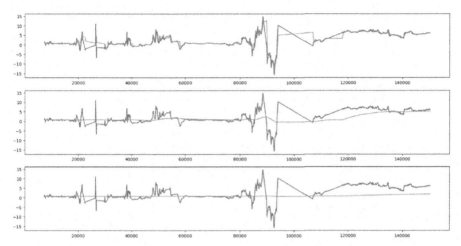

Fig. 6. Graphs of torque along the x axis (blue) and the filtered torque (orange) at 0.5 Hz, 0.01 Hz and 0.001 Hz filtering frequencies

4.2 Geometrical Transformation

We observed on the videos of the operation the needle is rotated often, so the tip of the needle and the handle have very different trajectories. We assumed that the needle is rigid, and we extrapolate the tip of the needle position from the measured handle position (where the IR reflectors are mounted).

In order to do so, we used the data provided by the OptiTrack. Using homogenous coordinates [7] and the OptiTrack documentation [8], we create a transformation matrix. Knowing the tetrahedron and the needle geometries, the OptiTrack sensor rotation and translation, we computed the needle tip position. From this we created a 3D plot of to check if the needles movement corresponds to the expectations (Fig. 7).

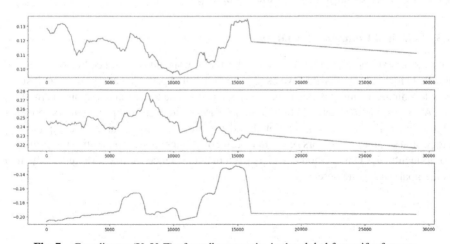

Fig. 7. Coordinates (X, Y, Z) of needle extremity in the global frame if reference

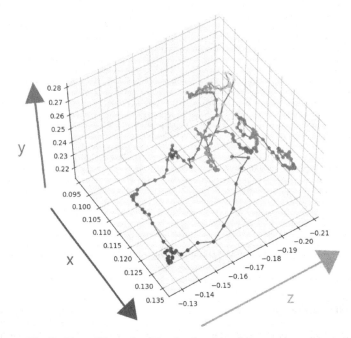

Fig. 8. Three 3D graph of the simulated needle's point's position

From the 3D plot and the variation along each axis, we can say that the variations are contained within the margins of length expected by the veterinarian which are between 5 to 6 cm (Fig. 8), considering that there are also parasitic lateral movements around the insertion axis and that the there is some margin around the actual needle insertion movement. To detect any error in this procedure, we created a function that computes, for each point of the needle tip calculated, the distance between this point and the handle. We checked this distance had no variation in length.

4.3 Principal Component Analysis

In order to extrapolate a one-dimensional direction profile for the needle and focus on insertion force along this axis, we used a method called principal component analysis to isolate a direction along which to project each point. The Principal Component Analysis (PCA) is a reduction method applied on a dataset to identify the principal components along which the data is differentiated and therefore distinguishable. It computes the eigenvalues and eigenvectors of the dataset's covariance matrix to identify the principal components [9]. This method is widely used in data science for linear regression, which is the goal we pursue here.

We implemented the PCA with the estimated point of insertion coordinates (cf. Section 4.2). To calculate this point, we found the first variation of the force and defined it as the needle point of insertion (Fig. 9 bottom).

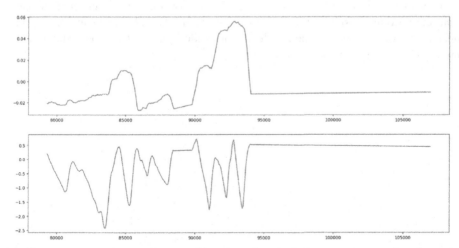

Fig. 9. (Top) Graph of the projected position along the principal component over time (bottom) Graph of the force insertion profile Fz along the needle axis over time

Most data treatment coding languages –mainly Python and Matlab – have built-in methods to implement the PCA. Here we used numpy.linalg.svd from NumPy, the fundamental package for scientific computing in Python. Once we implemented the PCA, we can directly plot the projected data and its variations during the timeframe (Fig. 9 top).

5 Discussion

5.1 PCA Limitations

As we can observe on (see Fig. 10), the results of the PCA are far from the expected results as we have variations on a length of almost twelve centimeters for the one-dimensional movement. Also, it would be expected for the needle to progress along the axis only to be pulled out when the procedure is done, hitting a plateau whenever the veterinarian stops to force the needle through a tissue that presents a resistance. Instead, this plot shows a needle that would be inserted then pulled back to be inserted further again. The problems to solve here are to use force, torque and position data together to create a more accurate insertion force profile, that compensates lateral movements of the needle.

5.2 Force Modifications

As we can observe on (see Fig. 9) that the force does not render the idea of three resistance plateaus but more of isolated peaks of exerted force from the veterinarian. This seems counter intuitive to a simple rectilinear movement, however, as previously said, the movement isn't rectilinear. Indeed, watching the video one can observe the operator moves the head of the needle around while trying to feel where and how to insert the needle at every step of the way.

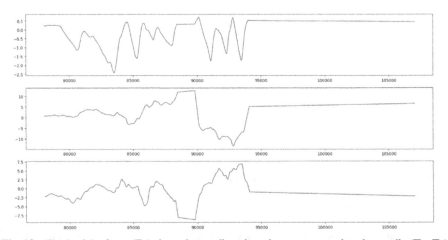

Fig. 10. Graph of the force (Fz) along the needle axis and torques normal to the needle (Tx, Ty) over time

If we look at the torque while the force is at a local minimum, we can see that a strong torque is measured most of the time (Fig. 10). This can be explained by the fact that the operator turns the needle while not moving. And this torque when impacting the force may cause this irregular profile. As a way of correcting those irregularity and creating a more accurate insertion force profile, we cancel the torque which would normally translate as:

$$(-Tx + Ty) \times l$$

where l is the needle's length.

By summing its opposite projection to the force:

$$F = Fz + (Tx - Ty) \times l$$

This operation gives us the following force profile (Fig. 11):

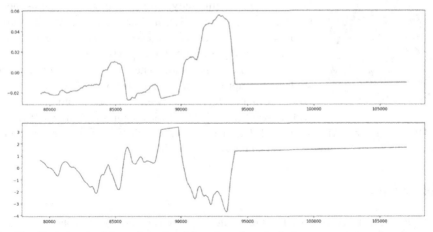

Fig. 11. (Top) Graph of the projected position along the principal component over time (bottom) Graph of the force with torque compensated over time.

When analysing the movements, we can set an approximative time gap of five seconds between the skin and the second resistance encountered and a fifteen seconds gap between the skin and the third resistance. Therefore, we have peaks of force that are coherent with the three timestamps of insertion and resistance met during the operation (as seen on Fig. 11). The incertitude on the data is the peak of force that indicates a traction of the needle by the skin. Although the epidural space has a negative pressure, the needle does not seem to be in the epidural space. This irregularity might be caused by a poor filtering of the torque, as we can see very high torque at this point.

6 Conclusion

Despite being able to measure force and torque on the needle with a highly sensitive sensor, performed by an expert anaesthesiologist able to compensate the tool weight variation, it was not straightforward to extract the force profile along the needle insertion axis. Nevertheless, the calculated force profile (Fig. 11) can be used to measure or train students haptic sensitivity, for example with simple 1D haptic devices.

We can further improve the correction to create the three proper peaks although we know data accuracy is counter balanced by expert perception: from previous studies [10], we know that, once inserted in a simulator, force profiles need to be tuned by experts before they trust the simulation and actually use it with trainees. The following step is then to check this first force profile with the experts.

7 Future Work

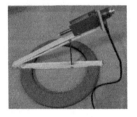
Based on this preliminary work, we have conducting new research on two tracks. The first one is measuring the same epidural procedure at the vet school with a cow, to get a force profile with a significant different physiology.

The second one is to build a new haptic device to train vet students to improve their performance with ligaments perforation. A first prototype has been built (see photo on the left) with an ATI Nano 17 6D force sensor and a FAULHABER linear DC servomotor LM1247-020-12. The displacement and the force feeling are smooth. It has to be evaluated now with vet students.

References

1. Halley, L.E., Riedesel, D.H.: Epidural Analgesia in the Dog, vol. 45. Iowa State University Veterinarian (1983)
2. Vaughan, N., Dubey, V.N., Wee, M.Y., Isaacs, R.: A review of epidural simulators: Where are we today? Med. Eng. Phys. 35(9), 1235–1250 (2013)
3. Neves, E.C.D., Pelizzari, C., Oliveira, R.S., Kassab, S., Lucas, K.D.A., Carvalho, Y.K.: 3D anatomical model for teaching canine lumbosacral epidural anesthesia. Acta Cir Bras. 35(6), e202000608 (2020). doi: https://doi.org/10.1590/s0102-865020200060000008. Epub 2020 Jul 13. PMID: 32667587; PMCID: PMC7357831
4. Jeong, S.-M., et al.: A proposal of a simple epidural simulator for training novice anesthesiologists. J. Anesth. 30(4), 591–595 (2016). https://doi.org/10.1007/s00540-016-2182-5
5. Lynch, C., Richards, C., Dorrington, P., Brousseau, E., Turner, M.: Measurement of forces from epidural catheter insertion. Anaesthesia 70, 1215–1216 (2015). https://doi.org/10.1111/anae.13214
6. HaCoE, Haptic perception measurement software for research. https://github.com/cyhd/HaCoE. Accessed 1 Oct 2021
7. John, F., Van Dam, A., McGuire, M., et al.: Computer graphics: Principles and practice. Chapter 11: Transformations in Three Dimensions, p. 263, Addison-Wesley (2014)
8. OptiTrack Documentation Wiki. https://v23.wiki.optitrack.com/index.php?title=OptiTrack_Documentation_Wiki. Accessed 1 Oct 2021
9. Gray, V.: Principal Component Analysis: Methods. Applications and Technology. Nova Science Publishers, Incorporated (2017)
10. Dumas, C., Coles, T., de Visser, H., Cao, C.G., Grimpen, F.: Haptic feedback tuning in colonoscopy simulation. In: 2016 IEEE International Conference on Systems, Man, and Cybernetics (SMC), pp. 004400–004404 (2016). doi: https://doi.org/10.1109/SMC.2016.7844923

Sensorless Force Approximation Control of 3-DOF Passive Haptic Devices

Maciej Łącki[1(✉)] and Carlos Rossa[2]

[1] Haply Robotics, Montreal, QC, Canada
maciej@haply.co
[2] Carleton University, Ottawa, ON, Canada
rossa@sce.carleton.ca

Abstract. Haptic devices using passive actuators such as brakes are intrinsically stable and offer greater transparency compared to traditional haptic devices actuated by electric motors. However, force control in passive haptic devices is particularly challenging and relies heavily on the use of force sensors, which can significantly increase the device's inertia and bandwidth. This paper proposes to use a nonlinear disturbance observer based on a Newtonian dynamic model of a 3-Degree-of-Freedom delta passive haptic device to estimate the force input of the user. The observer is tested using a series of simulations and the results confirm that the estimated input force closely matches the actual input force even when the system is subjected to unmodelled dynamics such as the brake's hysteresis.

Keywords: Force-feedback · Passive actuators · Force estimation

1 Introduction

Passive haptic devices offer unparalleled stability compared to traditional haptic devices actuated by electric motors. Their main limitation, however, is that multi-degree-of-freedom (DOF) passive haptic devices cannot generate forces in arbitrary directions. In fact, as discussed in [14], the portion of the workspace where forces can be displayed in any direction decreases rapidly with the number of degrees of freedom. To render forces outside of these regions, force approximation controllers need to be used. These control schemes typically attempt to eliminate the net force acting on the end-effector, perpendicular to the virtual surface, thereby ensuring that the device's end effector slides along the surface without penetrating it [3,15]. To balance the force at the end-effector these controllers require an accurate measurement of the user force input, which is typically achieved using a force sensor.

Force sensors are challenging to integrate in haptic devices as the added mass reduces transparency and measurement noise narrows the bandwidth [10]. An alternative to direct force sensing is to modify the force approximation scheme such that it considers the energy exchange between the virtual environment and the haptic device, eliminating the need for force measurement so long as the device moves [4]. However, in a stationary passive haptic device, the energy flow

S. Berretti and G.-M. Su (Eds.): ICSM 2022, LNCS 13497, pp. 381–394, 2022.
https://doi.org/10.1007/978-3-031-22061-6_28

is not observable and the direction of the user's force needs to be determined to control the device. This non-model-based controller is simple to integrate, however, it is difficult to adapt in multi-DOF. Thus, to void the use of a force sensor completely, the force input must be estimated.

Force estimation schemes generally use model-based disturbance observers to estimate an external force. In [11,12], the disturbance observer uses an inverse dynamic model with a low-pass filter to estimate the disturbances acting on the device. Expanding this observer to each joint of a multi-DOF manipulator allows for the estimation of the torque applied to each joint [18]. The observers presented in [11,12,18] base their structure on linearized models of highly nonlinear dynamics, which means that their stability is not guaranteed.

Nonlinear disturbance observers (NDOs) are a superior choice for robotic manipulators as they require no model linearization, no acceleration measurement, and are proven to be asymptotically stable. The first NDOs, introduced in [2], were limited to 2-DOF planar devices, however, their use quickly expanded to devices functioning in 3D space [13] followed by n-DOF manipulators in [19]. NDOs were previously used in haptic applications for closed-loop force [7] or impedance control [6] and in many other applications [17]. To the best of our knowledge, NDOs have not been used to estimate the input force in a multi-DOF haptic device.

This paper explores the possibility of using an NDO to estimate the force input in a passive haptic device. The preliminary analysis is based on the 3-DOF parallel device introduced in [15] and aims to prove, using simulations, that the force applied to a moving end-effector can be estimated, and that such an estimate is sufficient for force approximation control of the device.

To this end, Sect. 2 introduces the structure of the NDO, along with the dynamic model of the 3-DOF parallel passive device presented in [15]. Next, in Sect. 3 the observer is validated using a series of simulations aimed at proving the observer's ability to estimate various types of force inputs. The preliminary results are then evaluated in Sect. 4 to determine the feasibility of the presented approach in the context of controlling a multi-DOF passive haptic device without an accurate force measurement.

2   Force Observer Design

The NDO is model-based and thus it requires a dynamic model of the device. The device used to derive such a dynamic model is a 3-DOF haptic device with a modified Delta mechanism presented in [21].

2.1   Dynamic Model

The device we are considering uses particle brakes as actuators and employs a modified Delta kinematic structure as described in [21]. Let the position of each of the three actuated joints be $\boldsymbol{\theta} = [\theta_{11}\ \theta_{12}\ \theta_{13}]^{\mathrm{T}}$, the velocity be $d\boldsymbol{\theta}/dt = \dot{\boldsymbol{\theta}}$

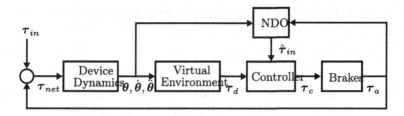

Fig. 1. The torque input estimate $\hat{\tau}_{in}$ of the NDO replaces the torque measurement τ_{in} as the input to the controller. The NDO requires measurements of the joint positions θ and velocities $\dot{\theta}$, along with the plant output estimate τ_a.

and acceleration be $d\dot{\theta}/dt = \ddot{\theta}$. Its dynamic model is given by the second order differential equation (Fig. 1)

$$\mathbf{M}(\theta)\ddot{\theta} + \mathbf{C}(\theta,\dot{\theta})\dot{\theta} + \mathbf{G}(\theta) = \tau_a + \tau_{in} \tag{1}$$

where $\{\tau_a, \tau_{in}, \mathbf{G}(\theta)\} \in \mathbb{R}^{3\times 1}$, $\{\mathbf{M}(\theta), \mathbf{C}(\theta,\dot{\theta})\} \in \mathbb{R}^{3\times 3}$ represent the applied torque, the input torque, the gravity vector, and the inertia and Coriolis matrices, respectively. The matrices in (1) are given using Newtonian dynamic analysis [21] as:

$$\mathbf{M}(\theta) = I_a\mathbf{I} + m(\mathbf{J}^{-1})^{\mathrm{T}}\,\mathbf{J}^{-1}$$

$$\mathbf{C}(\theta,\dot{\theta}) = c_d\mathbf{I} + m(\mathbf{J}^{-1})^{\mathrm{T}}\,\frac{d}{dt}(\mathbf{J}^{-1})$$

$$\mathbf{G}(\theta) = -ag\left(\frac{m_a}{2} + m_b\right)\cos(\theta) - m(\mathbf{J}^{-1})^{\mathrm{T}}\begin{bmatrix}0\\0\\g\end{bmatrix}$$

where $\mathbf{J} \in \mathbb{R}^{3\times 3}$ is the Jacobian matrix and

$$I_a = I_m + \frac{m_a a^2}{3} + m_b\,a^2$$

$$m = 3m_b + m_c$$

where m_a, m_b, and m_c, represent the mass of links a, b, and the end-effector; c_d is the viscous damping of the brake and g is the acceleration due to gravity. Note that there is no known closed form solution to $d/dt\,(\mathbf{J}^{-1})$ therefore this term must be approximated numerically [9]. The link lengths of the device are given in Table 1 while Table 2 summarizes their physical characteristics.

2.2 Nonlinear Disturbance Observer

The nonlinear disturbance observer, shown as a block diagram in Fig. 2, has the following form:

$$\dot{\mathbf{z}} = -\mathbf{L}(\theta,\dot{\theta})\mathbf{z} + \mathbf{L}(\theta,\dot{\theta})\left(\mathbf{C}(\theta,\dot{\theta}) + \mathbf{G}(\theta) - \tau_a - \mathbf{p}(\theta,\dot{\theta})\right) \tag{2a}$$

$$\hat{\tau}_d = \mathbf{z} + \mathbf{p}(\theta,\dot{\theta}) \tag{2b}$$

Table 1. Links (top row) of the haptic device and their corresponding lengths in millimetres (bottom row) of the 3-DOF passive haptic device from [15]

Link Lengths in millimeters								
a	b	c	d	e	f	g	r	s
60.0	102.5	14.4	13.0	13.0	25.0	27.9	36.6	27.2

Table 2. Physical characteristics of each link in the 3-DOF passive haptic device. I_m is the brakes' mass moment of inertia, c_d is the viscous damping of the actuator, m_a the mass of link a, m_b is the equivalent mass of link b, and m_c is the mass of the end-effector.

I_m (mm^4)	m_a (g)	m_b (g)	m_c (g)	c_d (Ns/m)
8.5	6.8	16	10	0.01

where $\mathbf{z} \in \mathbb{R}^{3 \times 1}$ represents internal observer states, $\mathbf{L}(\boldsymbol{\theta}, \dot{\boldsymbol{\theta}}) \in \mathbb{R}^{3 \times 3}$ is the observer gain matrix given as,

$$\mathbf{L}(\boldsymbol{\theta}, \dot{\boldsymbol{\theta}}) \, \mathbf{M}(\boldsymbol{\theta}) \ddot{\boldsymbol{\theta}} = \left[\frac{\partial \mathbf{p}(\boldsymbol{\theta}, \dot{\boldsymbol{\theta}})}{\partial \boldsymbol{\theta}} \; \frac{\partial \mathbf{p}(\boldsymbol{\theta}, \dot{\boldsymbol{\theta}})}{\partial \dot{\boldsymbol{\theta}}} \right] \begin{bmatrix} \dot{\boldsymbol{\theta}} \\ \ddot{\boldsymbol{\theta}} \end{bmatrix} \tag{3}$$

and the auxiliary function $\mathbf{p}(\boldsymbol{\theta}, \dot{\boldsymbol{\theta}})$ which is used to substitute acceleration measurements [2]. For a 3-DOF manipulator a possible formulation for the auxiliary variable is [6]

$$\mathbf{p}(\boldsymbol{\theta}, \dot{\boldsymbol{\theta}}) = c \begin{bmatrix} \dot{\theta}_{11} \\ \dot{\theta}_{11} + \dot{\theta}_{12} \\ \dot{\theta}_{11} + \dot{\theta}_{12} + \dot{\theta}_{13} \end{bmatrix} \tag{4}$$

giving the observer gain matrix as

$$\mathbf{L}(\boldsymbol{\theta}, \dot{\boldsymbol{\theta}}) = c \begin{bmatrix} 1\,0\,0 \\ 1\,1\,0 \\ 1\,1\,1 \end{bmatrix} \mathbf{M}(\boldsymbol{\theta})^{-1} \tag{5}$$

which can be proven to be asymptotically stable for a range of c, which represents a controllable observer gain [2,13].

NDOs are formulated on assumptions that disturbance varies relatively slowly and that all functions in the dynamic model of the device are smooth. As shown in [2], however, NDOs can estimate fast varying disturbances. In the context of passive haptic devices, it must also be assumed that no brake is stationary during the operation, as the states of a stationary brake are not observable.

3   Simulation Results

Since a human operator can apply a wide range of forces to the device at a wide range of frequencies, the simulations must test the ability of the observer to

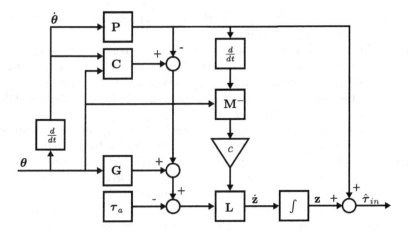

Fig. 2. The nonlinear disturbance observer estimates torque input $\hat{\tau}_{in}$ using angular position θ, velocity $\dot{\theta}$, and the torque applied by the brakes τ_a

estimate the input force in a variety of conditions. The testing scenarios 1 and 2, detailed below, validate the observer for cases where the device is not rendering forces to the user i.e., free motion. Scenarios 3 and 4, on the other hand, test the observer as the force sensing component in the force approximation scheme from [15]. Finally, scenarios 5 and 6 compare the observer and the controller performance with uncompensated brake hysteresis in the dynamic model of the device.

Scenario 1 - Free motion with constant vertical force input: In this scenario, the force applied to the end-effector has only a vertical Z component following a square wave pattern of ±400mN at a frequency of 1Hz. This scenario tests the NDO's ability to converge when subjected to an unsmooth force input.

Scenario 2 - Free motion with varying force input: The force simulates the user's input; it is smooth and varies in the X, Y, and Z axes. The magnitude of the applied forces in the Z axis is approximately 10N and 5N along the X and Y axes with a frequency of 3.5Hz. This scenario evaluates the observer's ability to estimate fast varying forces acting in arbitrary directions.

Scenario 3 - Rendering a displayable desired force in the vertical axis: Similar to the first scenario, the force input is a square wave in the Z axis with a magnitude of 400mN and a frequency of 2Hz, but the device attempts to render a force. The desired force is a sine wave with a magnitude of 100mN and a frequency of 1Hz. This scenario compares the output of the force approximation controller from [15] when using the actual and estimated force input.

Scenario 4 - Rendering partially displayable forces, with a varying force input: Like in Scenario 2 the force input simulates the user moving the device in the X, Y, axes and Z axes with a force magnitude of 10N in the Z axis, and 5N in the X and Y, all at a frequency of 3.5Hz. The desired force has a magnitude of 5N in the Z axis, and 2N in the X and Y axes all at a frequency of 0.5Hz.

This scenario compares the output of the force approximation control scheme when using the force input and its estimate.

Scenario 5 - Rendering a force in the vertical axis with uncompensated hysteresis: The force input and the desired force are the same as in scenario 3, however, the torque generated by the brake is subject to hysteresis. This scenario tests the ability of the observer to converge when the device model does not account for all device dynamics.

Scenario 6 - Rendering a varying force, with a varying force input and uncompensated hysteresis: The force input and the desired force are the same as in scenario 4, however, the torque generated by the brakes is subject to hysteresis. This scenario tests the performance of the NDO and force control scheme from [15] when the force input is not the only disturbance in the system.

The simulations are conducted using MATLAB Simulink 2020a running at 1kHz sampling frequency. The device dynamics obey the equations given in Sect. 2.1 but the gravity is omitted in the simulations such that the only force acting on the device is the force input of the user. The brake hysteresis in scenarios 5 and 6 is modelled using the Presiach hysteresis model described below.

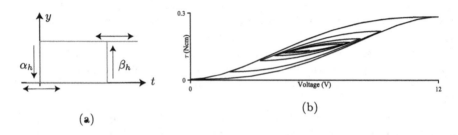

(a)

(b)

Fig. 3. (a) shows the hysteresis loop of a nonideal relay with excitation thresholds of α and β while (b) shows the hysteresis curve for the simulation imitates the curve of the actual particle brakes.

3.1   Hysteresis Model

The Preisach model is constructed using the model of a nonideal relay hysteron, see Fig. 3(a), with a square hysteresis loop described by α_h and β_h excitation thresholds such that,

$$y(t) = \begin{cases} 1 & \text{if } t \leq \alpha_h \\ y(t-1) & \text{if } \alpha_h < t < \beta_h \\ 0 & \text{if } t \geq \beta_h \end{cases} \tag{6}$$

where $y(t)$ is the output of the relay for input t, and $y(t-1)$ represents the previous output value. The model assumes that the magnetization properties of an object are a result of integrating the magnetization of many independent hysterons $\gamma_{\alpha_h, \beta_h}$ modelled in (6), such that

$$y(t) = \iint\limits_{\alpha_h \geq \beta_h} \mu(\alpha_h, \beta_h) \, \gamma_{\alpha_h, \beta_h} \, d\alpha_h \, d\beta_h \tag{7}$$

where $\mu(\alpha_h, \beta_h)$ is the Preisach density function [5,20]. The analytical solution to (7) does not exist and, thus, its implementation is discretized into a finite number of hysterons. The shape of the resulting hysteresis loop is defined by

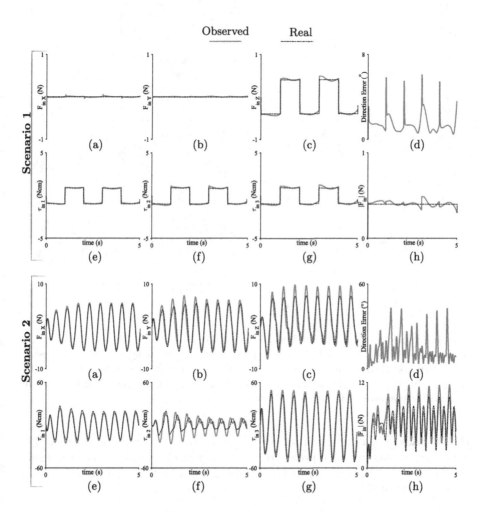

Fig. 4. The results for scenarios 1 and 2 where (a), (b), (c) show the forces input and its estimate in the X, Y, and Z axes, the torque inputs for the three joints are in (e), (f), (g). The angle between the actual force input and the estimate is shown in (d) and the magnitude difference in (h).

388   M. Łącki and C. Rossa

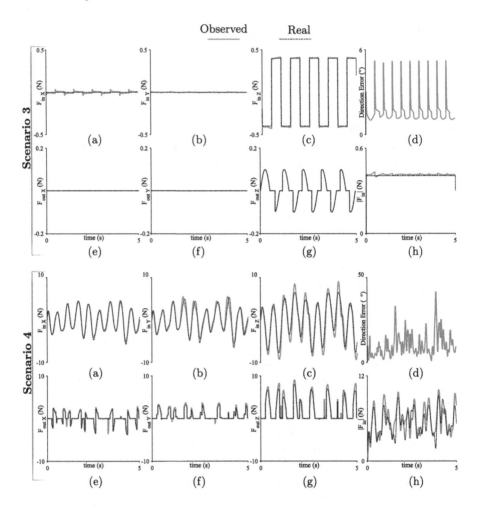

Fig. 5. The results for scenarios 3 and 4 where (a), (b), (c) show the forces input and its estimate in the X, Y, and Z axes and the force calculated by the controller using the actual and the estimated force input in the X, Y, and Z axes is shown in (e), (f), (g). The angle between the actual force input and the estimate force input is shown in (d) and the magnitude difference is given in (h).

the Preisach density function, however, finding such a function is difficult [20]. The Presach function is approximated as a matrix of coefficients created using a series of linear operations, thus obtained model results in the hysteresis loop shown in Fig. 3(b).

3.2 Results

The results for scenarios 1 and 2 are given in Fig. 4. Each set of results compares the force input and its estimate in the X, Y, and Z axes in (a), (b), (c), respec-

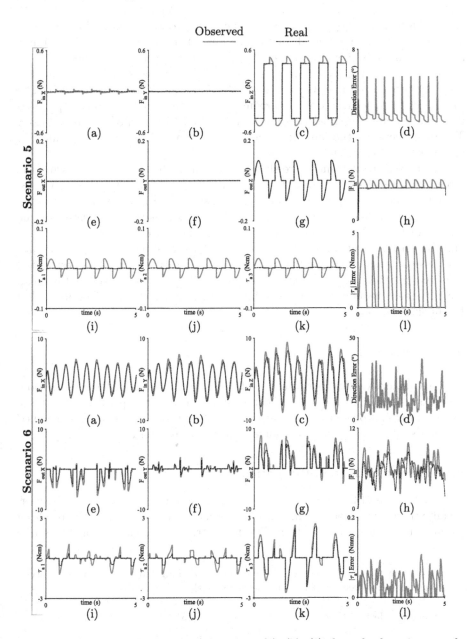

Fig. 6. The results for scenarios 5 and 6 where (a), (b), (c) show the force input and its estimate in the X, Y, and Z axes and the force calculated by the controller using the actual and estimated force input in the X, Y, and Z axes is shown in (e), (f), (g). The angle between the actual force input and the estimate is shown in (d) and the magnitude difference in (h). The effects of hysteresis on torque output of each brake is shown in (i), (j), and (k). Lastly, (l) shows the normalized error between the ideal applied torque and the actual applied torque.

tively. Subfigures (e), (f), (g), compare the actual torque input and its estimate at each of the three actuated joints. The angle between the actual force input and its estimate is shown in (d), while (h) compares their magnitudes.

Scenario 3 and 4 results are presented in Fig. 5. Subfigures (a), (b), and (c) show the force input and its estimate in the X, Y, and Z axes, respectively. The force calculated by the force approximation scheme using the force input and its estimates are compared in (e), (f), and (g), one for each of the spatial axes. The angle between the force input and its estimate is shown in (d), while (h) compares their magnitude.

Lastly, scenarios 5 and 6 are shown in Fig. 6, with (a), (b), (c) showing the force input and its estimate in the X, Y, and Z axes; (e), (f), (g) compare the output of the force approximation scheme using force input and its estimate; (i), (j), (k) compare desired torque and the hysteresis-uncompensated torque output of the brakes, while (l) shows the normalized error between the ideal and the actual torque applied to the brakes. The angle between the force input and its estimate is shown in (d). Finally (h) compares the actual and estimated force magnitudes. The mean and maximum direction error as well as the mean magnitude errors between the force input and its estimate are given in Table 3 for all six scenarios.

4   Discussion

First let us analyze the performance of the NDO in terms of force estimation, then discuss how errors in the force estimate affects the performance of the force approximation control scheme, and finish with the discussion of unmodelled hysteresis and its impact on the force estimation and control.

4.1   Force and Torque Estimation

In scenarios 1 and 2 the device is subjected only to the force input of the user. Ideally, there should be no difference between the input force and the resulting torque, and their estimates. In reality, however, the nonlinearity of the system

Table 3. The mean and maximum direction error along with the mean magnitude error between the force input and its estimate for the six scenarios.

Scenario	Mean direction error (deg)	Max direction error (deg)	mean magnitude error (%)
1	1.46	6.07	1.88
2	13.55	42.93	7.86
3	1.62	5.2	1.92
4	10.31	41.5	8.61
5	1.60	5.35	11.57
6	13.78	36.96	11.15

creates a discrepancy between the observer approximation and the torque inputs leading to a difference in both the magnitude and direction.

In the first scenario, a vertical force is applied to the end-effector, and since the device is in the centre of its workspace, that is $\theta_{11} = \theta_{12} = \theta_{13}$, the torque applied to each brake is the same. The results from Fig. 4 show that the estimated torques differ depending on the brake; the torque estimate at the first joint, shown in (e), is accurate while the estimates for joints two and three, see (f) and (g), exhibit some overshoot. This type of inaccuracy is intrinsic to the NDO as similar patterns can be seen in results obtained in [2,13]. These errors in the torque estimate translate to errors in the estimated force, as shown in (a), (b), and (c). From (d) and (f) it is evident that when the force input changes direction there is a delay in the response of the observer resulting in the angle and magnitude difference between the force input and its estimate. The direction error after the instantaneous change in the force direction is initially $6°$, but it quickly converges to below $2°$. The magnitude of the force also shows a large initial spike but likewise, it converges to approximately 2%. On average the direction error of the force estimate is $2°$ and the magnitude error is 1.5%. Note, however, that this is the worst-case scenario for the NDO as the disturbance is not a smooth function. Despite the unfavourable conditions the observer can still provide an accurate force estimate.

The force input in the second scenario is a smooth sinusoidal function acting in all three axes. The observer estimates the torque applied to each of the three joints with varying results. For instance, the estimate for the first and third joints, shown in Fig. 4(e) and (g), closely tracks the torque input, however, joint two, shown in (f), differs from the actual input torque. This error, however, does not seem to significantly affect the force estimate shown in (a), (b) and (c). Like in scenario 1, the error in the estimate is the greatest when the force direction changes rapidly, however, since the input is smooth the spikes in the direction error, shown in (d), and magnitude, shown in (h), are far less pronounced. The maximum direction error is $43°$ but on average the error is only $14°$ while the average magnitude error is about 8%.

4.2 Force Approximation Using Estimated Forces

Scenarios 3 and 4 incorporate the force approximation scheme from [15] and compares forces calculated by the controller using the input force measurement and its estimate. Ideally, the force output calculated by the controller based on the input force estimate should be the same as the one calculated using the actual force input. The results for scenarios 1 and 2, however, show that the estimated force may have a different direction and magnitude than the actual force. In scenario 3, the force input and the desired force act in the vertical direction and thus rendering the force does not require force approximation. On the other hand, in scenario 4 both the force input and the desired force act in all three directions meaning that there will be instances where force will need to be approximated.

Similar to the previous two scenarios, the observer estimates the force input with an average direction error of 2° and 10° for scenarios 3 and 4, respectively and an average magnitude error of 2% and 9%, respectively. The controller output using the true and estimated force input are similar for the two scenarios, as shown in Fig. 5(e), (f), and (g) for each of the three axes. Notably, the output based on the observed force input is higher when the force estimate is used. The cause of this anomaly is likely the overestimation of the force input magnitude which can be observed in (h). Such overestimation changes the balance of forces at the end-effector which may lead to stiction. The direction error, on the other hand, does not seem to have a major effect on the output of the controller.

4.3   Effects of Unmodelled Dynamics

Scenarios 5 and 6 test the sensitivity of the NDO to unmodelled dynamics, such as the brake hysteresis. Since the user is no longer the sole contributor of disturbance in the system the resulting force estimate is subjected to both the hysteresis and the user input.

The results in Fig. 6 (i), (j), and (k) show a significant difference between the output of the brake with and without unmodelled hysteresis. In both scenarios 5 and 6, the torque output by the brake is significantly smaller than the desired torque and the error is most pronounced when rending low torques, likely due to the highly nonlinear hysteresis loop.

The resulting force input estimate, shown in (a), (b), and (c), causes a noticeable increase in the error when compared with Fig. 5. In scenario 5 the error in the direction of the force input is nearly the same as in scenario 3, which is to be expected as the torque applied to each brake is equal. However, both the mean and the maximum magnitude errors in scenario 5, respectively 12% and 35%, are higher than what is observed in scenario 3, that is 2% and 6%. In scenario 6, on the other hand, the mean direction error is higher than in scenario 4, 14° compared with 10°. The magnitude error is also higher in scenario 6 compared with scenario 5, respectively 11% and 8%, which should increase the force output error of the controller.

In scenario 5 there is no difference in the output of the controller using the force input and its estimate, as shown in (e), (f), and (g). In scenario 6, on the other hand, the error in the force input estimate is noticeably higher than that in scenario 5. This indicates that unmodeled dynamics of the device can further promote stiction and, thus, must be compensated to guarantee the best performance. Despite the exaggerated effects of modelled hysteresis, the results clearly show that the NDO is versatile and it can provide an estimate of the force input.

5   Conclusions

Control schemes for passive haptic devices depend on force measurements and thus typically require force sensors. To reduce the reliance on the force sensor, we propose the use of a nonlinear disturbance observer (NDO) for sensorless force estimation. The proposed approach is tested using a dynamic simulation of the 3-DOF parallel passive device from [15]. The results show that the NDO can estimate the force input with a magnitude error lower than 9% and the direction error of 13° or less. The preliminary tests show that the force estimate obtained using the NDO can be used in force approximation control schemes as a replacement for a force sensor, though the concept requires further experimental validation.

It remains to be seen how unmodelled dynamics affect the force estimation in experimental tests. For large magnitudes of the input force, the disturbances caused by hysteresis, friction, and unmodeled dynamics become insignificant, meaning that the observed disturbance should closely match the user force input. On the other hand, when the force input is small these effects may dominate the response resulting in less accurate force approximation. An accurate Lagrangian model of the 3-DOF passive device proposed in [21] may improve accuracy at a cost of increasing computational cost [1,16]. Modelling particle brakes using Presiach model [20] or a more sophisticated approach like Boc-Wen Model [8] can further improve the accuracy of the NDO.

The NDO can only estimate forces when the device is in motion; if brakes stop moving the system is no longer observable. In situations where the brakes do not move, for instance, if the virtual environment imposes high viscous damping, a force measurement will still be required. The control of stationary brakes, however, requires only the measurement of the force input direction which can be estimated using low fidelity force sensors such as strain gauges placed directly on the links of the device. As a result, it may be possible to improve performance and reduce the cost of passive haptic devices, making them a more viable option in many haptic applications.

References

1. Briot, S., Khalil, W.: Dynamics of Parallel Robots. Springer, Cham (2015). https://doi.org/10.1007/978-3-319-19788-3
2. Chen, W.H., Ballance, D.J., Gawthrop, P.J., O'Reilly, J.: A nonlinear disturbance observer for robotic manipulators. IEEE Trans. Ind. Electron. **47**(4), 932–938 (2000)
3. Cho, C., Song, J.B., Kim, M.: Design and control of a planar haptic device with passive actuators based on passive force manipulability ellipsoid (FME) analysis. J. Robot. Syst. **22**(9), 475–486 (2005)
4. Cho, C., Song, J.B., Kim, M.: Energy-based control of a haptic device using brakes. IEEE Trans. Syst. Man Cybern. **37**(2), 341–349 (2007)
5. Diep, B.T., Vo, N., Le, T.D.: Hysteresis identification of bidirectional magneto-rheological actuators employing Preisach model. In: Applied Mechanics and Materials, vol. 889, pp. 355–360. Trans Tech Publ. (2019)
6. Flores-Mendez, J.D., Schiøler, H., Madsen, O., Bai, S.: Impedance control and force estimation of a redundant parallel kinematic manipulator. In: Gusikhin, O., Madani, K. (eds.) ICINCO 2017. LNEE, vol. 495, pp. 174–191. Springer, Cham (2020). https://doi.org/10.1007/978-3-030-11292-9_9

7. Gupta, A., O'Malley, M.K.: Disturbance-observer-based force estimation for haptic feedback. J. Dyn. Syst. Meas. Control **133**(1) (2011)
8. Ismail, M., Ikhouane, F., Rodellar, J.: The hysteresis Bouc-Wen model, a survey. Arch. Comput. Methods Eng. **16**(2), 161–188 (2009)
9. Karbasizadeh, N., Zarei, M., Aflakian, A., Masouleh, M.T., Kalhor, A.: Experimental dynamic identification and model feed-forward control of Novint Falcon haptic device. Mechatronics **51**, 19–30 (2018)
10. Katsura, S., Matsumoto, Y., Ohnishi, K.: Analysis and experimental validation of force bandwidth for force control. IEEE Trans. Ind. Electron. **53**(3), 922–928 (2006)
11. Komada, S., Nomura, K., Ishida, M., Hori, T.: Robust force control based on compensation for parameter variations of dynamic environment. IEEE Trans. Ind. Electron. **40**(1), 89–95 (1993)
12. Komada, S., Ohnishi, K.: Force feedback control of robot manipulator by the acceleration tracing orientation method. IEEE Trans. Ind. Electron. **37**(1), 6–12 (1990)
13. Korayem, M.H., Haghighi, R.: Nonlinear disturbance observer for robot manipulators in 3D space. In: Xiong, C., Huang, Y., Xiong, Y., Liu, H. (eds.) ICIRA 2008. LNCS (LNAI), vol. 5314, pp. 14–23. Springer, Heidelberg (2008). https://doi.org/10.1007/978-3-540-88513-9_3
14. Lacki, M., Rossa, C.: On the feasibility of multi-degree-of-freedom haptic devices using passive actuators. In: 2019 IEEE/RSJ International Conference on Intelligent Robots and Systems (IROS), pp. 7288–7293. IEEE (2019)
15. Lącki, M., Rossa, C.: Design and control of a 3 degree-of-freedom parallel passive haptic device. IEEE Trans. Haptics **13**(4), 720–732 (2020)
16. Merlet, J.P.: Parallel Robots, vol. 128, 2nd edn. Springer, Netherlands (2000). https://doi.org/10.1007/978-94-010-9587-7
17. Mohammadi, A., Marquez, H.J., Tavakoli, M.: Nonlinear disturbance observers: design and applications to Euler Lagrange systems. IEEE Control Syst. Mag. **37**(4), 50–72 (2017)
18. Murakami, T., Yu, F., Ohnishi, K.: Torque sensorless control in multidegree-of-freedom manipulator. IEEE Trans. Ind. Electron. **40**(2), 259–265 (1993)
19. Nikoobin, A., Haghighi, R.: Lyapunov-based nonlinear disturbance observer for serial n-link robot manipulators. J. Intell. Robot. Syst. **55**(2–3), 135–153 (2009)
20. Stakvik, J.Å., Ragazzon, M.R., Eielsen, A.A., Gravdahl, J.T.: On implementation of the Preisach model: Identification and inversion for hysteresis compensation. Model. Identif. Control: Nor. Res. Bull. **36**(3), 133–142 (2015). https://doi.org/10.4173/mic.2015.3.1
21. Stamper, R.E.: A three degree of freedom parallel manipulator with only translational degrees of freedom. Ph.D. thesis, University of Maryland (1997)

Passive Haptic Learning as a Reinforcement Modality for Information

Connor Giam[(✉)], Joseph Kong, and Troy McDaniel

Arizona State University, Mesa, AZ 85212, USA
connorgiam@gmail.com

Abstract. This paper outlines the experimental procedure are results of a project focused on furthering the understanding of Passive Haptic Learning (PHL) for future exploration. PHL is the use of haptic signals to assist in the reinforcement of previously learning information without the need for subjects to actively focus on memorization of content. In this experiment, it was proposed to attempt to utilize PHL while providing subjects as little connecting details as possible between memorization targets and haptic inputs to assess if PHL still applied. This experiment found that without any connecting information the phenomenon known as PHL does not occur and may even be partially detrimental to memory retention based on subject responses.

Keywords: Multimedia and education · Multi-modal integration · Human-machine interaction and human factors

1 Introduction

Passive Haptic Learning (PHL) is the use of haptic signals to assist in the memorization of information over an extended period without the need for subjects to actively focus on attempting to memorize target content. PHL was first conceptualized as a means of improving ability to learn difficult skills like stenography, also known as shorthand, which suffered from extremely high failure rates due to the difficulty of memorization required to be successful. Since then, several experiments have been run to begin to explore the phenomenon of PHL including teaching subjects how to play minor piano songs and teaching the beginnings of Braille. These previous experiments have shown consistently better performance from subjects who underwent PHL than those who did not, suggesting that PHL had a positive correlation with the learning of the presented content.

PHL is a memory reinforcement technique used to improve user retention of knowledge over time by associating the haptic stimuli with a desired piece of information. Despite the word learning being in the name, PHL does not actively teach entirely new information, but instead assists in making sure newly learned or introduced information is better reinforced. Subjects are usually introduced to the information independent of the passive learning methods. This is reflected in the methods of which PHL is applied in most experiments with the process taking form in two phases with an evaluation after each phase to assess user knowledge. The first phase, known as the training phase, is comprised of a directed learning session where subjects are actively taught desired content to attempt to commit to memory. This can be the timings and keys to a piano song, the various patterns and buttons associated with Braille, the necessary associations of randomized keys to a keyboard, or any other intentional input. This phase is followed by an evaluation where subjects are to display their initial understanding of the previously taught content. The results of this evaluation create a form of base line to see what subjects have gained from the initial training to be compared to later. The following phase is the passive training phase, an extended period where subjects are periodically subjected to haptic stimuli associated with the previously learned information. It is to be noted that during this passive training period, subjects are intentionally not to be focused on memorizing the information the passive haptic signals are trying to reinforce. They are instead usually given distractions in the forms of trying to type things, respond to emails, or otherwise be unfocused while the signals passively continue in the background. The final phase of the PHL process is a final evaluation where subjects are again tested like after the initial training to see their retention of the initially presented information.

At the time of this experiment, the limitations of PHL are relatively unexplored as most experiments to date have focused on associating the haptic signals to a linked motor motion usually in the form of pressing a button. The focus of the experiment in this paper is to attempt to apply PHL with as little linking information as possible while still retaining the main phases utilized by other experiments. This is to be done by attempting to have subjects memorize randomized pairs of words with a haptic signal associating each individual word pair, but still utilizing the same initial and passive phases used by other experiments. By taking away associated physical motion and instead attempting to memorize randomized word pairs with the assistance of associated haptic signals, the experiment attempts to strip PHL down to as close to a minimum possible to see if the effect is still present in any capacity with minimal connection of the haptic signals to the target information.

2 Related Works

In previous works, Passive Haptic Learning has been largely focused on assisting individuals in passively learning or memorizing patterns of buttons to press or definitions to specific motor actions. Notably in these experiments was *PianoTouch: A Wearable Haptic Piano Instruction System For Passive Learning of Piano Skills*, an experiment that used PHL to attempt to teach individuals basic songs [1]. In this experiment, subjects were taught how to play two songs with both audio and haptic inputs over the course of about 10–15 min before a longer passive learning session with one song repeated with haptic stimulus and the other with just audio. The results of the study showed individuals made 2.75 to 7.25 less errors in playing the audio and haptic reinforced song versus the solely audio reinforced song during a post training evaluation with a 95% confidence. This indicated a proof of concept that individuals could actively learn through haptic inputs even if their focus was not explicitly devoted to attempting to learn or memorize new content.

A similar result was gained from a later experiment, *Passive Haptic Learning of Braille Typing*, that attempted to teach individuals the basics of Braille through PHL [2]. In this experiment, subjects were asked to perform an initial Braille reading test with before entering a passive learning phase. In this phase, the experimental group would hear audio of words with a haptic stimulus while control groups would have neither input before both groups would conduct a retest on Braille reading. The results of these experiments showed an average increase of 32.85% in Braille writing ability with a 72.5% increase in ability to recognize and read Braille when utilizing PHL techniques. This was significant compared to the control group, which saw a 2.73% decrease in writing ability and 22.4% increase in reading ability. This experiment showed that PHL had potential for more than just memorization, as the Braille reading tests did not have combinations of previously memorized content, but instead independent pieces of information that utilized skills that had been learned over the previous passive sessions.

While many experiments show a bright future for PHL, it is important to be careful of experimental design, as showcased in the study *Reevaluating Passive Haptic Learning of Morse Code* [3]. This study was conducted with the intent to replicate the actions of a previous study that utilized Google Glass vibrotactile actuators and PHL to teach subjects how to understand Morse Code. While the previous study showed extremely high learning ability, this study attempted to recreate the results while excluding any opportunities for accidental active learning by participants. While the results did show that PHL had a minor positive effect, the effect was significantly lower than the recorded impact from the previously questioned experiment. This reevaluation experiment marks an important caution as to the potential limitations of PHL and begins the exploration of where some of the boundaries of PHL may lie.

3 Proposed Method

As previously mentioned, the purpose of this experiment is to attempt to apply PHL with as little connected information as possible to see if the effect still appears. In previous experiments, subjects are placed into an initial active learning session with the intent of being actively exposed and engaged by desired memorization content before undergoing a long, passive training session before a final evaluation. This experiment seeks to conduct a similar method but differs in what information is attempting to be memorized as well as its association with the provided haptic signals. In this experiment, subjects will be asked to memorize randomized, unrelated pairs of words with a unique, but unrelated haptic code assigned to each pair. In previous experiments, most haptic signals were used to directly influence users to associate a piece of information with either an action such as pressing a key/button, or a meaning such as the haptics directly translating to intentional words or definitions. By removing both the relevance of the memorization targets as well as any possible meanings behind the haptic signal's association to the target information, the experiment hopefully seeks to remove all potential associations to see if the PHL phenomenon still occurs in minimalistic conditions.

3.1 Apparatus

The apparatus of this experiment included a singular glove with one vibration motor at each knuckle excluding the thumb totaling 4 vibration motors. The glove selected was a simple cloth glove that provided minimal impedance of motion. To house the motors, additional small pockets were sewn on to the knuckles of each finger to allow for targeted haptic signaling. The vibration motors were mini vibration motors utilized in devices such as mobile phones and tablets. These motors were connected to a micro-controller which was programmed to output both the haptic sequences as well as the associated audio outputs. Audio outputs utilized a pair of over-the-ear headphones. These headphones were non-noise cancelling as to allow subjects to still be able to hear their surroundings during the test and minimize apparatus interference. The microcontroller featured two buttons which acted as the user interface with pre-programmed navigation of various modes that were used throughout the experiment. The entire apparatus together is shown in Fig. 1.

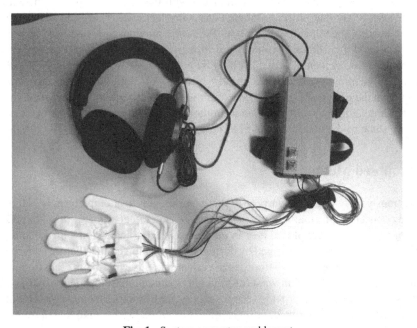

Fig. 1. System apparatus and layout

3.2 Target Information

Experimental subjects were asked to memorize two sets of 10 individual word pairs, with each word pair being given a unique haptic code for that memorization set. The word pairs selected were gathered from a randomized word generator and featured words with exactly two syllables. The reasoning behind this was to minimize the chances that longer or shorter words would inadvertently increase or decrease the difficulty of memorization. Aside from the number of syllables, no other factors were considered in selecting word pairs, and any perceived correlations were not intentionally placed. The associated haptic codes were comprised of 3 digit codes with numbers ranging from 1 to 4. These numbers correlated to individual fingers on the haptic glove and when a specific word pair was played, the associated code would be played in tandem. Vibrations played in this manner were played individually with a duration of 100 ms and a 250 ms delay between each signal. This allowed subjects to clearly identify the code being played without the code taking too long to play. Figures 2 and 3 show the initial information page subjects were given at the start of their initial training sessions.

Group A Word Pairs

Below is a listing of 10 word-pairs. Each word-pair features a pair of two-syllable words with no intentional association. Over the course of initial training, your goal is to memorize these pairs and their associated haptic signal.

At the end of this session there will be an evaluation to see how many you initially remember.

Pair No.	Word 1	Word 2	Haptic Code
1	Jockey	Struggle	124
2	Desert	Coma	231
3	Afford	Statement	143
4	Research	Chicken	341
5	Discount	Morsel	421
6	River	Elapse	243
7	Orange	Aware	321
8	Craftsman	Zero	412
9	Suntan	Entry	134
10	Format	Delay	214

Fig. 2. Group A word pairs

Group B Word Pairs

Below is a listing of 10 word-pairs. Each word-pair features a pair of two-syllable words with no intentional association. Over the course of initial training, your goal is to memorize these pairs and their associated haptic signal.

At the end of this session there will be an evaluation to see how many you initially remember.

Pair No.	Word 1	Word 2	Haptic Code
1	Water	Confine	432
2	Support	Limit	123
3	Transform	Morale	213
4	Sandwich	Captain	234
5	Tribute	Illness	312
6	Letter	Meaning	134
7	Export	Whisper	413
8	Damage	Virtue	324
9	Donor	Begin	341
10	Dairy	Council	421

Fig. 3. Group B word pairs

3.3 Study

The experiment included two main phases with evaluations after each phase. Test subjects shown how to wear and use the apparatus at the start of the experiment and are asked to wear the entire apparatus for the duration of the experiment. Initially, subjects will be given a 15-min training period where they are given the groups of word pairs and haptic sequences they are intended to memorize. During this period, they are also instructed to use the apparatus to assist their memorization, which can provide the associated audio and haptic signals. At the end of this period or when subjects tell a research assistant that they are ready, test subjects are given an initial evaluation. These evaluations feature a series of fill-in-the-blank questions that ask subjects to write the correct alternate word in each pair. Figure 4 shows an example of one of the evaluation papers.

Once the test subject has completed the initial memory evaluation they begin a passive training phase. During this phase, subjects are given a 30 min period where they are instructed to focus their attention on completing a series of puzzles while intermittent signals are played reflecting the word pairs from the earlier training. If subjects are in the experimental group, they will hear the first word from each word pair matched with

Participant ID#: _____

Initial Memory Evaluation A

Below is a listing of 10 word-pairs in a randomized order. Your goal is to fill in the word pairs correctly according to how you remember them. An example is shown below:

Word Pairs: Apple – Blanket, Jingle – Bucket, Wooden – Swiftly

Apple	___Blanket___
___Swiftly___	Wooden
___Bucket___	Jingle

Fill in the blanks for the word pairs below. Note that order of the pairs and order of words in pairs are not consistent.

Research	_____
Discount	_____
_____	Desert
Orange	_____
Jockey	_____
_____	Format
Afford	_____
_____	Suntan
River	_____
_____	Craftsman

Fig. 4. Memory evaluation form example

its associated haptic code while control groups will only hear the audio of the first word. Subjects are instructed to fully focus on the puzzles and ignore the additional signals with the objective being to complete the maximum number of puzzles with options to skip puzzles that prove too difficult. This experiment utilized the puzzle game, *Rush Hour*, a puzzle game where players are to attempt to slide cars in linear slots to allow a red car to leave the board. Figure 5 shows a picture of the game. This game was selected as it required minimal motor skill to play and did not impede audio or haptic inputs while still proving mentally challenging as no subjects managed to complete all the provided puzzles in the allotted time.

After the end of the passive training session subjects are given a follow up evaluation to see how much of the target content they managed to retain. This examination is similar to the initial evaluation with the exact same instructions, but with the positions of each pair shifted so that subjects prove memorization of the actual pairs and not just a pattern.

Fig. 5. Rush hour puzzle game

Once subjects have completed the entire process with their given condition, they then did the entire experiment a second time with an alternate condition of either control or experimental.

At the end of both runs of the experiment, subjects were asked to share their thoughts through a post experiment survey. These questions were mostly rated on a scale of 1 to 5 based on subjective opinion with one free response question at the end for subjects to explain any additional strategies or notes they wanted to share. These questions included subject perception of difficulty in memorizing the words, impressions on performance on the first set, impressions on performance on the second set, scale of how distracted they felt during the passive period, and how disruptive they perceived the device.

4   Results

The results of this study showed a notable lack of impact of PHL on users reflected by both statistical results as well as by user responses. With the addition of PHL on each subject's first run on the experiment, memory retention averaged about 89.6%. On the second run without PHL applied, memory retention averaged 90.0%. This shows a difference of 0.4% of initially learned material retained over the 30-min passive training period. Over the course of the experiment's 8 participants, the addition on PHL had a Cohen's d between −1.020 and 0.404 with a 95% confidence. Because this effect range includes zero, this implies that the true mean difference is likely not impacted by the addition of PHL in this form for memory retention.

4.1 Distraction Task

Over the course of the passive training phases, subjects were asked to attempt multiple puzzles from the boardgame, *Rush Hour*. The success rate of each individual was recorded to see if the addition or exclusion of PHL was having any effect on their overall ability to be successful at solving the puzzles. Puzzles were not repeated between individual runs of the experiment regardless of experimental group, and both times subjects went through the distraction task they were given an equal number of puzzles from each difficulty level ranging from beginner to expert as per provided by the game itself. With PHL applied, subjects averaged 10 puzzles completed over the 30-min period with an average of 1.5 puzzles skipped, while without PHL they averaged 11 puzzles completed with 1.3 puzzles skipped. The difference of 1 additional puzzle completed and 0.2 less puzzles is well within a 95% confidence interval, with the Cohen's d between −1.415 and 0.129 for the completed puzzles and −0.494 and 0.909 for the puzzles skipped. This shows that it is unlikely that the PHL had an impact on subject ability to complete the distraction task.

4.2 User Feedback

User feedback reflected an impression of consistently better performance between the first and second experimental runs. All users reported between a 2 and 3 for perceived difficulty in memorizing the words with an average of 2.6, implying that users did not find the words provided excessively difficult to memorize. However, the average perception of success for the first set of words was 3.5, versus the second run with a perceived success of 4.5. This shows that must subjects believed consistently that they performed better on the second set of words than the first, regardless of PHL inclusion. Subjects varied slightly on perception of how distracted they felt during the passive training sessions but averaged around 3.8 and for the most part felt the experimental apparatus slightly disruptive with an average response of 2.8.

The most important results were the user notes, as most subjects left detailed notes on the various strategies they utilized to remember the word pairs. The vast majority of these responses included notes on how the first attempt at the experiments were not as focused, but in the second attempt they utilized a unique strategy to help them memorize better. Several of these responses included the use of phrases, sentences, or imagined catch phrases with the word pairs in mind or attempting to focus solely on the audio and ignoring the haptic inputs entirely. One subject noted that the addition of trying to memorize the haptic signals along with the word pairs proved more difficult for them than just trying to memorize the words alone. Only one subject noted a positive impression of the haptic patterns helping them remember the target content.

5 Discussion

The results of the experiment show a consistent trend of the addition of PHL having low or negligible effect at improving subject ability to retain information past their own personal strategies. One major difference in the data is in the difference in number of words memorized between the first and second experiments for all subjects. On the first set of words, subjects successfully initially memorized an average of 7.9 of the 10 individual word pairs, while during the second set they memorized 9.8 of the 10 pairs. This is reflected in the post experiment survey data as on average subjects perceived that they did worse on the first set of words than the second. Despite this difference in total memorized content, the amount of information retained between the initial and post evaluations for both the first and second sets of word pairs remained consistent at close to 90%. Conjoined with the subject responses stating that many of them developed specialized memorization techniques through the first set of words and applied them on the second set suggests that the observed difference is likely linked to subjects developing independent strategies rather than the inclusion or exclusion of PHL having an effect.

The PHL applied in this experiment was intended to be used as a memory reinforcement technique for assisting subjects in retaining new information. Unlike other PHL experiments, this experiment sought to completely remove as much connecting information as possible from the nature of the signals to the target information, resulting in the only linking information being the fact that each haptic code of 3 digits is unique to the pair of words attempting to be memorized. Instead of observing PHL benefiting users as an additional modality to remind them during the passive training session, it was instead noted by users that the additional information to memorize proved in some cases more confusing as they not only felt the need to memorize the word pairs, but also having to memorize the associated code to gain any benefits from the haptic inputs. This suggests that for the PHL phenomenon to properly be applied, there must be some intentional connection between the nature of the haptic signals and the intended information the users are attempting to learn. As PHL was applied in this experiment, it had little to no effect on the memorization or retention abilities of any of the test subjects over the course of the experiment regardless of word group or order.

6   Conclusion and Future Works

This experiment sought to attempt to apply PHL with the bare minimum information between the target information and nature of the provided haptic stimulus. By changing the nature of the memorization targets from a motor response to an information pairing and substituting the haptic signals to correlate to a numeric code, the experiment focused on applying PHL purely by using it as an association between one set of information to another. This application proved unsuccessful as the PHL was concluded to not have any effect on subject ability to retain the provided information. This implies that at least some additional information must be provided linking the nature of the haptic signals to the target information or the target information may not be in a form which PHL would have an effect. While the experiment did fail to apply the PHL effect in this manner, it did show a barrier to the technique's versatility and provides a potential start to future experiments that will further the understanding of what the PHL phenomenon is, and which factors are most crucial in applying it.

Moving forward to future experiments, main objectives would be to begin the process of exploring the limits of the PHL phenomenon as well as the metrics necessary to apply it most effectively for future application in education. This experiment proved that it is likely that PHL requires more than just the presence of haptic signals unique to the desired paired information attempting to be memorized. Some of the more major questions revolve around what kinds of information PHL requires to function or what combinations of connecting information are the most impactful. For instance, is it necessary for PHL to require association to a physical motion for it to take effect, or what other factors matter?

One of the comments from the subject feedback mentioned the difficulty of having to memorize both the information pair as well as the associated haptic signal. Unlike previous PHL experiments, where the signals were directly linked to the physical motion of pressing an associated button, this experiment required subjects to also memorize an associated code. It is believed that this may have added to the complexity of the target information, and that if it could be prevented, the experiment may have seen more positive application of PHL. Future experiments should focus on continuing to diagnose the various examples of PHL for trends and attempting to isolate the individual elements that cause it to take effect.

References

1. Huang, K., Do, E.Y.-L., Starner, T.: PianoTouch: A wearable haptic piano instruction system for passive learning of piano skills. In: 2008 12th IEEE International Symposium on Wearable Computers, pp. 41–44. IEEE (2008)
2. Seim, C., Chandler, J., DesPortes, K., Dhingra, S., Park, M., Starner, T.: Passive haptic learning of braille typing. In: Proceedings of the 2014 ACM International Symposium on Wearable Computers, pp. 111–118, ACM (2014)
3. Pescara, E., Polly, T., Schankin, A., Beigl, M.: Reevaluating passive haptic learning of Morse code. In: Proceedings of the 23rd International Symposium on Wearable Computers, pp. 186–194, ACM (2019)

Industrial

Lighting Enhancement Using Self-attention Guided HDR Reconstruction

Shupei Zhang[1]([✉]), Kangkang Hu[2], Zhenkun Zhou[2], and Anup Basu[1]

[1] University of Alberta, Edmonton, AB T6R 2E8, Canada
{shupei2,basu}@ualberta.ca
[2] UAHJIC, University of Alberta, Edmonton, AB T6R 2E8, Canada

Abstract. Computational photography has become an increasingly popular technique for capturing images in high contrast scenes. Current imaging systems solve this problem by capturing a set of images with different exposure settings and then reconstructing a final image. However, this approach cannot solve the problem of revealing or predicting details in already-captured images. Convolutional neural networks (CNNs) can address this problem to some extent, but existing single image lighting enhancement methods based on deep learning suffer from CNNs' limited receptive field and thus cannot yield the optimal results. To overcome this problem, we propose a self-attention based learning strategy inspired by high dynamic range (HDR) reconstruction process to reconstruct a properly exposed image from a single input image. Specifically, we leverage the self-attention mechanism to model the interdependencies between different locations and help reduce the local color artifacts during reconstruction. Furthermore, we adapt the idea of a generative adversarial network (GAN) and design a custom HDR loss function to achieve better image quality. We compare our method with several other recent image enhancement methods using several full-reference and non-reference image quality assessment methods. Experimental results show that our approach can produce images with better details in both over-exposed and under-exposed areas, and thereby outperform existing methods.

Keywords: Attention mechanism · Generative adversarial networks · Image enhancement

1 Introduction

Dynamic range in an image or video is the ratio between the largest and smallest values of luminance. High dynamic range (HDR) can reveal more details, especially in the darkest and brightest areas, than low dynamic range (LDR). HDR content can offer a better viewing experience because it is closer to human perception of the scene. Human eyes can adapt to a wide range of luminance levels by controlling the pupil and having two types of photoreceptors that work

Supported by NSERC and UAHJIC.

in both bright and dark environments. This is a result of adaptation to the large range of illumination values exhibited by natural scenes [11]. However, a large dynamic range causes most conventional imaging systems to get either overexposed or underexposed. Unfortunately, devices that are capable of HDR image and video capture are still rare, not to mention that it is impossible to capture already existing images and videos again. Thus, developing HDR enhancement algorithms for images and videos has become a necessity.

Currently, many researchers are using deep learning to enhance images. An and Lee used a deep convolutional neural network (CNN) to reconstruct the radiance map from raw Bayer images [1]. Marnerides et al. used three CNN branches to process local, large pixel neighborhood and global information [9]. Gabriel et al. proposed a UNet-like CNN for HDR reconstruction [2]. They trained their CNN on a simulated HDR dataset created from a subset of the MIT Places dataset to make the model more generalizable. RetinexNet is a deep learning model with a decomposition net and an enhancement net that is intended for low-light enhancement [15]. Deep SR-ITM is a joint super-resolution and inverse tone-mapping framework that boosts the contrast and details of images [5]. Yifan et al. proposed EnlightenGAN, which does not require paired HDR-LDR images for training [4]. EnlightenGAN also utilizes adversarial learning.

A problem with these approaches is that the sizes of the convolutional kernels are relatively small and not enough to leverage distant or global dependencies in the images. Thus, artifacts usually appear in their reconstructions when there is high contrast or tinted light source in the scene.

In this paper, we propose a new neural network model combining the self-attention mechanism, adversarial training, and customized loss function inspired by the HDR reconstruction process to enhance over or under-exposed images and address the problems mentioned above. We also conduct several ablation tests to demonstrate the effectiveness of our proposed method. Our contributions are listed below:

- This is the first work to utilize self-attention mechanism to model long-distance dependencies across different regions in images for lighting enhancement. This mechanism helps reduce artifacts and boosts the output image quality.
- We design a new HDR loss function inspired by the characteristics of HDR images and the HDR reconstruction process. We show that this loss function can alleviate the color shift/artifacts in the output images.
- We compare our work with several state-of-the-art methods utilizing objective tests to show that our proposed method outperforms all other existing methods.

2   Proposed Method

As shown in Fig. 1, we adapt a UNet-like CNN with a self-attention mechanism as the generator, and a pre-trained ResNet as the discriminator. Several custom loss functions are used together with the discriminator to guide the training process of the generator.

2.1 Generator

The generator has an encoder-decoder structure similar to UNet [13]. Between the second and third upsampling modules we introduce a self-attention module to eliminate local color artifacts.

2.2 Self-attention

Traditional CNNs can only model relatively local features due to their limited reception field. This prevents CNNs from capturing dependencies across the entire image and creates some local color artifacts in our experiments. The self-attention mechanism is one approach to solve this problem [16].

The self-attention module in our model is complementary to the convolution layers and helps capture cues from all positions in the image to reduce artifacts introduced by only using convolutions. This attention module works differently from convolutional kernels. The parameters in convolutional kernels cannot change after the training process, which means that the features they can extract are fixed. However, the attention module computes the attention map based on the input. There is no fixed rule to decide which features are related to another one, so the attention module is more flexible than the convolutional layers.

The input image features x of shape (C, H, W) is first transformed into two feature matrices, $f(x)$ and $g(x)$, to calculate the attention map, where C represents the image feature channels, H and W are feature height and width respectively, and $f(x) = W_f x$, $g(x) = W_g x$. Here, we use a 1×1 convolutional kernel with an output channel of $C/8$ to reduce memory utilization. Thus, $f(x)$, $g(x)$ and $h(x)$ are matrices of size $(C/8, H, W)$. They are then reshaped to size $(C/8, N)$ to convert the 2D features into 1D features, where $N = H \times W$.

The attention map is obtained by multiplying the transpose of $f(x)$ with $g(x)$ and then applying the softmax function to the result:

$$\beta_{j,i} = \frac{\exp(s_{ij})}{\sum_{i=1}^{N} \exp(s_{ij})}, \text{ where } s_{ij} = f(x_i)^T g(x_j). \tag{1}$$

$\beta_{j,i}$ denotes the extent to which the model attends to the ith location when synthesizing the jth region. This is the key to the self-attention mechanism, as it enables the model to find the relationship between any two locations in the entire image. The final output is calculated using the following formula:

$$o = v(h(x)\beta^T), \text{ where } h(x) = W_h x \text{ and} \\ v(x) = \gamma h(x) + x \tag{2}$$

h represents a convolution layer using an 1×1 kernel, and v is a linear transformation with a learnable parameter γ. The input features are added back to the final output.

The self-attention layer is applied after the second upsampling block of the generator.

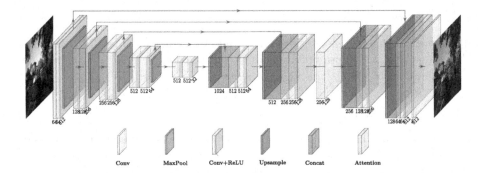

Fig. 1. Generator structure.

2.3   Discriminator

We use a pre-trained ResNet-18 as our discriminator for its simplicity and relatively good performance. It was originally trained for classification, but in the training process it is also trained to classify good and bad HDR reconstructions. The classifier layer in the original structure is replaced with a linear transformation layer: $y = xA^T + b$, where x and y are the input and output of this layer, A is the weight matrix of size $(1, N)$ and b is a scalar bias. N is the channel size of the output from the last average pooling layer. Thus, the final result is a scalar indicating the quality of the HDR reconstruction from the generator. We use the simple minimax GAN loss function:

$$E_x\left[\log\left(D(x)\right)\right] + E_z\left[\log\left(1 - D(G(z))\right)\right], \qquad (3)$$

where $D(x)$ is the discriminator's estimate of the probability that the HDR reconstruction x (x is from the dataset) is good, E_x is the expected value over all good reconstructions, $G(z)$ is the generator's output when the given image is z, $D(G(z))$ is the discriminator's estimate of the probability that a reconstruction from the generator is good, E_z is the expected value over all generated reconstructions. The generator tries to minimize this loss while the discriminator tries to maximize it.

2.4   Custom Loss Functions

We also design some other custom loss functions to help train the network.

Feature Preserving Loss. The image features are similar in the original image and its HDR reconstruction, except that the HDR version will have better contrast and details. Thus, if both images are sent to a classifier network the image features extracted should be similar. In order to preserve image features in our enhancement process, we adopt ResNet to extract image features from the input and output of the generator and build a loss function based on the mean square error between these features.

Only the first six ResNet basic blocks are used in our implementation. Two image feature tensors are obtained by feeding the original image and its HDR counterpart to the partial ResNet. The feature preserving loss is the mean square error between these two tensors:

$$L_{FP}(\boldsymbol{I}, \boldsymbol{I'}) = \frac{1}{N} \sum_{i=1}^{N} \left(\varphi_i(\boldsymbol{I}) - \varphi_i(\boldsymbol{I'}) \right)^2, \tag{4}$$

where I and I' are the original image and the corresponding generator output, N is the number of elements in the image features and φ denotes the partial ResNet feature extractor.

HDR Loss. HDR images usually have better contrast, more saturated colors and correct exposure. Thus, we combine these three metrics to develop a HDR loss function. This loss function helps the generator produce images that have more HDR image characteristics.

Contrast. The contrast of an image is calculated by applying a Laplacian filter to the input. This operation highlights regions with rapid intensity change, which are regions with high contrast levels. By using this metric we ensure that the details in images are revealed.

The Laplacian $L(x, y)$ of an image with pixel intensity $I(x, y)$ is given by:

$$L(x, y) = \frac{\partial^2 I}{\partial x^2} + \frac{\partial^2 I}{\partial y^2} \tag{5}$$

For simplicity we implement this using convolution. The convolutional kernel is a discrete approximation of the Laplacian function:

$$k_L = \begin{bmatrix} 0 & -1 & 0 \\ -1 & 4 & -1 \\ 0 & -1 & 0 \end{bmatrix} \tag{6}$$

The contrast score map S_c is obtained by convolution: $S_c = I * k_L$, where I is the input image and $*$ denotes the convolution operation.

Saturation. Saturation at position (i, j), $S_s(i, j)$, is calculated by converting the image from RGB to HSV color space using the following formulae:

$$\begin{aligned} C_{max}(i, j) &= \max\left(R_{i,j}, G_{i,j}, B_{i,j}\right) \\ C_{min}(i, j) &= \min\left(R_{i,j}, G_{i,j}, B_{i,j}\right) \\ S_s(i, j) &= C_{max}(i, j) - C_{min}(i, j). \end{aligned} \tag{7}$$

Exposure. HDR images need to have correct exposure so that the image is neither too dark or too bight. We use the maximum value among the three color channels as the exposure value:

$$S_e(i, j) = \max\left(R_{i,j}, G_{i,j}, B_{i,j}\right) \tag{8}$$

Finally, the HDR loss is calculated using the following formulae:

$$L_{HDR}(\boldsymbol{I}, \boldsymbol{I'}) = \sum \left[S(\boldsymbol{I}) - S(\boldsymbol{I'}) \right]^2, \text{ where}$$
$$S(\boldsymbol{I}) = S_c(\boldsymbol{I})S_s(\boldsymbol{I})S_e(\boldsymbol{I}) \text{ and} \tag{9}$$
$$S(\boldsymbol{I'}) = S_c(\boldsymbol{I'})S_s(\boldsymbol{I'})S_e(\boldsymbol{I'})$$

The training procedure is described in Algorithm 1. z and y refer to the original images and ground truth HDR reconstructions in the dataset.

3 Experiments

In the training process we use 3340 image pairs from the HDR+ Burst Photography Dataset. We use the remaining 280 non-synthetic images pairs as the test data.

Algorithm 1: Training procedure using GAN and custom loss functions.

---
1. Initialize the parameters in the generator G and the discriminator D;
while *maximum training steps not reached* **do**
  2. Load one batch of training data (z, y) from the dataset;
  Stage 1: Train discriminator
  3. Generate the output $G(z)$ by passing z to the generator;
  4. Get the response for the ground truth and the output from the discriminator, $D(y)$ and $D(G(z))$;
  5. Use gradient descent to maximize $D(y) - D(G(z))$;
  Stage 2: Train generator
  6. Calculate the total loss $L_{total} = L_{FP} + L_{HDR}$;
  7. Use gradient descent to maximize $D(G(z))$ and minimize L_{total};
end

Table 1. Image Quality Comparison. Scores in red indicate the best performance, scores in blue indicate the second-best performance. Metrics used in this table are described in [6,8,10,14]

IQAs	PSNR	SSIM	MS-SSIM	NIQE	HDR-VDP-3	HIGRADE1	HIGRADE2
Input	14.9391	0.4806	0.675	3.7306	6.5258	−1.1556	−1.1911
LIME	14.2289	0.3995	0.6801	4.3187	6.9466	−0.5466	−0.4578
HDR-CNN	14.8821	0.4521	0.6242	2.7261	6.6668	−0.2866	−0.0032
EnlightenGAN	12.5575	0.4258	0.6796	3.079	6.2393	0.1201	0.2906
HDR-ExpandNet	15.4712	0.4573	0.6832	3.2965	6.8556	−1.095	−0.748
RetinexNet	12.203	0.3121	0.5249	7.8457	5.2298	−0.0579	0.2514
Deep-SR-ITM	14.3853	0.4937	0.6724	4.2377	6.7546	−0.0894	−0.0659
Proposed	17.0074	0.4837	0.7011	3.1442	7.2566	0.2741	0.3346

3.1 Ablation Study

We train our model in several different settings using various techniques mentioned in the last section. This helps us understand the function of each part in the model. The settings we use are listed in Table 2. Note that Model 6 is the one we use to report our final performance.

Table 2. Model settings.

Model	GAN	Self-attention	MSE loss	Feature Preserving loss	HDR loss
1			*		
2					*
3	*				
4	*				*
5	*			*	*
6	*	*		*	*

These 6 models are tested using NIQE and two HIGRADE metrics, with the results listed in Table 3. From the scores of Models 1–3, it is clear that when only HDR loss or GAN is used, the model performs worse than the baseline model. But the scores of Model 4 indicate that those techniques work better when combined together. Model 5 improves the NIQE score and HIGRADE1 score to 4.5651 and 0.1154, respectively, with the help of the feature preserving loss. Model 6 increases the performance considerably, indicating that the self-attention module is a key part of the model.

Table 3. Model performance.

Model	NIQE	HIGRADE1	HIGRADE2
1	4.8311	−0.2557	−0.2222
2	5.1346	−0.2615	−0.2946
3	5.0736	0.0428	0.0171
4	4.8205	0.0642	−0.0164
5	4.5651	0.1154	0.0736
6	3.1442	0.2741	0.3346

3.2 Performance Comparison

Comparisons are made between LIME [3], HDR-CNN [2], EnlightenGAN [4], HDR-ExpandNet [9], RetinexNet [7], Deep-SR-ITM [5] and our proposed method. HDR-CNN and HDR-ExpandNet originally produce HDR images with linear luminance levels, so the output of those two methods is tone-mapped using the Reinhard curve [12].

Original	LIME	HDR-CNN	EnlightenGAN	Proposed

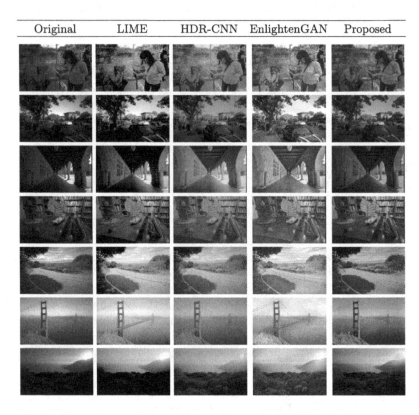

Fig. 2. Comparing performance of various methods. (Color figure online)

Some example output images are shown in Fig. 2. The original images and output images of some compared methods are shown in various columns. RetinexNet's output contain many unnatural colors and noise. The output of Deep-SR-ITM has a strong blue tinge, and HDR-ExpandNet produces some images that look washed out. Compared to LIME, HDR-CNN, EnlightenGAN, and our method, the other methods perform worse in terms of image quality.

For the first two test images, EnlightenGAN seems to be the best, followed by our proposed method. Both methods are able to reveal details in dark areas, while LIME is not able to recover dark areas in the second image. For yellowish images, like Images 3 and 4, LIME, HDR-CNN, and EnlightenGAN appear to ignore this problem and produce more yellow-orangish images, but our method is able to identify the problem and produce more reasonable colors. For images that include large parts of sky in them, e.g., Images 2, 5, and 6, the proposed method appears to be better at recovering the colors of the sky and details on the clouds. Furthermore, for images with large contrast or relatively high brightness, such as Images 6 and 7, LIME introduces some noise into the output, while HDR-CNN's and EnlightenGAN's output contain color artifacts. In comparison, our method can successfully recover details without introducing noise and artifacts.

We use both Full Reference Image Quality Assesements (FR-IQAs) and No Reference Image Quality Assessments (NR-IQAs) to measure the image quality of the above-mentioned algorithms. Note that for NIQE, a lower score is better, while for all other metrics, a higher score is better. The results are listed in Table 1.

For the four FR-IQAs, all the models generate relatively low scores. For example, an image with good quality usually gets a PSNR higher than 40 dB when compared to its reference, but all models score lower than 20 dB. The same situation occurs in evaluations using SSIM, MS-SSIM, and HDR-VDP-3. We believe the reason for this is that the output images are very different from the reference images from the dataset. The reference images are synthesized using multiple images with different exposure settings, but all compared models produce their output based on only one input image. Thus, the reconstructions are very different since less information is available.

However, the results still show that our model can produce the output with the highest quality. Our PSNR score shows that the output image of our model has a 1.6 times higher signal-to-noise ratio than the output of the model in the second place. The proposed method comes in second place in the SSIM test, but only falls short by a small margin. As for MS-SSIM and HDR-VDP-3, the proposed method ranks first and leads by a large margin.

For the NIQE test, the proposed method performs worse than HDR-CNN and EnlightenGAN, but is much better than the other methods. For HIGRADE-1 and HIGRADE-2, our model scores the highest.

In summary, the proposed model performs better than all other methods for 5 of the 7 IQAs we used in the comparisons. This shows that our model can effectively enhance SDR images and attain good image quality.

4   Conclusion

We proposed an encoder-decoder model with adversarial training, self-attention mechanism and customized loss functions to enhance incorrectly exposed images. To the best of our knowledge, we are the first to combine adversarial training with a self-attention mechanism to improve reconstruction quality. This helps our network exploit the location interdependency and reduce artifacts. Objective comparisons between our proposed model and several other state-of-the-art methods were conducted. These comparisons demonstrated that our model performs better in terms of image naturalness and the ability to adapt to different lighting conditions.

References

1. An, V.G., Lee, C.: Single-shot high dynamic range imaging via deep convolutional neural network. In: 2017 Asia-Pacific Signal and Information Processing Association Annual Summit and Conference (APSIPA ASC), pp. 1768–1772, December 2017. https://doi.org/10.1109/APSIPA.2017.8282319
2. Eilertsen, G., Kronander, J., Denes, G., Mantiuk, R.K., Unger, J.: HDR image reconstruction from a single exposure using deep CNNs. ACM Trans. Graph. **36**(6), 178 (2017). https://doi.org/10.1145/3130800.3130816
3. Guo, X., Li, Y., Ling, H.: LIME: low-light image enhancement via illumination map estimation. IEEE Trans. Image Process. **26**(2), 982–993 (2017). https://doi.org/10.1109/TIP.2016.2639450
4. Jiang, Y., et al.: EnlightenGAN: Deep Light Enhancement Without Paired Supervision. arXiv:1906.06972 [cs, eess], June 2019
5. Kim, S.Y., Oh, J., Kim, M.: Deep SR-ITM: joint learning of super-resolution and inverse tone-mapping for 4K UHD HDR applications. In: 2019 IEEE/CVF International Conference on Computer Vision (ICCV), pp. 3116–3125. IEEE, Seoul, Korea (South), October 2019. https://doi.org/10.1109/ICCV.2019.00321
6. Kundu, D., Ghadiyaram, D., Bovik, A.C., Evans, B.L.: No-reference quality assessment of tone-mapped HDR pictures. IEEE Trans. Image Process. **26**(6), 2957–2971 (2017). https://doi.org/10.1109/TIP.2017.2685941
7. Land, E.H.: The Retinex theory of color vision. Sci. Am. **237**(6), 108–129 (1977)
8. Mantiuk, R., Kim, K.J., Rempel, A.G., Heidrich, W.: HDR-VDP-2: a calibrated visual metric for visibility and quality predictions in all luminance conditions. ACM Trans. Graph. **30**(4), 40:1–40:14 (2011). https://doi.org/10.1145/2010324.1964935
9. Marnerides, D., Bashford-Rogers, T., Hatchett, J., Debattista, K.: ExpandNet: a deep convolutional neural network for high dynamic range expansion from low dynamic range content. Comput. Graph. Forum **37**(2), 37–49 (2018). https://doi.org/10.1111/cgf.13340
10. Mittal, A., Soundararajan, R., Bovik, A.C.: Making a "completely blind" image quality analyzer. IEEE Sig. Process. Lett. **20**(3), 209–212 (2013). https://doi.org/10.1109/LSP.2012.2227726
11. Narwaria, M., Perreira Da Silva, M., Le Callet, P.: HDR-VQM: an objective quality measure for high dynamic range video. Sig. Process.: Image Commun. **35**, 46–60 (2015). https://doi.org/10.1016/j.image.2015.04.009
12. Reinhard, E., Stark, M., Shirley, P., Ferwerda, J.: Photographic tone reproduction for digital images. In: Proceedings of the 29th Annual Conference on Computer Graphics and Interactive Techniques, SIGGRAPH 2002, pp. 267–276. Association for Computing Machinery, New York, July 2002. https://doi.org/10.1145/566570.566575
13. Ronneberger, O., Fischer, P., Brox, T.: U-Net: Convolutional Networks for Biomedical Image Segmentation. arXiv:1505.04597 [cs], May 2015
14. Wang, Z., Bovik, A., Sheikh, H., Simoncelli, E.: Image quality assessment: from error visibility to structural similarity. IEEE Trans. Image Process. **13**(4), 600–612 (2004). https://doi.org/10.1109/TIP.2003.819861
15. Wei, C., Wang, W., Yang, W., Liu, J.: Deep Retinex Decomposition for Low-Light Enhancement. arXiv:1808.04560 [cs], August 2018
16. Zhang, H., Goodfellow, I., Metaxas, D., Odena, A.: Self-attention generative adversarial networks. In: International Conference on Machine Learning, pp. 7354–7363. PMLR, May 2019

MoCap Trajectory-Based Animation Synthesis and Perplexity Driven Compression

Guanfang Dong$^{(\boxtimes)}$ and Anup Basu

University of Alberta, Edmonton, Canada
{guanfang,basu}@ualberta.ca

Abstract. We design a 3D Motion Capture Animation Synthesis and Compression pipeline that allows reproducing people's movement in a 3D environment and compressing it at ultra-low bitrates. The method deploys a stage-wise strategy. A surveillance video is used as the input to obtain the movement trajectory by adopting an object detection algorithm. The acquired trajectory is lifted to 3D space by spatial transformations. Based on the reference MoCap animation, an adaptive animation synthesis algorithm processes the input through position and rotation correction; frame interpolation and deletion; trajectory smoothing. Finally, a perceptually adaptive compression algorithm driven by perplexity is applied for MoCap animation compression. Experimental results demonstrate that our animation synthesis method eliminates artifacts, such as foot-sliding, while ensuring accurate position and smooth transition. Our proposed compression algorithm can achieve results that are perceptually similar compared to the PCA-based benchmark, while requiring only about one-tenth of the storage space.

Keywords: Animation · Multi-modal integration

1 Introduction

Motion capture is the process of translating human or object motion into a digital model [1]. Motion capture technology is used extensively in virtual reality, games, animation, and film production [2]. Utilizing recent motion capture equipment for data acquisition, many companies and organizations have created large MoCap databases. For example, in our experiments, we use the CMU motion capture database from Carnegie Mellon University [3]. This open-source database has 2650 different motion capture animations. However, using the existing animations in the database to match motions of characters is challenging. In addition, efficient storing and compiling MoCap data are also important. We propose a stage-wise solution to these two challenges for surveillance videos. This approach not only utilizes and improves state-of-the-art motion detection and trajectory restoration techniques, but it also introduces a new motion capture compression method based on perplexity estimation.

Figure 1 describes our Trajectory-based Animation Synthesis and Perplexity Driven Compression pipeline. The input is a static surveillance video. Static

© The Author(s), under exclusive license to Springer Nature Switzerland AG 2022
S. Berretti and G.-M. Su (Eds.): ICSM 2022, LNCS 13497, pp. 419–429, 2022.
https://doi.org/10.1007/978-3-031-22061-6_31

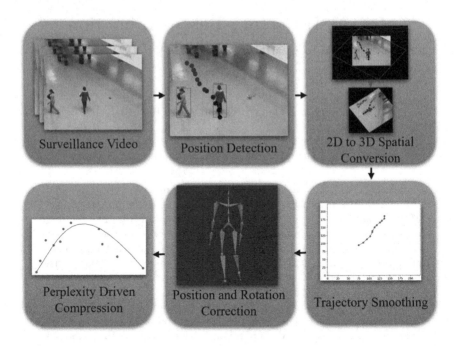

Fig. 1. The pipeline of MoCap animation synthesis and compression.

means the camara is not freely moving. A state-of-the-art learning-based method, PP-YOLO, detects the position of pedestrians at a specific sampling rate. Following this, the camera's viewport is converted to an absolute position in 3D space via four anchor points. The converted result can also be regarded as the top view of the camera's viewport. With the transformation matrix derived from the view transformation, we lift the pedestrian trajectory to the same absolute position in the real 3D space. Based on the existing sampled trajectory, we use Catmull-Rom curve fitting to interpolate the trajectory and smooth the position. According to the above processing, we synthesize the corresponding MoCap animation based on the smoothed pedestrian trajectory. An adaptive animation synthesis algorithm is proposed. The adaptive animation speed control system in the algorithm ensures that artifacts like foot sliding are avoided while the referenced animation is unmodified. Based on the hierarchical relationship of the skeleton in MoCap, we analyzed the error propagation pattern under this system. Finally, a perplexity estimation formula considering the relative motion amplitude, skeleton length, and MoCap hierarchical relationships were designed. This formula simulates the human understanding of perception. We adopt different compression rates based on the perplexity value to eliminate spatio-temporal redundancy.

Although we borrow some methods from previous work in object position detection and trajectory smoothing, our proposed MoCap animation position, rotation self-correction algorithms and the perplexity-based MoCap animation data compression algorithm effectively refine our pipeline. The main contributions of our work are:

1. We build a MoCap animation synthesis algorithm whose input is based on the position of a moving object at fixed time points. This algorithm automatically calculates the ratio of step size to distance. Thus, the algorithm can interpolate frames or remove redundant frames as required. The adaptive algorithm ensures that the results generated are free from undesirable artifacts, such as foot-sliding and misalignment.
2. We design a perceptually adaptive MoCap animation compression algorithm based on joint perplexity. This algorithm analyzes the hierarchical structure of the MoCap skeleton to determine how distance to the *root* affects human perception. This information along with the amplitude of the joint during motion are used to predict joint perplexity. Compression rates are varied across joints based on perplexity estimates. The final result ensures lower impact on human perception for high compression rates.

2 Related Work

An earlier project [4] considered the surveillance video of an Edmonton metro station to reconstruct MoCap data, and also proposed using specialized hardware. They used optical flow and Haar cascade classifiers to detect moving objects, followed by placing MoCap models in desired positions. Their result demonstrate the ability of tracing moving people. However, given the limitation of optical flow and Haar cascade classifiers, their methods have difficulty detecting people with exaggerated or slow movement. Also, occlusion and posters can also decrease the accuracy of moving object tracking. In comparison our proposed methods utilizes state-of-the-art recognition algorithms, which are robust to different types of scenes. Besides that, Cheng et al. proposed a perceptually driven method for compressing MoCap data in crowd scenes [5]. They first cut MoCap data into several segments, and use the wavelet-based SPIHT method to encode MoCap data. They use the lower compression rates for closer models. However, this work only demonstrates compression results from a fixed viewing angle. In our method, the camera can be freely rotated and translated and the compression rate can be adaptively adjusted in real time. Firouzmanesh et al. proposed a wavelet-based perceptually guided compression algorithm [6]. However, their approach ignores the influence of the MoCap hierarchy. Also, their algorithm cannot handle spatial redundancy removal very well.

3 Methodology

Figure 1 shows our pipeline for MoCap Animation synthesis and Compression. In this section, we will elaborate on the details of each part.

3.1 Experimental Data Preparation

The surveillance video used in this experiment is the Atrium video provided by Atousa Torabi and Guillaume-Alexandre-Bilodeau [7]. This video provides a surveillance view inside a building; including movement of multiple pedestrians, reflection of light on the floor, overlapping of pedestrians, and appearance of disturbing objects. We choose this video since these occurrences challenge the robustness of our algorithm.

3.2 Surveillance Video Sampling and Moving Object Position Detection

The frame rate of the experimental video is 10 frames per second, with a resolution of 800×600. This frame rate and resolution is less computationally demanding than a high-resolution, high frame-rate video. However, to further reduce the computational cost, we pre-process the video in two steps. First, we sample the video twice in one second, since human walking speed is only about $1.4\,\mathrm{m/s}$ [8]. Second, even if we use the most advanced deep learning-based methods to detect pedestrians, there is no guarantee that we can accurately identify the pedestrians under each frame. Thus, if one sampled frame is undetected, two adjacent frames (before and after the targeted frame) are used for pedestrian identification. If these two frames are detected, we assume that the previous frame was undetected due to an error. The position of the pedestrian in the previous frame will be obtained from the average of the pedestrian positions in the two current frames.

We apply PaddlePaddle-YOLO (PP-YOLO) to detect the position of pedestrians [9]. After the PP-YOLO detection, the identified pedestrians are marked by the bounding box. Since humans walk upright, we consider the coordinates at the bottom of the bounding box to represent the pedestrian's position in 2d space. In the 3d MoCap space, the y-axis represents the height of the model. Thus, we need to determine the position of the x- and z-axes to match the 2d position and transform the 2d coordinates of the object from the camera's top view. This step would be difficult without the height and view angle of the surveillance camera. Fortunately, our surveillance video has a clear boundary wall on the left. Using this wall, we can roughly outline the capture range of the surveillance video. Then, based on the four anchor points (rectangular floor area), we use the least squares method to find the homogeneous transformation matrix. The least squares method minimizes the back-projection error between two set of anchor points. The back-projection error is given below:

$$\sum_i \left(x_i' - \frac{h_{11}x_i + h_{12}y_i + h_{13}}{h_{31}x_i + h_{32}y_i + h_{33}} \right)^2 + \left(y_i' - \frac{h_{21}x_i + h_{22}y_i + h_{23}}{h_{31}x_i + h_{32}y_i + h_{33}} \right)^2$$

Here, h_{11}, h_{12} etc. belong to a matrix H that calculates the transformation between the source and target planes. x, y belong to the anchor points of original plane and x', y' belong to the anchor points of the target plane.

Figure 2 shows the result of moving object position detection and 2d to 3d view conversion. Now, we standardize the size of 3d space in terms of MoCap units. As the result, the size of the 3d space will be as follows:

$$\text{StandardizedSize} = \left(\frac{(w, h)}{\Delta_v} \right) \times \Delta_m$$

Here, w and h are image width and height. Δ_v is a one step pixel used in the 3d space. Δ_m is the one step distance in the MoCap sample data (derived from the MoCap coordinate system of the referenced animation). This allows us to convert the pedestrian position to a position in the MoCap coordinate system.

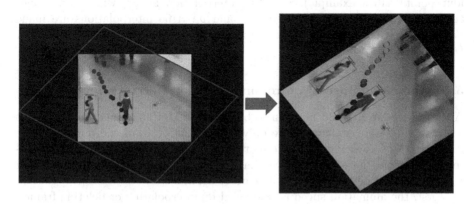

Fig. 2. Position detection and 2d to 3d view conversion.

3.3   Motion Analysis and Trajectory Interpolation

Next, we need to obtain the referenced animations from the CMU MoCap database. At present, we classify a pedestrian's movement into two categories: walking and waiting. To classify trajectory, we first calculate the velocity of a pedestrian in a specific segment. The current pedestrian trajectory can be viewed as a set of positions $\{p_1, p_2, \ldots, p_n\}$ with a sampling rate of twice a second. Thus, for one detected position $p_i = (x_i, y_i)$, the velocity will be $V_i = \frac{((x_i - x_{i+1}) + (y_i - y_{i+1}))^2}{t}$. Here, x_i and y_i belong to the current detected positions and x_{i+1} and y_{i+1} belong to next detected positions. t is the time interval. In the current setting, t is 0.5s since we detect video twice a second. Then, depending on the velocity, we set a threshold to identify the status of the pedestrian as waiting or walking.

The current trajectory is 2 FPS, as we detect it twice per second. However, the animation in the CMU database is 120 FPS. Therefore, we need to interpolate the existing trajectory to match the referenced MoCap animation. Based on the need for trajectory smoothing during interpolation, we apply the Catmull-Rom curve fitting to construct an accurate and smoothed trajectory. In contrast to the Bezier curve [10], the Catmull-Rom curve is guaranteed to pass through all points [11]. The formula for the Catmull-Rom curve is given by:

$$\mathbf{p}(s) = \begin{bmatrix} 1 & u & u^2 & u^3 \end{bmatrix} \begin{bmatrix} 0 & 1 & 0 & 0 \\ -\tau & 0 & \tau & 0 \\ 2\tau & \tau - 3 & 3 - 2\tau & -\tau \\ -\tau & 2 - \tau & \tau - 2 & \tau \end{bmatrix} \begin{bmatrix} \mathbf{p}_{i-2} \\ \mathbf{p}_{i-1} \\ \mathbf{p}_i \\ \mathbf{p}_{i+1} \end{bmatrix}$$

In this equation, τ is the tension factor controlling the degree of bending of the interpolation curve at the control point. We choose 0.5 as the preset tension factor. $p_{i-2}, p_{i-1}, p_i, p_{i+1}$ are the four control points (we take four points as an example). u is the interpolation factor, which determines the interpolated position between p_{i-1} and p_i. If 60 interpolations are made between p_{i-1} and p_i, u will be symmetrically sampled 60 times between 0 and 1 ($u = \{0, 1/59, 2/59, \ldots, 1\}$).

3.4   Animation Speed Control and Rotation Correction

Although we have the 3d position to build a new MoCap animation, the result will contain severe artifacts if we only change the position coordinates of the candidate MoCap animation. These artifacts include foot-sliding and misalignment. To avoid these artifacts, we match the animation speed to the walking velocity and rotate the model to its correct position.

First, the animation speed is controlled by interpolating or deleting frames. As the base animation reference, we first analyze the properties of *walking* animation from the CMU database. The *walking* model takes 1.25 s and 20 MoCap units for one step (units are derived from the MoCap coordinate system). Considering that our sampling FPS is 2, the time between two control points of the trajectory is 0.5 s. Thus, for standardized animation, in 0.5 s the model walks $\frac{20}{1.25} \div 2 = 8$ units. Since the FPS of the CMU MoCap database is 120, a 0.5 s sample requires 60 frames. We regard these information as prior-knowledge. Different animations may have different data. However, our proposed interpolation and deletion method can be extended to other base animations by changing corresponding parameters above. The modification is simply a process of changing the preset values for step length, frame sampling rate and time taken for a step after the initial analysis of the referenced animation.

After the first step, we calculate the distance between the two trajectory control points. Based on the ratio of the distance to the standardized distance, we calculate the number of frames that should be extracted from the referenced animation (*walking* in our case). The frames number can be calculated as $N_f = \mathrm{ceil}\left(\frac{d_p}{d_s} \times N_s\right)$. d_p is the distance between two points. d_s and N_s are standardized distance and standardized number of frames. For *walking* animation, these two values are 8 and 60. The obtained animation segments can be matched with the walking distance between the trajectory control points. We then interpolate or delete frames to reach 60 FPS to ensure the same frame sampling rate as the referenced animation. The above process helps us to raise or lower the stride rate after changing the speed, leading to the elimination of artifacts like sliding feet.

Next, we need to correct the model's direction. The model's rotation for MoCap data along the z-axis is controlled by ry in *root* DOF. To calculate the correct rotation angle, we compare the position between two points in the interpolated trajectory. The angle is $\arctan\left(\frac{y_{i+1}-y_i}{x_{i+1}-x_i}\right)$. We replace the angle value to obtain the correct position. During the actual experiments, we reverse the sign to obtain the true rotational direction when angle closes to the 180 °C (Euler angle) due to the ASF/AMC MoCap coordinate system definition.

The pseudo-code for the MoCap animation synthesis algorithm is given below.

Algorithm 1. MoCap Animation Synthesis Based on Trajectories

Require: Trajectory: $\{x_i, y_i\}$
    Base MoCap Animation References: *Walking, Standing*
Ensure: FPS for *walking* $= 120$
    Approximate seconds for one step $= 1.25$
    Trajectory Sample rate $= 2$ per second
    Calculate the MoCap distance for *Walking* $= 8$
    (All input can vary by different base MoCap Animation)
    while trajectory not finished **do**
        Current and next location $= \{x_i, y_i\}, \{x_{i+1}, y_{i+1}\}$
        Current velocity $V_i = \frac{\left(\left(x_i-x_{i+1}\right)+\left(y_i-y_{i+1}\right)\right)^2}{t}$
        if $V_i <$ velocity threshold **then**
            Set all frames as *Standing* frames
        else
            Calculate the MoCap distance d_p units for two locations
            Slicing the *Walking* frames as $N_f = \text{ceil}\left(\frac{d_p}{d_s} \times N_s\right) \cdot d_p$
            Interpolating or uniformly deleting frames into desired frames ($120/2 = 60$)
        end if
        Set new $\{x, y\}$ batches by Catmull-Rom curve fitting result
        Set new rotation by angle calculated from $\arctan\left(\frac{y_{i+1}-y_i}{x_{i+1}-x_i}\right)$
        Write new frame batches to the new MoCap file
    end while

3.5 Perceptually Adaptive MoCap Compression Algorithm

The previous steps ensure that for a pedestrian trajectory, we synthesize a matching MoCap animation from the referenced animations. However, because MoCap animation sequences are redundant in both the temporal and spatial domains, there is a considerable amount of room for animation sequences to be compressed. Specifically, spatial redundancy refers to the motion of joints that are not independent and can be predicted by other joints. Temporal redundancy indicates that the motion trajectory of a joint can be predicted by a fitted curve. Thus, we design a perceptually adaptive MoCap compression algorithm to remove both redundancies.

Our compression algorithm imitates human perception to adopt different compression rates for different joints. MoCap's skeleton architecture has a rigorous tree-like hierarchy structure. Therefore, when there is an error in the parent joint, the error will be propagated to the child joints at the same time. In addition to this, the length of the bones between the parent and child joints also affects the propagation of errors. The above analysis can yield the compression rate for different joints should match their importance to ensure human perception. A better way to solve this problem is to calibrate the compression rates of different joints by trial and error. This method compares the errors before and after compression to determine the importance of the joints. Eventually, the compression rates for different joints will converge to an equilibrium point balancing the error rate and the compression size. However, this repeats unnecessary compression and decompression steps. This approach is less practical in both compression efficiency and specific scenarios such as real-time transmission. Therefore, for the aspects of MoCap skeletons and animation sequences that have the greatest impact on human perception, we propose a joint perplexity estimation formulation.

Three variables determine the perplexity of a given joint. They are the hierarchical level, bone length, and joint movement amplitude. The perplexity is a value between 0 and 1. We set perplexity of the *root* to 1, which means that *root* has the greatest impact on the perception and should set the lowest compression rate. To calculate joint perplexity, first, we split a MoCap animation sequence into several sequences with an equal number of frames. Usually the segments are between 16 and 64 frames. For each segment, we first calculate the amplitude of motion for the root as $\gamma_{\text{root}} = \sum_{i=1}^{k-1} |\text{sum}(DOF_i - DOF_{i+1})|$. The remaining joints are compared to the root. The perplexity is given by following formula:

$$P_i = \frac{\log(\gamma_i - \gamma_{root} + e) + d_i + (10 - h_i)}{c}$$

Here, P_i is the perplexity for joint i. γ_i is the amplitude. d_i is the bone length. h_i is the hierarchical level. The reason for $(10 - h_i)$ is that there are 10 hierarchical levels for joints in MoCap, the closer the joints to the 10th level, the less perplexity they should be. c is a constant used to control the compression rate. Higher c means higher compression rate.

After calculating the perplexity for each joint, we quantify the joint perplexity into five categories. They are $\{p < 0.6, 0.6 < p < 0.7, 0.7 < p < 0.8, 0.8 < p < 0.9, p > 0.9\}$. In each category, their DOFs (degree of freedom) are extended to form a 2D matrix. Thus, the matrix size is $k \times (\text{concat}(DOF_1, \ldots, DOF_n))$. k is the segment size. n is the total number of DOFs, which is the sum of DOFs for each joint. For this matrix, we apply PCA (Principal Components Analysis) to remove spatial redundancy. For each category, we select different number of components to meet the variance value. In our implementation, the variances are $\{v = 0.5, v = 0.6, v = 0.7, v = 0.8, v = 0.95\}$.

After removing spatial redundancy, we remove the temporal redundancy for the disassembly matrix after PCA downscaling by fitting cubic Bezier curves. For each segment, we use the first, the one third, the two third and the end points as four control points to generate the curve. Then, for each point in a segment, we calculate the $L2$ error. If the error is larger than a threshold, we add this point as a new Bezier curve control point. We repeat the process until the error for every point is less than the threshold. Furthermore, we apply quantization and Huffman coding to further reduce the MoCap file size.

4   Experiment Results

Our experiment synthesize a robust MoCap animation by manipulating an existing referencing MoCap animation. Our new animation follows the trajectories of pedestrians correctly. Besides, MoCap animation has the same speed and step frequency as pedestrians. Artifacts such as sliding-foot, misalignment are mostly eliminated by the animation speed and rotation direction control algorithms.

For the MoCap compression part, our proposed algorithm achieves high compression rates without affecting the main structure of a model. Except for the animations we created, we also test several MoCap animations from the CMU database. To test for the error in the compressed files, we calculate the Euclidean distance between the original and compressed data for each of the five categories ($\{p < 0.6, 0.6 < p < 0.7, 0.7 < p < 0.8, 0.8 < p < 0.9, p > 0.9\}$). Then, since different categories have different number of joints, we standardize errors by the following formula:

$$\text{Error}_{1\sim5} = \frac{(M_c - M_o)^2}{N_{\text{dof}} \times N_{\text{frames}}}$$

Here, M_c and M_o are the compressed and original data. N_{dof} and N_{frames} are the number of dof in the category and number of frames. The compression result are given in Table 1.

By comparing the original file size and the baseline Naive PCA, we can have the following conclusions. Since our generated MoCap animations have more redundant information in the temporal and spatial scales, our compression algorithm can compress 90 times while ensuring almost the same human perceptual quality. For animations with more complex motions, more joints have a higher degree of perplexity due to the greater relative amplitude of motion between them. To ensure consistency in perception, smaller compression rates are adaptively deployed. Thus, for the animation data in the CMU MoCap database, the compression rate of our algorithm is around 40–50 times. Compared to naive PCA, our compression algorithm can achieve almost similar perceptual results with a tenth of the size of the former. This advantage is even more significant for animations with similar motions.

Figure 3 shows the trend of compression errors for five different perplexity categories. We can clearly see that the compression error decreases as the perplexity level increases. The perplexity degree represents the importance of

Table 1. MoCap compression result

Animation	Original size	Compressed size	Naive PCA size	Error 1	Error 2	Error 3	Error 4	Error 5
1.amc*	931 KB	**11 KB**	135 KB	0.0016	0.0007	0.0020	0.0005	0.0008
2.amc*	1863 KB	**22 KB**	203 KB	0.0005	0.0003	0.0004	0.0002	0.0002
3.amc*	2203 KB	**25 KB**	278 KB	0.0013	0.0007	0.0011	0.0007	0.0008
4.amc*	930 KB	**11 KB**	123 KB	0.0011	0.0004	0.0009	0.0004	0.0005
Walk.amc	389 KB	**9 KB**	107 KB	0.0013	0.0005	0.0008	0.0006	0.0005
Dance.amc	818 KB	**17 KB**	144 KB	0.0034	0.0007	0.0015	0.0009	0.0001
Basketball .amc	560 KB	**13 KB**	126 KB	0.0014	0.0016	0.0010	0.0007	0.0006
Martialart.amc	2442 KB	**53 KB**	575 KB	0.0162	0.0048	0.0013	0.0007	0.0014

1. *amc files are generated by our experiments.
2. Error 1–5 corresponds to the errors of five categories.

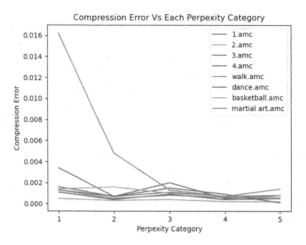

Fig. 3. Line graph of compression errors for different perplexity categories.

joints in human perception. In this way, we can conclude that joints important to human perception adaptively receive smaller compression rates. This also demonstrates the correctness of our compression algorithm.

We provide a video[1] demonstrating the model matching and compression. For the compression part, the right one is the compressed animation.

5   Conclusion

We designed a complete pipeline for surveillance video MoCap animation synthesis and perplexity driven compression. We first detected the positions of pedestrians in a 2D space. Then, we converted the 2D space into a 3D MoCap space. We interpolated and further smoothed the trajectories. The position and rotation were corrected to avoid artifacts, such as foot-sliding and misalignment. Finally, we performed a novel perplexity-based perceptually adaptive compression algorithm to reduce bandwidth requirements. In our experiments, we demonstrated

---

[1] https://youtu.be/saXfRFzYyew.

application and superiority in MoCap animation synthesis and ultra-low bitrate animation compression. Experimental results also showed that our method synthesizes MoCap animations in a manner consistent with human perception while requiring very low storage.

References

1. Browning, R.C., et al.: Effects of obesity and sex on the energetic cost and preferred speed of walking. J. Appl. Physiol. **100**(2), 390–398 (2006)
2. Barry, P.J., Goldman, R.N.: A recursive evaluation algorithm for a class of Catmull-Rom splines. ACM SIGGRAPH Comput. Graph. **22**(4), 199–204 (1988)
3. Cheng, I., Firouzmanesh, A., Basu, A.: Efficient interactive visualization of crowd scenes on mobile devices. In: SIGGRAPH Asia 2014 Mobile Graphics and Interactive Applications, pp. 1–6 (2014)
4. Firouzmanesh, A., Cheng, I., Basu, A.: Perceptually motivated real-time compression of motion data enhanced by incremental encoding and parameter tuning. In: Eurographics (Short Papers) (2013)
5. Gross, R., Shi, J.: The CMU Motion of Body (MOBO) Database. The CMU Motion of Body (MoBo) Database, Carnegie Mellon University, 13 June 2001. https://www.ri.cmu.edu/publications/the-cmu-motion-of-body-mobo-database/
6. Jodoin, J.-P., Bilodeau, G.-A., Saunier, N.: Urban tracker: multiple object tracking in urban mixed traffic. In: IEEE Winter Conference on Applications of Computer Vision. IEEE (2014)
7. Long, X., et al.: PP-YOLO: an effective and efficient implementation of object detector. arXiv preprint arXiv:2007.12099 (2020)
8. Lv, N., et al.: A survey on motion capture data compression algorithm. In: Proceedings of the 2nd International Conference on Big Data Technologies (2019)
9. Noonan, D.P., et al.: A stereoscopic fibroscope for camera motion and 3D depth recovery during minimally invasive surgery. In: 2009 IEEE International Conference on Robotics and Automation. IEEE (2009)
10. Farin, G.: Algorithms for rational Bézier curves. Comput. Aided Des. **15**(2), 73–77 (1983)
11. Tan, S., et al.: Crowd visualization on low bandwidth mobile devices based on video analysis. In: SIGGRAPH ASIA 2016 Mobile Graphics and Interactive Applications, pp. 1–5 (2016)

Hyperspectral Image Denoising Based on Dual Low-Rank Structure Preservation

Mingpei Tang[1], Chengcai Leng[1(✉)], and Irene Cheng[2]

[1] Northwest University, Xi'an 710127, China
tangmingpei@stumail.nwu.edu.cn, ccleng@nwu.edu.cn
[2] University of Alberta, Edmonton T6G 2E8, AB, Canada
locheng@ualberta.ca

Abstract. The research on hyperspectral images (HSIs) has attracted the attention of many scholars in recent years due to its wide applications, e.g., soil and mineral properties analysis. A main challenge in hyperspectral imaging is the various noise disturbances encountered during the image analysis process. Therefore, denoising of HSIs is an initial but essential step to facilitate further analytics steps. In the literature, many denoising methods based on low-rank representations have been applied in the study of HSIs. This paper proposes a structure-preserving method based on dual low-rank, and we demonstrate that our proposed method achieves better denoising results (qualitatively and quantitatively) on both simulated and real datasets, compared to related work.

Keywords: Image processing · Hyperspectral image · Low-rank representation · Structure preservation

1 Introduction

Hyperspectral images (HSIs) are data obtained by imaging spectrometers, which simultaneously contain two-dimensional spatial information and one-dimensional spectral information. HSIs have the characteristics of high resolution, wide spectral range and large amount of data. They are used in many applications, such as medical diagnosis [1], agricultural inspection [2], radar monitoring [3] and so on. Since HSIs are captured by instruments, various noise interferences, such as Gaussian noise, impulse noise, etc., are inevitably incorporated during the imaging process. Therefore, denoising of HSIs is necessary in order not to adversely affect the desired application results. The bands of HSIs have high similarities, so it is natural to apply low-rank representations (LRR) to HSI data. Zhang et al. [4] proposed HSI restoration method based on the low-rank matrix recovery (LRMR), which regards the spectral information of each pixel as a one-dimensional vector. The three-dimensional HSI data is then converted into two-dimensional matrix data, and solved by the "Go Decomposition" (GoDec) algorithm. Based on LRMR, He et al. [5] considered the noise distribution in HSI to be band-dependent and proposed a noise-adjusted iterative low-rank matrix approximation (NAILRMA). However, the denoising methods based on LRR only consider the properties of the HSI spectral dimension and ignore the spatial dimension information.

In recent years, many HSI denoising methods considered both spatial dimension and spectral dimension information, and have achieved good results. Among them, the denoising methods based on total variation (TV) can preserve edge information well, while removing noise. For example, Yuan *et al.* [6] proposed a denoising method that can adaptively adjust the denoising strength of different bands on the one hand, and consider the denoising strength of different attribute regions in the spatial dimension on the other hand. The solution is obtained by using the split Bregman iteration algorithm. Wei *et al.* [7] put the TV regularization term and the nuclear norm into a denoising framework, using the former to explore the spatial structure and the latter to exploit the low-rank properties of the spectrum.

In this paper, we propose a dual low-rank structure-preserving (DLRSP) HSI denoising model, which can simultaneously explore spatial structure and spectral information. The main contributions of this paper are as follows:

(1) Both spectral and spatial similarity are exploited with dual low-rank representation, and the DLRSP model is proposed by combining spatial structural information.
(2) Linearized alternating direction method with adaptive penalty (LADMAP) algorithm is used to solve the proposed DLRSP model, and experiments are conducted on both the simulated and real datasets, to demonstrate the effectiveness of our method.

The rest of this paper is organized as follows. Section 2 reviews the background computation on LRR and dual LRR. Section 3 proposes DLRSP and presents the corresponding optimization algorithm. In Sect. 4, we describe our experimental study and analysis. Section 5 concludes the work.

2   Background

2.1   Low-Rank Representation (LRR)

Suppose n samples x_i form a data matrix $X = [x_1, x_2, \cdots, x_n] \in \mathbb{R}^{m \times n}$, when given a dictionary $A \in \mathbb{R}^{m \times n}$, each sample x_i can be represented as a linear combination of dictionary atoms, that is,

$$x_i = Az_i, (i = 1, \cdots, n) \tag{1}$$

where z_i is the representation of x_i. If $Z = [z_1, z_2, \cdots, z_n] \in \mathbb{R}^{n \times n}$, then there is matrix form:

$$X = AZ. \tag{2}$$

When desiring the learned representation Z is the lowest rank, a low-rank representation model is given as [8]:

$$\min_{Z,E} rank(Z) + \gamma \|E\|_l \text{ s.t.} X = AZ + E \tag{3}$$

where $E \in \mathbb{R}^{m \times n}$ represents noise and error, $\| \cdot \|_l$ is a specific regularization, and γ is a regularization parameter. Since $rank(\cdot)$ is non-convex, the kernel norm $\| \cdot \|_*$ is usually used to approximate $rank(\cdot)$. And when the original data matrix X is selected as the dictionary, i.e., $A = X$, the LRR model is equivalent to:

$$\min_{Z,E} \|Z\|_* + \gamma\|E\|_l \ \text{s.t.} X = XZ + E \tag{4}$$

Compared with the robust principal component analysis (RPCA) [9], LRR is not constrained by the fact that the data comes from a single subspace and is applicable to a wider range.

2.2   Dual Low-Rank Representation (DLRR)

When $X = XZ$, it is easy to understand that Z contains the row space information of the original data matrix X. Similarly, we consider the matrix W to make $X = WX$ true, then W contains the column space information of X. Without considering noise, X can be expressed as [10]:

$$X = XZ + WX \tag{5}$$

Based on the above dual matrix factorization, the following DLRR model is considered:

$$\min_{Z,E} \|Z\|_* + \|W\|_* + \gamma\|E\|_l$$
$$\text{s.t.} X = XZ + WX + E \tag{6}$$

where γ is the regularization parameter used to balance E and the first two low-rank terms.

3   Problem Formulation

3.1   HSI Noise Degradation Model

Suppose $\mathcal{X} \in \mathbb{R}^{n_{row} \times n_{col} \times b}$ is a HSI, where $n = n_{row} \times n_{col}$ represents the number of spatial pixels, and b represents the number of bands. HSI will inevitably be disturbed by noise, such as Gaussian noise, impulse noise, Poisson noise, stripe noise, dead lines etc., Gaussian noise affects all bands of HSI, while the rest of the noise only disturbances a few of bands, so this type of noise is called sparse noise. The observation model of HSI can be written as follows:

$$\mathcal{X} = \mathcal{C} + \mathcal{N} + \mathcal{E} \tag{7}$$

where $\mathcal{C}$, $\mathcal{N}$ and $\mathcal{E} \in \mathbb{R}^{n_{row} \times n_{col} \times b}$ represent the clean HSI, Gaussian noise and sparse noise, respectively. The purpose of HSI denoising is to recover clean $\mathcal{C}$ from the original HSI $\mathcal{X}$.

Let X, C, N and $E \in \mathbb{R}^{n \times b}$ be the Casorati matrix of $\mathcal{X}, \mathcal{C}, \mathcal{N}$ and $\mathcal{E}$, respectively. Then the matrix form of (7) is:

$$X = C + N + E \tag{8}$$

Considering the dual matrix factorization into (8), then we have:

$$X = XZ + WX + N + E \tag{9}$$

where $Z \in \mathbb{R}^{b \times b}$ and $W \in \mathbb{R}^{n \times n}$. Therefore, the purpose of denoising becomes to recover Z and W from X.

3.2 Proposed DLRSP

Consider the structural information of Z and W. In fact, Z and W can be regarded as two coefficient matrices that reconstruct X, and then Z and W are decoupled in the absence of noise, so we have:

$$X = XZ \tag{10}$$

$$X = WX \tag{11}$$

According to [11] and [12], Z and W should have the same structure as coefficient matrices, then substituting $X = Z$ and $X = W$ into (10) and (11) respectively:

$$Z = ZZ = Z^2 \tag{12}$$

$$W = WW = W^2 \tag{13}$$

The above properties are called the structure preservation of Z and W. Based on this, we propose a dual low-rank representation based on structure preservation. The specific model is as follows:

$$
\begin{aligned}
&\min_{Z,W,E} \|Z\|_* + \|W\|_* + \alpha \left\|Z - Z^2\right\|_F^2 + \beta \|W - W^2\|_F^2 + \gamma \|E\|_{2,1} \\
&\text{s.t} \quad X = XZ + WX + E, \quad I_b Z = I_b, \quad I_n W = I_n
\end{aligned}
\tag{14}
$$

We choose $l_{2,1}$-norm to constrain the noise E, because the $l_{2,1}$-norm is more robust. The two constraints $I_b Z = I_b$ and $I_n W = I_n$ are to prevent $Z = \mathbf{0}_{b \times b}$ and $W = \mathbf{0}_{n \times n}$. In addition, in order to make better use of the similarity between adjacent pixels and reduce the complexity of the algorithm, the above optimization problem will be solved in patch units. In particular, let the blocksize be k and the stepsize be s.

3.3 LADMAP Algorithm

In this section, we use the LADMAP algorithm [13] to solve the proposed model. Unlike the alternating direction method of multipliers (ADMM) algorithm, here we do not need to introduce additional auxiliary variables to complete the solution of (14).

434   M. Tang et al.

First we get the Lagrangian function of (14) as follows:

$$\mathcal{L}(Z,W,E,Y,\mu) = \|Z\|_* + \|W\|_* + \alpha \left\|Z - Z^2\right\|_F^2 + \beta \left\|W - W^2\right\|_F^2 + \gamma \|E\|_{2,1}$$
$$+ \frac{\mu}{2}\left\|X - XZ - WZ - E + \frac{Y_1}{\mu}\right\|_F^2 + \frac{\mu}{2}\left\|I_b Z - I_b + \frac{Y_2}{\mu}\right\|_F^2$$
$$+ \frac{\mu}{2}\left\|I_n W - I_n + \frac{Y_3}{\mu}\right\|_F^2$$

(15)

where $Y = [Y_1, Y_2, Y_3]$ are the Lagrange multipliers, and μ is the penalty parameter.

Then we denote the sum of all quadratic terms in the Lagrangian function as $\mathcal{Q}(\cdot)$, called smooth function:

$$\mathcal{Q}(Z,W,E,Y,\mu) = \alpha \left\|Z - Z^2\right\|_F^2 + \beta \left\|W - W^2\right\|_F^2 + \frac{\mu}{2}\left\|X - XZ - WZ - E + \frac{Y_1}{\mu}\right\|_F^2$$
$$+ \frac{\mu}{2}\left\|I_b Z - I_b + \frac{Y_2}{\mu}\right\|_F^2 + \frac{\mu}{2}\left\|I_n W - I_n + \frac{Y_3}{\mu}\right\|_F^2$$

(16)

Using $\nabla_Z \mathcal{Q}(Z)$ and $\nabla_W \mathcal{Q}(W)$ to represent the gradient of the smooth function with respect to Z and W, so we have:

$$\nabla_Z \mathcal{Q}(Z) = -\mu\left(X^\top\left(X - XZ - WX - E + \frac{Y_1}{\mu}\right)\right) + \mu I_b^T\left(I_b Z - I_b + \frac{Y_2}{\mu}\right) + 2\alpha T_Z$$

(17)

$$\nabla_W \mathcal{Q}(W) = -\mu\left(\left(X - XZ - WX - E + \frac{Y_1}{\mu}\right)X^T\right) + \mu I_n^T\left(I_n W - I_n + \frac{Y_3}{\mu}\right) + 2\beta T_W$$

(18)

where $T_Z = Z - Z^T - ZZ^T + Z^2 Z^T - Z^T Z + Z^T Z^2$ and $T_W = W - W^T - WW^T + W^2 W^T - W^T W + W^T W^2$.

Next, we update each variable.

Updating Z: fixing W and E, Z can be updated as follows:

$$Z_{k+1} = \operatorname*{argmin}_{\mathrm{rank}(Z)\leq r_1} \|Z\|_* + \langle\nabla_Z \mathcal{Q}(Z_k), Z - Z_k\rangle + \frac{\eta\mu_k}{2}\|Z - Z_k\|_F^2$$
$$= \operatorname*{argmin}_{\mathrm{rank}(Z)\leq r_1} \|Z\|_* + \frac{\eta\mu_k}{2}\left\|Z - \left(Z_k - \frac{\nabla_Z \mathcal{Q}(Z_k)}{\eta\mu_k}\right)\right\|_F^2$$

(19)

Then, we have:

$$Z_{k+1} = D_{\frac{1}{\eta\mu_k}}(A_Z)$$

(20)

where $A_Z = Z_k - \frac{\nabla_Z \mathcal{Q}(Z_k)}{\eta\mu_k}$ and $D_{\frac{1}{\eta\mu_k}}(\cdot)$ is the singular value operator.

Updating W : similar to Z, we have:

$$W_{k+1} = \operatorname*{argmin}_{\mathrm{rank}(W)\leq r_2} \|W\|_* + \frac{\eta\mu_k}{2}\left\|W - \left(W_k - \frac{\nabla_W \mathcal{Q}(W_k)}{\eta\mu_k}\right)\right\|_F^2$$

(21)

Let $B_W = W_k - \frac{\nabla_W \mathcal{Q}(W_k)}{\eta \mu_k}$, we have:

$$W_{k+1} = D_{\frac{1}{\eta \mu_k}}(B_W) \tag{22}$$

Updating E : fixing Z and W, E can be updated as: footnotesize

$$
\begin{aligned}
E_{k+1} &= \underset{E}{\operatorname{argmin}} \gamma \|E\|_{2,1} + \frac{\mu_k}{2} \left\| X - XZ_k - W_k X - E + \frac{Y_{1k}}{\mu_k} \right\|_F^2 \\
&= \underset{E}{\operatorname{argmin}} \frac{\gamma}{\mu_k} \|E\|_{2,1} + \frac{1}{2} \left\| E - \left(X - XZ_k - W_k X + \frac{Y_{1k}}{\mu_k} \right) \right\|_F^2
\end{aligned}
\tag{23}
$$

Then the $l_{2,1}$ minimization operator can be used to obtain E.

Updating Y : fixing Z, W and E, Lagrange multipliers $Y = [Y_1, Y_2, Y_3]$ can be updated as:

$$
\begin{aligned}
Y_{1k+1} &= Y_{1k} + \mu_k \left(X - XZ_k - W_k X - E_k \right) \\
Y_{2k+1} &= Y_{2k} + \mu_k \left(I_b Z_k - Z_k \right) \\
Y_{3k+1} &= Y_{3k} + \mu_k \left(I_n W_k - W_k \right)
\end{aligned}
\tag{24}
$$

Therefore, we define Algorithm 1.

Algorithm 1. DLRSP

Require: $X \in \mathbb{R}^{n \times b}$, α, β, γ, r_1, r_2, s, k
Ensure: Z, W
    Initialization: $Z = eye(b,b)$, $W = eye(n,n)$, $E = \mathbf{0}$, $Y = \mathbf{0}$, $\mu_0 = 0.01$, $\mu_{max} = 10^6$, $\eta = \|X\|_F^2 + 1$, $\rho = 1.1$, $\varepsilon_1 = 10^{-6}$, $\varepsilon_2 = 10^{-4}$
    Repeat until convergence:
    1.Update Z by (20)
    2.Update W by (22)
    3.Update E by (23)
    4.Update Y by (24)
    5.Update the penalty parameter μ by

$$\mu_{k+1} = \min\left(\rho \mu_k, \mu_{\max} \right)$$

    6.Check the convergence condition:

$$\|X - XZ - WX - E\|_F / \|X\|_F > \varepsilon_1$$

$$\mu_k \max\left(\sqrt{\eta} \|Z_{k+1} - Z_k\|, \sqrt{\eta} \|W_{k+1} - W_k\|, \|E_{k+1} - E_k\| \right) / \|X\|_F > \varepsilon_2$$

---

4 Experimental Studies

In this section, we report the experimental studies on both simulated and real datasets, and compare our results with related work, in order to validate the effectiveness of our proposed method. We select six classical (state-of-the-art) contrast methods: low rank tensor approximation (LRTA) [14], block-matching 4D filtering (BM4D) [15], parallel factor analysis (PARAFAC) [16], LRMR [4], NAILRMA [5] and non-local low-rank factorization (GLF) [17]. Mean peak signal to noise ratio (MPSNR) and mean structural similarity (MSSIM) are used as evaluation metrics. In our experiments, we use the regularization parameters $\alpha = \beta = 0.1$, $\gamma = 0.01$, blocksize $k = 20$, stepsize $s = 10$, rank upper bounds $r_1 = r_2 = 4$.

4.1   Experiments on Simulated Data

In order to demonstrate the robustness of our method, when using simulated data, we add more noises to the Pavia City Center dataset [18] in the experiments. The six noises are described as follows:

Case 1: Add Gaussian noise with zero-mean and variance 0.001 to all bands, and then add impulse noise with noise intensity 0.01.

Case 2: Add the same Gaussian noise and impulse noise as in Case 1 to all the bands, and then randomly select 20% of the bands to add vertical stripe noise, where the number of stripe noises accounts for 20%–30% of the total width of the image.

Case 3: Add the same Gaussian noise and impulse noise as in Case 1 to all the bands, and then randomly select 40% of the bands to add vertical stripe noise, where the number of stripe noises accounts for 40%–50% of the total width of the image.

Case 4: Add Gaussian noise with zero-mean and variance 0.05 to all bands, and then add impulse noise with noise intensity 0.1.

Case 5: Add the same Gaussian noise and impulse noise as in Case 4 to all the bands, and then randomly select 20% of the bands to add vertical stripe noise, where the number of stripe noises accounts for 20%–30% of the total width of the image.

Case 6: Add the same Gaussian noise and impulse noise as in Case 4 to all the bands, and then randomly select 40% of the bands to add vertical stripe noise, where the number of stripe noises accounts for 40%–50% of the total width of the image.

Table 1 shows the denoising results of related work and our DLRSP on the Pavia City Center dataset, where the best result under the current noise case is shown in bold, and the second best is underlined. It can be seen from Table 1 that when the intensity of Gaussian noise and impulse noise is small (Case 1–3), the denoising results of PARAFAC, LRMR, GLF and DLRSP are good, but our DLRSP method is the best. When the intensity of Gaussian noise and impulse noise is relatively large (Case 4–6), except our DLRSP, the other methods have poor denoising results. This shows that regardless of the noise intensities or types, the proposed DLRSP has a good denoising capability on the Pavia City Center dataset.

Table 1. Denoising results on the Pavia City Center dataset.

Noise case	Index	LRTA	BM4D	PARAFAC	LRMR	GLF	DLRSP
Case 1	MPSNR	26.515	24.418	35.069	35.308	39.596	**40.696**
	MSSIM	0.761	0.828	0.928	0.937	0.969	**0.976**
Case 2	MPSNR	25.590	23.833	31.865	34.361	37.586	**40.607**
	MSSIM	0.713	0.779	0.853	0.923	0.953	**0.976**
Case 3	MPSNR	24.271	22.936	29.316	32.352	35.762	**39.535**
	MSSIM	0.642	0.701	0.754	0.883	0.919	**0.970**
Case 4	MPSNR	24.379	24.254	25.074	25.188	26.945	**35.901**
	MSSIM	0.626	0.589	0.669	0.707	0.827	**0.944**
Case 5	MPSNR	23.589	24.295	24.581	25.118	26.943	**35.231**
	MSSIM	0.568	0.586	0.654	0.704	0.825	**0.938**
Case 6	MPSNR	21.744	23.870	23.314	24.865	26.678	**33.165**
	MSSIM	0.453	0.565	0.589	0.693	0.812	**0.893**

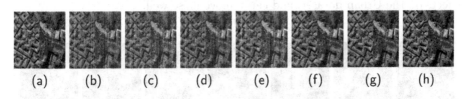

(a)   (b)   (c)   (d)   (e)   (f)   (g)   (h)

Fig. 1. False-color denoising results in Case 2 of the Pavia City Center dataset. (a)Original image (R:90, G:52, B:8). (b) Noisy image. (c) LRTA. (d) BM4D. (e) PARAFAC. (f) LRMR. (g) GLF. (h) DLRSP. (Color figure online)

As an example, Fig. 1 is the denoising results in Case 2, which shows that although the noise intensity is small, there is still obvious noise residual after LRTA, BM4D or PARAFAC denoising. For the LRMR and GLF, the denoised image is cleaner, but there is still a small amount of stripes residual. DLRSP has the best denoising effect: all kinds of noise are removed.

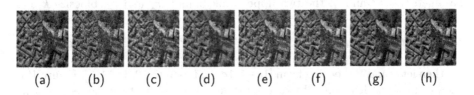

(a)   (b)   (c)   (d)   (e)   (f)   (g)   (h)

Fig. 2. False-color denoising results in Case 4 of the Pavia City Center dataset. (a)Original image (R:96, G:47, B:4). (b) Noisy image. (c) LRTA. (d) BM4D. (e) PARAFAC. (f) LRMR. (g) GLF. (h) DLRSP. (Color figure online)

Another example is illustrated in Fig. 2, which is the denoising results in Case 4. It can be seen from Fig. 2 that when Gaussian noise and impulse noise of larger intensity are added, only the denoised images of GLF and DLRSP have no obvious residual noise. However, the loss of detail information by GLF is noticeable, while DLRSP can remove noise and keep detail information better.

4.2 Experiments on Real Data

In this section, we report experimental results on the Urban dataset, and compare six other methods with our proposed DLRSP. Since the real dataset does not have clean images as ground truth, the denoising effect of each method cannot be judged by the evaluation index used for simulated data. Subjective visual evaluation is used.

Figure 3 shows the denoising results of the 109th band in the Urban dataset [19]. As can be seen from Fig. 3, LRTA and BM4D almost do not remove noise, while the PARAFAC noise removal is not clean, and there are obvious noise residues. The three methods, LRMR, GLF and DLRSP, have similar denoising effects on the Urban dataset, and can remove noise well.

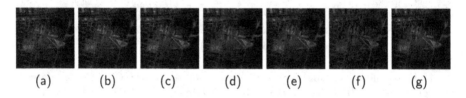

(a)   (b)   (c)   (d)   (e)   (f)   (g)

Fig. 3. Denoising results of the Urban band 109. (a) Original image. (b) LRTA. (c) BM4D. (d) PARAFAC. (e) LRMR. (f) GLF. (g) DLRSP.

4.3 Parametric Value Analysis

In this section, we explain the rationale behind the parameter values used. Taking Case 1 in the simulated data experiments as an example, we perform parametric analysis of each parameter in the proposed DLRSP, namely the regularization parameters α, β and γ, the two upper rank bounds r_1 and r_2, blocsize k and stepsize s.

Regularization Parameters α, β and γ: We first fix the regularization parameter $\gamma = 0.01$, the rank upper bounds $r_1 = r_2 = 4$, blocksize $k = 20$ and stepsize $s = 10$, then let α and β choosing from $[1e-5, 1e-4, 1e-3, 1e-2, 1e-1, 1]$. Figure 4(a) shows the MPSNR results obtained for different α and β. It can be seen that no matter what the values of α and β are, the MPSNR is similar, so in this paper we fix $\alpha = \beta = 0.1$. Then we select γ from $[0.005, 0.0075, 0.01, 0.0125, 0.015]$ on the basis of fixing the rest of the parameters. Figure 4(b) shows the denoising results obtained by different γ. It can be seen that when $\gamma = 0.01$, the MPSNR is the highest, so in this paper we fix $\gamma = 0.01$.

Upper Rank Bounds r_1 and r_2: Fixing the remaining parameters, let r_1 and r_2 be selected from $[1, 2, 3, 4, 5, 6, 7, 8, 9]$, the results are shown in Fig. 4(c) and 4(d). It can be seen from the figure that the MPSNR of r_1 and r_2 tends to be stable after 3, so we fix $r_1 = r_2 = 4$ in this paper.

Blocksize k and Stepsize s: Fixing the remaining parameters, we select blocksize k from $[10, 20, 30, 40, 50]$, and get Fig. 4(e). As can be seen from the Fig. 4(e), when $k = 20$, the MPSNR is the highest, so in this paper we fix $k = 20$. Similarly, selecting s from $[4, 8, 10, 15, 20]$, we get Fig. 4(f). Although the MPNR is the highest when $s = 4$, considering the time cost, we choose $s = 10$ in our experiments.

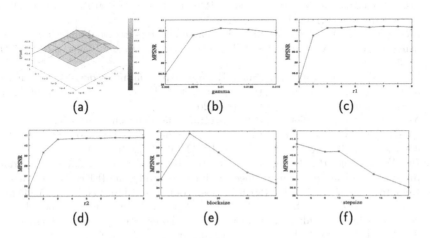

Fig. 4. Display of parameter selection results. (a) Regularization parameters α and β. (b) Regularization parameter γ. (c) Upper rank bound r_1. (d) Upper rank bound r_2. (e) Blocksize k. (f) Stepsize s.

5   Conclusion

In this paper, we propose a HSI denoising method based on dual low-rank structure preservation. The dual low-rank representation is combined with the structural properties of the representation coefficients. Our model not only reflects the similarity between spectra, but also utilizes the spatial result information. To validate the robustness of our method, we conduct experiments on both simulated and real datasets, and compare our results with related work. Experimental results demonstrate that our method outperform others when using both simulated and real datasets.

Acknowledgement. This work is supported by the National Natural Science Foundation of China under Grant No. 61702251, the Natural Science Basic Research Plan in Shaanxi Province of China under Program No. 2018JM6030, in part by the Natural Sciences and Engineering Research Council of Canada, and Youth Academic Talent Support Program of Northwest University under Grant No. 360051900151.

References

1. Fei, B., Lu, G., Halicek, M.T., et al.: Label-free hyperspectral imaging and quantification methods for surgical margin assessment of tissue specimens of cancer patients. In: 2017 39th Annual International Conference of the IEEE Engineering in Medicine and Biology Society (EMBC), pp. 4041–4045 (2017). https://doi.org/10.1109/EMBC.2017.8037743
2. Xiaohe, G., Yansheng, D., Kun, W.: Mapping farmland organic matter using HSI and its effects of land-use types. In: 2012 First International Conference on Agro-Geoinformatics, pp. 1–4 (2012). https://doi.org/10.1109/Agro-Geoinformatics.2012.6311723
3. Gu, Y., Wang, Q.: Discriminative graph-based fusion of HSI and LiDAR data for urban area classification. IEEE Geosci. Remote Sens. Lett. **14**(6), 906–910 (2017). https://doi.org/10.1109/LGRS.2017.2687519
4. Zhang, H., He, W., Zhang, L., Shen, H., Yuan, Q.: Hyperspectral image restoration using low-rank matrix recovery. IEEE Trans. Geosci. Remote Sens. **52**(8), 4729–4743 (2014)
5. He, W., Zhang, H., Zhang, L., Shen, H.: Hyperspectral image denoising via noise-adjusted iterative low-rank matrix approximation. IEEE J. Sel. Topics Appl. Earth Observ. Remote Sens. **8**(6), 3050–3061 (2015)
6. Yuan, Q., Zhang, L., Shen, H.: Hyperspectral image denoising employing a spectral-spatial adaptive total variation model. IEEE Trans. Geosci. Remote Sens. **50**(10), 3660–3677 (2012). https://doi.org/10.1109/TGRS.2012.2185054
7. He, W., Zhang, H., Zhang, L., Shen, H.: Total-variation-regularized low-rank matrix factorization for hyperspectral image restoration. IEEE Trans. Geosci. Remote Sens. **54**(1), 178–188 (2016). https://doi.org/10.1109/TGRS.2015.2452812
8. Liu, G., Lin, Z., Yan, S., Sun, J., Yu, Y., Ma, Y.: Robust recovery of subspace structures by low-rank representation. IEEE Trans. Pattern Anal. Mach. Intell. **35**(1), 171–184 (2013). https://doi.org/10.1109/TPAMI.2012.88
9. Wright, J., Ganesh, A., Rao, S., et al.: Robust principal component analysis: exact recovery of corrupted low-rank matrices via convex optimization. In: Advances in Neural Information Processing Systems, 22 (2009)
10. Liu, G., Yan, S.: Latent low-rank representation for subspace segmentation and feature extraction. In: 2011 International Conference on Computer Vision, pp. 1615–1622 (2011)
11. Roweis, S.T., Saul, L.K.: Nonlinear dimensionality reduction by locally linear embedding. Science **290**(22), 2323–2326 (2000)
12. Qiao, L., Chen, S., Tan, X.: Sparsity preserving projections with applications to face recognition. Pattern Recognit. **43**(1), 331–341 (2010)
13. Lin, Z., Liu, R., Su, Z.: Linearized alternating direction method with adaptive penalty for low-rank representation. In: Proceedings Advances in Neural Information Processing Systems (2011)
14. Renard, N., Bourennane, S., Blanc-Talon, J.: Denoising and dimensionality reduction using multilinear tools for hyperspectral images. IEEE Geosci. Remote Sens. Lett. **5**(2), 138–142 (2008)
15. Maggioni, M., Katkovnik, V., Egiazarian, K., Foi, A.: Nonlocal transform-domain filter for volumetric data denoising and reconstruction. IEEE Trans. Image Process. **22**(1), 119–133 (2013)
16. Liu, X., Bourennane, S., Fossati, C.: Denoising of hyperspectral images using the PARAFAC model and statistical performance analysis. IEEE Trans. Geosci. Remote Sens. **50**(10), 3717–3724 (2012)

17. Zhuang, L., Bioucas-Dias, J.M.: Hyperspectral image denoising based on global and non-local low-rank factorizations. In: IEEE International Conference on Image Processing (ICIP), pp. 1900–1904 (2017)
18. Hyperspectral Images. http://www.ehu.es/ccwintco/index.php/Hyperspectral_Remote_Sensing_Scenes
19. Hyperspectral Images. http://www.tec.army.mil/hypercube

SimFormer: Real-to-Sim Transfer with Recurrent Restoration

Yingnan Ma[✉], Fan Yang, Xudong Li, Chen Jiang, and Anup Basu

Department of Computing Science, University of Alberta, Edmonton, Canada
ma4@ualberta.ca

Abstract. Real-to-Sim transfer is a popular research topic in robotics. Utilizing a simulated environment, the development processes can achieve lower costs and make the testing process easier. In addition, after the Real-to-Sim transfer, the simulated environment can lower the texture effect and light effect, which can be further applied to other computer vision tasks, such as robot grasping. Differing from artistic style transfer, Real-to-Sim transfer has higher accuracy requirements for content preservation. In this paper, we utilize the transformer to solve the Real-to-Sim transfer. We creatively design the restoration stage to preserve the content information. We also propose the restoration loss function. After these improvements, our architecture can achieve better performance on light removal, content preservation, and feature embedding.

1 Introduction

Style transfer has been a popular research topic in the computer vision field. It aims to render photographs with specific given styles. Style transfer can be explored in the [6,7] robotic field. Specifically, it can be applied to Real-to-Sim and Sim-to-Real transfer, which transfer photo-realistic scenes into simulated scenes [18] or vice versa. Differing from artistic style transfer, Real-to-Sim transfer needs special style images like indoor scenes to synthesize simulated scenes. It can also be seen as a domain adaptation problem, where gaps exist between the real and simulated data. We focus on the transfer from real to simulated data since the volume of the realistic content image data enables us to achieve different kinds of transfer as long as the specific style is determined.

In this paper, we propose the SimFormer architecture to solve the Real-to-Sim transfer. We optimize the transformer by recurrent restoration and design the restoration loss function to ensure the learning process. Utilizing the restoration stage, the content features can be better preserved so that the simulated results can be further applied to realistic tasks, such as object tracking [15] and robotic grasping. In the proposed architecture, we utilize SANet [20] as the first-stage backbone to achieve the transformation. The self-attention mechanism can ensure the basic feature embedding between style features and content features.

The original version of this chapter was revised: For all authors, a second, incorrect affiliation had been stated. This has been removed. The correction to this chapter is available at https://doi.org/10.1007/978-3-031-22061-6_35

© The Author(s), under exclusive license to Springer Nature Switzerland AG 2022, corrected publication 2023
S. Berretti and G.-M. Su (Eds.): ICSM 2022, LNCS 13497, pp. 442–452, 2022.
https://doi.org/10.1007/978-3-031-22061-6_33

In the second stage, we creatively involve the restoration process for the transformer to preserve the content information. After this improvement, our architecture can achieve better performance on light removal, content preservation, and feature embedding.

The reminder of this paper is organized as below: The next section reviews state-of-the-art approaches for style transfer. Section 3 introduces our proposed SimFormer architecture and related loss functions. Data arrangement and experimental results are summarized in Sect. 4. We compare the results of the proposed method with five state-of-the-art approaches and evaluate the performance using both qualitative and quantitative measures.

2 Related Work

In robotic researches, learning a simulator from real-world input has attracted attention recently. Some classical methods use a simulator, such as Unity or Maya, which chooses a set of features and characteristics to present the simulation. However, classical methods have many limitations. For example, the represented features require an expert-level designer during the simulation process. In addition, if a parameter of the simulator changes, the whole process needs to be re-designed from the beginning.

In addition to classical solutions, many computer vision methods have also been applied to solve the Real-to-Sim problem. The deep neural network (DNN) was originally used, with a supervised DNN to map features from Domain A to Domain B. Gatys et al. proposed a CNN-based image style transfer framework [1]. It demonstrated that both the content and the style of images could be represented by CNNs independently. Later on, Li et. al. embedded a pair of feature transforms, namely the whitening and coloring transform (WCT), to the image reconstruction network [2]. Huang and Belongie proposed an adaptive instance normalization (AdaIN) layer to align the content features with the style features [3]. However, these methods fail to perform well in detailed texture areas.

In addition to DNN, generative adversarial networks (GANs) [2] was also used in Real-to-Sim as it is simply an unsupervised transformation from Domain A to Domain B, which could be seen as a domain adaptation problem [3,4]. For example, the pixel-level domain adaptation model GraspGAN was proposed to solve the domain adaptation problem and was applied to grasp novel objects from raw RGB images [5]. Similarly, CycleGAN was proposed for unpaired image-to-image translation between domains, which involves two GANs. One is to adapt from the source to the target domain and the other is to adapt from the target to the source [6]. For instance, CycleGAN maps an image from the simulator to a realistic image or from a realistic image to the simulation. However, GANs have many limitations. It is hard to know where it will end up after applying the transformation from Domain A to Domain B. Also, the output from GANs may contain unexpected patterns that might be good for the human eye but will result in unsatisfactory output for classification or [17] detection tasks.

Recently, the transformer was utilized for solving style transfer. The transformer was first proposed to solve natural language processing (NLP) problems like machine translation [8]. With the success of transformers applied in NLP tasks, recently many vision transformers have been developed to solve the problems in computer vision, including object detection and semantic segmentation [9–12]. For the style transfer task, SANet [20] firstly involved the self-attention mechanism to learn the mapping between content features and style features. Afterward, Deng et al. [13] proposed the StyTr^2 architecture, which involves the content-aware positional encoding (CAPE). It split the images into blocks, and utilized a linear projection to obtain image sequences. Later on, Chen extended SANet's architecture by involving contrastive learning and internal-external learning [14] for style transfer. However, these methods still suffer from some artifacts, such as discord patterns [16] and detail loss.

3 Proposed Method

We propose a novel transformer architecture to solve Real-to-Sim transfer. Differing from artistic style transfer, the Real-to-Sim transfer has higher content preservation requirements. Existing style transfer approaches usually produce the halation and curve the edges, which cannot satisfy the requirements for Real-to-Sim transfer. To solve this problems, we propose the SimFormer, which can optimize the Real-to-Sim transfer with recurrent restoration. We also design the restoration loss function to support the restoration stage. The overview of the proposed architecture is shown in Fig. 1.

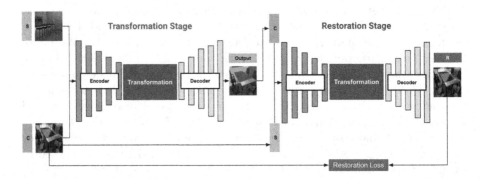

Fig. 1. The overview of the proposed architecture. In the first stage, S represents the simulated images, which provide style information. C represents the real images, which provide the content information. In the second stage, the stylized outputs are utilized to provide content information, and real images in the first stage are utilized to provide style information.

3.1 Network Architecture

As shown in Fig. 1, our proposed architecture involves two stages. The first stage is extended from SANet [20], which is utilized for the transformation from real to simulated images. The second stage is designed for recurrent restoration, which can transfer stylized outputs to real images. In Real-to-Sim transfer, two types of data are involved. The real images are utilized to provide the content information and the simulated images are utilized to provide the style information. For the first stage, the architecture [20] contains an encoder, a transformation module, and a decoder. Encoder E is a pre-trained VGG-19 network [21], which is used to extract image features. VGG-19 contains 5 convolution blocks. For each block, the kernel size is set to 3 × 3 and the kernel number is set as 64, 128, 256, 512, and 512 respectively. Between convolution blocks, max pooling is used to decrease the feature size. With the increase of depth, convolution blocks can obtain high-level features. After the feature maps are obtained from the encoder, the transformation module T is utilized for feature embedding. We employ SANet [20] as the transformation module. SANet is a style attentional network that can flexibly match the style features onto the content features. It takes content feature maps F_c and style feature maps F_s as inputs. It normalizes these feature maps and transforms them into two feature spaces k and g to calculate the attention between each other, as shown below [20].

$$F_{cs}^i = \frac{1}{C(F)} \sum_{\forall j} exp(k(\bar{F}_c^i)^T g(\bar{F}_s^j)) h(F_s^j) \tag{1}$$

With the self-attention mechanism, the SANet module can appropriately embed a style pattern for each position of the content feature maps by mapping correspondence between the content and style feature maps. After the embedded feature map is obtained from the transformer, decoder D is utilized to transform embedded feature maps into stylized images, where D is a reversed VGG-19 network.

After the first stage, the simulated result can be obtained. However, Real-to-Sim transfer needs to preserve the content when removing the textures [19] and light effects. To improve the accuracy of Real-to-Sim transfer, we design the recurrent restoration as the second stage, which is shown in Fig. 1. With the utilization of the restoration process, the proposed architecture can achieve better performance on feature embedding and content preservation. In the first stage, real images C are utilized to provide content information, and simulated images S are used to provide style information. After feature embedding, the stylized results O should share the same content as the original real images and share the same style as the original simulated images. However, due to the limitations of widely-used content loss [3] and style loss, the content feature cannot be preserved well. Thus, we propose recurrent restoration as the second stage. In the second stage, we utilize the stylized results O as content images to provide content information and utilize the real images C in the first stage as style images to provide style information. After the transformation in the second stage, we can transform the stylized results into the real images in the first stage and get the restored images R. In theory, a restored image R should be the same as the original real image C since they share the same content and style information. However, due to the limitations of content preservation, the restored images can be slightly different from the original real images. The difference between R and C is due to the difference in content information. Thus, we propose the restoration loss to concentrate on the difference between R and C. With the same style information, the restoration loss can avoid the obstruction from style features so that our architecture can learn from the difference to achieve better content preservation. With the use of style loss, the stylized results O do not need to be the same as the content images C since the style loss also minimizes the difference of style feature between stylized results O and style images S.

3.2   Loss Function

To ensure the performance of the proposed architecture, we design the restoration loss function $\mathcal{L}_r$ to achieve content preservation. We also involve state-of-the-art loss functions for style transfer to guarantee basic transformation performance. The final loss function $\mathcal{L}$ can be summarized as below, where adaptive hyperparameters are included.

$$\mathcal{L} = \lambda_1 \mathcal{L}_c + \lambda_2 \mathcal{L}_s + \lambda_3 \mathcal{L}_{identity} + \lambda_4 \mathcal{L}_r + \lambda_5 \mathcal{L}_{adv} + \lambda_6 \mathcal{L}_{s-contra} + \lambda_7 \mathcal{L}_{c-contra} \quad (2)$$

We involve the content loss $\mathcal{L}_c$ and style loss $\mathcal{L}_s$. Content loss calculates the difference of content features between the output images I_{cs} and input content images I_c. Style loss calculates the difference of style features between the output images I_{cs} and input style images I_s. The equations [3] are shown below, where ϕ denotes the i_{th} layer of the VGG-19 network. μ denotes the mean value of feature maps and σ represents the standard deviation of feature maps.

$$\mathcal{L}_c = \|\phi_{relu4\_1}(I_{cs}) - \phi_{relu4\_1}(I_c)\|_2 + \|\phi_{relu5\_1}(I_{cs}) - \phi_{relu5\_1}(I_c)\|_2 \quad (3)$$

$$\mathcal{L}_s = \sum_{i=1}^{L} \|\mu(\phi_i(I_{cs})) - \mu(\phi_i(I_s))\|_2 + \|\sigma(\phi_i(I_{cs})) - \sigma(\phi_i(I_s))\|_2 \quad (4)$$

We also involve the identity loss $L_{identity}$ proposed by SANet [20], which can achieve the identity mapping when the content information and the style information are obtained from the same image. Identity loss can improve the preservation of content structure and style characteristics. The equation [20] is shown below.

$$\mathcal{L}_{identity} = \lambda_{identity1}(\|I_{cc} - I_c\|_2 + \|I_{ss} - I_s\|_2)$$
$$+ \lambda_{identity2} \sum_{i=1}^{L} (\|\phi_i(I_{cc}) - \phi_i(I_c)\|_2 + \|\phi_i(I_{ss}) - \phi_i(I_s)\|_2) \quad (5)$$

In addition, we include contrastive content loss $\mathcal{L}_{c-contra}$ and contrastive style loss $\mathcal{L}_{s-contra}$, which are proposed by the IEContraAST [22] architecture. For each content image, it provides two different style images, which can obtain two results $s_i c_j$ and $s_y c_j$ that share the same content with different styles. Similarly, for each style image, it matches two different content images, which can obtain two results $s_i c_j$ and $s_i c_x$ that share the same style with different content. Contrastive loss measures the association among several stylized results to learn the stylization-to-stylization relations.

$$\mathcal{L}_{c-contra} = -log(\frac{exp(l_c(s_i c_j)^T l_c(s_y c_j)/\tau)}{exp(l_c(s_i c_j)^T l_c(s_y c_j)/\tau) + \sum exp(l_c(s_i c_j)^T l_c(s_m c_n)/\tau)}) \quad (6)$$

$$\mathcal{L}_{s-contra} = -log(\frac{exp(l_s(s_i c_j)^T l_s(s_i c_x)/\tau)}{exp(l_s(s_i c_j)^T l_s(s_i c_x)/\tau) + \sum exp(l_s(s_i c_j)^T l_s(s_m c_n)/\tau)}) \quad (7)$$

In addition to contrastive learning [23], we also involve [22] the adversarial loss $\mathcal{L}_{adv}$ to learn human-aware style information with the GAN [24] architecture. We make the transformation stage as the generator $\mathbb{G}$. We send the original simulated images as real data and the stylized images as fake data to the discriminator $\mathbb{D}$. The adversarial loss $\mathcal{L}_{adv}$ is shown below, where E, T, and D represent the encoder, transformation module, and decoder, respectively.

$$\mathcal{L}_{adv} = \mathbb{E}[log(\mathbb{D}(I_s))] + \mathbb{E}[log(1 - \mathbb{D}(D(T(E(I_c), E(I_s)))))] \quad (8)$$

In addition to the above state-of-the-art loss functions, we also design the restoration loss $\mathcal{L}_r$ to achieve better content preservation. The restored loss $\mathcal{L}_r$ can avoid the obstruction from style features since the restored image I_r and content image I_c have almost the same style features. The difference between the restored image I_r and content image I_c is due to the loss of content information. Thus, to calculate the difference of content features between I_r and I_c, we take I_r into the encoder to get the restored feature map and calculate the difference between the restored feature map and the original content feature map. The equation of the restoration loss $\mathcal{L}_r$ is summarized below, where ϕ denotes the layer of VGG-19 network and μ denotes the mean value of feature maps.

$$\mathcal{L}_r = \|\phi_{relu4\_1}(I_r) - \phi_{relu4\_1}(I_c)\|_2 + \|\phi_{relu5\_1}(I_r) - \phi_{relu5\_1}(I_c)\|_2 \quad (9)$$

By utilizing the restoration loss $\mathcal{L}_r$, the restoration stage can learn from the difference between restored images and original content images to achieve better content preservation.

4   Experimental Results

4.1   Dataset

In Real-to-Sim transfer, two types of data are involved. For real images, we take MS-COCO [25] to provide the content information. For simulated images, we create our own dataset to provide simulated style information, which contains 4500 images. The sim data is captured from simulated scenes, which contains less texture and light effect than real images. In this section, all the comparative experiments are trained on the same dataset.

4.2   Evaluation and Comparison

To demonstrate the performance of the proposed architecture, we make comparisons with several state-of-the-art methods, including IEContraAST [22], AdaIN [3], LST [26], SANet [20] and Gatys et al. [1]. We evaluate the Real-to-Sim transfer performance both qualitatively and quantitatively. The quantitative and qualitative results are shown in Table 1 and Fig. 2 separately.

Quantitative Comparisons. For the Real-to-Sim quantitative evaluation, we utilize Learned Perceptual Image Patch Similarity (LPIPS) as the [22] evaluation metric, which is widely used as the evaluation metric for image-to-image translation. To demonstrate the stability and consistency of the transfer task, we expect a lower LPIPS result. We apply the LPIPS metric on every consecutive frame in order to demonstrate the coherence of the video. The LPIPS value for original realistic video data is 0.235. After the Real-to-Sim transfer, the simulated video obtained by IEContraAST can reach 0.200 for the LPIPS value. With the improvement of the proposed restoration process, our proposed architecture can reach 0.195, which improves the stability by 2.5%. The result of LST is a bit higher than IEContraAST, which is 0.201. AdaIN, SANet, and Gatys et al. have worse consistency than the original video. Gatys et al. returns the highest LPIPS value of 0.286. AdaIN and SANet have similar results with a value of 0.273 and 0.276, respectively. Overall, our proposed architecture can achieve better stability and consistency for Real-to-Sim transfer.

Table 1. LPIPS results for different methods.

	Inputs	Ours	IE ContraAST	AdaIN	LST	SANet	Gatys et al.
LPIPS	0.235	**0.195**	0.200	0.273	0.201	0.276	0.286

Qualitative Comparison. In addition to quantitative evaluation, we also apply qualitative evaluation to demonstrate the improvement of our proposed method. As shown in Fig. 2, the first row represents the original real scene. The simulated results from our proposed architecture are shown in the second row. The third row reveals the Real-to-Sim transfer results from the original architecture. The remaining rows show the Real-to-Sim transfer results by other state-of-the-art approaches. From the below comparisons, we can see that our proposed architecture can achieve better performance on feature embedding, light removal, and content preservation. To observe the improvement more clearly, we draw boxes on the results of the proposed architecture and the original IEContraAST architecture. From the regions in green boxes, we can see that the original architecture can remove the light effects slightly. However, our proposed architecture can better remove the light reflection on the desks, which achieves better performance for the removal of light effects. From the pink boxes, we can observe that the original model performs badly for feature embedding. It wrongly embeds

Input

Ours

IEContraAST

AdaIN

LST

SANet

Gatys et al.

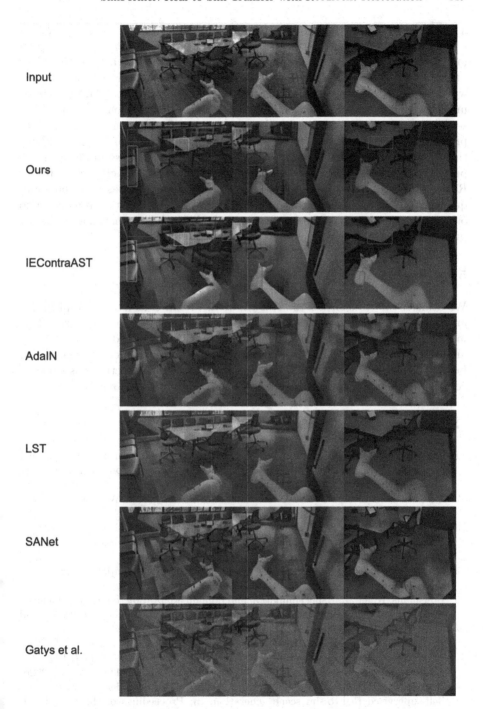

Fig. 2. Qualitative comparisons on Real-to-Sim transfer. The first row shows the input real images. The remaining rows show the results after Real-to-Sim transfer by different methods. The above experiments are based on the same sim image.

features when regions are near the borders and corners. However, our proposed method can avoid these inaccurate embedding. From the red boxes, the original model produces unwanted halation, which makes the regions around the edges blurred. Our proposed architecture can better avoid the halation near edges, which results in better performance for content preservation. AdaIN produces unwanted white patterns near the floor and desks. In addition, it cannot preserve the edges well near the highlight region. LST performs well on content preservation but it can only remove the light effect slightly. SANet strengthens the shadow effect without the removal of the light effect. It also produces unexpected red patterns near the edges and corners. Gatys et al. performs badly for Real-to-Sim transfer with blurred and foggy results. Overall, based on the above qualitative comparison, we can see that our proposed architecture is superior to other state-of-the-art methods on feature embedding, light removal, and content preservation.

5 Conclusion

We utilized the transformer to solve the Real-to-Sim transfer. In the proposed architecture, we proposed a novel restoration stage and designed the restoration loss function. Based on qualitative and quantitative comparisons, our proposed architecture can achieve Real-to-Sim transfer with better light effect removal and better feature embedding performance. Our proposed architecture has superior content preservation performance based on an analysis of the results.

References

1. Gatys, L.A., Ecker, A.S., Bethge, M.: Image style transfer using convolutional neural networks. In: Proceedings of the IEEE Conference on Computer Vision and Pattern Recognition, pp. 2414–2423 (2016)
2. Li, Y., et al.: Universal style transfer via feature transforms. In: Advances in Neural Information Processing Systems, 30 (2017)
3. Huang, X., Belongie, S.: Arbitrary style transfer in real-time with adaptive instance normalization. In: Proceedings of the IEEE International Conference on Computer Vision, pp. 1501–1510 (2017)
4. Csurka, G.: Domain adaptation for visual applications: a comprehensive survey. arXiv preprint arXiv:1702.05374 (2017)
5. Bousmalis, K., et al.: Using simulation and domain adaptation to improve efficiency of deep robotic grasping. In: 2018 IEEE International Conference on Robotics and Automation (ICRA), pp. 4243–4250. IEEE, May 2018
6. Rao, K., Harris, C., Irpan, A., Levine, S., Ibarz, J., Khansari, M.: Rl-CycleGAN: reinforcement learning aware simulation-to-real. In: Proceedings of the IEEE/CVF Conference on Computer Vision and Pattern Recognition, pp. 11157–11166 (2020)
7. Prakash, A., Debnath, S., Lafleche, J.F., Cameracci, E., Birchfield, S., Law, M.T.: Self-supervised real-to-sim scene generation. In: Proceedings of the IEEE/CVF International Conference on Computer Vision, pp. 16044–16054 (2021)
8. Vaswani, A., et al.: Attention is all you need. In: Advances in Neural Information Processing Systems, 30 (2017)

9. Carion, N., Massa, F., Synnaeve, G., Usunier, N., Kirillov, A., Zagoruyko, S.: End-to-end object detection with transformers. In: Vedaldi, A., Bischof, H., Brox, T., Frahm, J.-M. (eds.) ECCV 2020. LNCS, vol. 12346, pp. 213–229. Springer, Cham (2020). https://doi.org/10.1007/978-3-030-58452-8_13

10. Dai, Z., Cai, B., Lin, Y., Chen, J.: UP-DETR: unsupervised pre-training for object detection with transformers. In: Proceedings of the IEEE/CVF Conference on Computer Vision and Pattern Recognition, pp. 1601–1610 (2021)

11. Zheng, S., et al.: Rethinking semantic segmentation from a sequence-to-sequence perspective with transformers. In: Proceedings of the IEEE/CVF Conference on Computer Vision and Pattern Recognition, pp. 6881–6890 (2021)

12. Wang, Y., et al.: End-to-end video instance segmentation with transformers. In: Proceedings of the IEEE/CVF Conference on Computer Vision and Pattern Recognition, pp. 8741–8750 (2021)

13. Deng, Y., Tang, F., Pan, X., Dong, W., Ma, C., Xu, C.: StyTr^2: unbiased image style transfer with transformers. arXiv preprint arXiv:2105.14576 (2021)

14. Chen, H., et al.: Artistic style transfer with internal-external learning and contrastive learning. In: Advances in Neural Information Processing Systems, 34 (2021)

15. Qian, K., Zhou, J., Xiong, F., Zhou, H., Du, J.: Object tracking in hyperspectral videos with convolutional features and kernelized correlation filter. In: Basu, A., Berretti, S. (eds.) ICSM 2018. LNCS, vol. 11010, pp. 308–319. Springer, Cham (2018). https://doi.org/10.1007/978-3-030-04375-9_26

16. Mukherjee, S., Valenzise, G., Cheng, I.: Potential of deep features for opinion-unaware, distortion-unaware, no-reference image quality assessment. In: McDaniel, T., Berretti, S., Curcio, I.D.D., Basu, A. (eds.) ICSM 2019. LNCS, vol. 12015, pp. 87–95. Springer, Cham (2020). https://doi.org/10.1007/978-3-030-54407-2_8

17. Liu, C., Cheng, I., Basu, A.: Synthetic vision assisted real-time runway detection for infrared aerial images. In: Basu, A., Berretti, S. (eds.) ICSM 2018. LNCS, vol. 11010, pp. 274–281. Springer, Cham (2018). https://doi.org/10.1007/978-3-030-04375-9_23

18. Avila, M., Ponce, P., Molina, A., Romo, K.: Simulation framework for load management and behavioral energy efficiency analysis in smart homes. In: McDaniel, T., Berretti, S., Curcio, I.D.D., Basu, A. (eds.) ICSM 2019. LNCS, vol. 12015, pp. 497–508. Springer, Cham (2020). https://doi.org/10.1007/978-3-030-54407-2_42

19. Lugo, G., Hajari, N., Reddy, A., Cheng, I.: Textureless object recognition using an RGB-D sensor. In: McDaniel, T., Berretti, S., Curcio, I.D.D., Basu, A. (eds.) ICSM 2019. LNCS, vol. 12015, pp. 13–27. Springer, Cham (2020). https://doi.org/10.1007/978-3-030-54407-2_2

20. Park, D.Y., Lee, K.H.: Arbitrary style transfer with style-attentional networks. In: Proceedings of the IEEE/CVF Conference on Computer Vision and Pattern Recognition, pp. 5880–5888 (2019)

21. Simonyan, K., Zisserman, A.: Very deep convolutional networks for large-scale image recognition. arXiv preprint arXiv:1409.1556 (2014)

22. Chen, H., et al.: Artistic style transfer with internal-external learning and contrastive learning. In: Advances in Neural Information Processing Systems, 34 (2021)

23. Kang, M., Park, J.: ContraGAN: contrastive learning for conditional image generation. In: Advances in Neural Information Processing Systems, 33, 21357–21369 (2020)

24. Arjovsky, M., Chintala, S., Bottou, L.: Wasserstein generative adversarial networks. In: International conference on machine learning, pp. 214–223. PMLR, July 2017

25. Lin, T.-Y., et al.: Microsoft COCO: common objects in context. In: Fleet, D., Pajdla, T., Schiele, B., Tuytelaars, T. (eds.) ECCV 2014. LNCS, vol. 8693, pp. 740–755. Springer, Cham (2014). https://doi.org/10.1007/978-3-319-10602-1_48
26. Li, X., Liu, S., Kautz, J., Yang, M.H.: Learning linear transformations for fast image and video style transfer. In: Proceedings of the IEEE/CVF Conference on Computer Vision and Pattern Recognition, pp. 3809–3817 (2019)

A Novel Method Based on Spatio-Frequency Analysis for HFSWR Ship Detection

Wandong Zhang$^{(\boxtimes)}$, Q. M. Jonathan Wu, Jiayuan Wang, and Zeng Li

University of Windsor, Windsor, Canada
{zhang11q,jwu,wang621,li1kj}@uwindsor.ca

Abstract. The high-frequency surface wave radar (HFSWR) detects ship targets in the exclusive economic zone (EEZ) effectively. Most of the existing ship target detection algorithms for HFSWR process the spatial domain features. However, the ship target is usually concealed and interfered with various clutters and background noise in the Doppler spectrum. In this paper, an efficient ship target detection approach based on spatio-frequency analysis (SFA) and extreme learning machine (ELM) is proposed. The algorithm subsumes two successive phases: Phase I - ship target coarse detection using discrete wavelet transform (DWT) and Phase II - ship target fine detection via a classifier. Experimental results on a challenging ship target RD image dataset demonstrate the effectiveness and efficiency of the proposed method.

Keywords: High frequency surface wave radar · Range-doppler image · Extreme learning machine

1 Introduction

The HFSWR plays an irreplaceable place in sea surface ship target surveillance and EEZ monitoring [1–4]. It operates in a high-frequency band ranging from 3–30 MHz, where the surface wave propagation mode is capable of beyond-the-horizon ship targets detection [5]. However, the data processing of the HFSWR signal is more complex than the traditional sensors since various interference co-exists with the vessel signal. The ship target detection using HFSWR is therefore a formidable and challenging task. Figure 1 depicts a sample of an HFSWR range-Doppler (RD) image with denoted ship targets, the clutters and background noise.

The traditional ship target detection algorithms for HFSWR can be categorized into two families: the constant false alarm rate (CFAR)-based algorithms and the RD image-based strategies. The CFAR-based algorithms [6–9] discriminate the interference and targets by comparing the power of cell under test (CUT) and the chosen threshold. Concurrently, the RD image processing-based methods [2,10,11] exploit the morphological characteristics including the geometrical feature and intensity distribution to separate targets from strong sea

© The Author(s), under exclusive license to Springer Nature Switzerland AG 2022
S. Berretti and G.-M. Su (Eds.): ICSM 2022, LNCS 13497, pp. 453–461, 2022.
https://doi.org/10.1007/978-3-031-22061-6_34

454   W. Zhang et al.

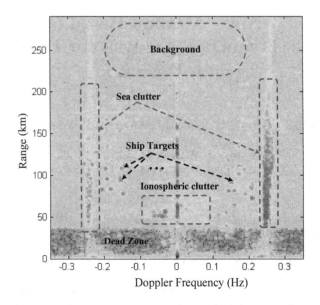

Fig. 1. A sample of an RD image.

clutter and the blind zones. Recently, Zhang et al. [12] proposed an advanced solution using Haar-like features and a simple classifier to detect ship targets in RD images. Despite its high ship target detection performance, it still has several drawbacks explained as follows.

1.1 Non-hierarchical Approach

The proposed ship target detection strategy in [12] utilizes the ELM on the whole RD image. However, this learning algorithm is inefficient as some of the regions in an RD image are relatively easy to be identified. In this paper, we consider that the removal of easily identified regions before having a complicated detection structure can improve the ship target detection performance. For example, one can speed up ship target detection by first removing the background regions from an RD image. Thus, it motivates us to develop a novel coarse-to-fine model that recognizes the potential region of interest in the first stage and localizes the ship targets in the second stage.

1.2 Spatial Information Dependant

The second drawback in [12] is that it only utilizes spatial information to detect ship targets from RD images. However, this is inefficient. For example, from Fig. 1, one can steadily find that the morphological characteristics of ship targets and clutters are significantly different in the frequency domain [2]. The ship targets in RD images can be considered as "peak point" regions, which belong to high-frequency components after having row-wise/column-wise DWT. On the other hand, the sea and ionospheric clutters appear as strips in an RD image. Thus, these components are low-frequency components in the range direction or Doppler frequency direction, respectively.

1.3   Contributions of the Proposed Scheme

To solve the above-mentioned limitations, we propose a novel RD image-based ship target detection algorithm, which is depicted in Fig. 2. To solve the first limitation, we propose a coarse-to-fine ship target detection algorithm. As for the second one, we utilize a spatio-frequency analysis method to identify the potential region of interest. Specifically, the proposed model is composed of two phases: Phase I - Spatio-frequency analysis for ship target coarse detection, which aims to remove the low-frequency components in an RD image, and Phase II - The ELM for ship target fine detection.

2   The Proposed Method

The proposed strategy is a coarse-to-fine method based on spatio-frequency analysis and ELM, which is shown in Fig. 2. It contains two learning phases elaborated as follows.

2.1   Phase I: Spatio-Frequency Analysis for Ship Target Coarse Detection

The main objective of the first phase is to detect the potential region of interest from the raw RD image. From Fig. 1, we can find the following characteristics.

First, the sea clutter has a ridge structure in the range direction, i.e., vertical axis. However, the ionospheric clutter appears as a strip in the Doppler shift along the horizontal axis. On the other hand, due to the local dominant amplitude and the effect of the window function in the processing, one can find that the movement of the ship target in a period always results in presenting strong isolated peaks in several adjacent cells [13]. Thus, the morphological characteristics of ship targets and clutters are significantly different. From the view of image processing, the ship targets can be considered as high-frequency components in the frequency domain; The sea clutter and ionospheric clutter are low varying information in the row-wise direction and column-wise direction, respectively. Thus, these two types of clutters are the low-frequency components in the range direction and Doppler frequency direction. In other words, the grayscale of background regions and ship targets have huge differences in the spatial domain. In this paper, the one-dimensional DWT is utilized to capture the frequency domain details.

Second, Fig. 2 indicates that the energy level of the ship targets is of a remarkable higher value compared to that of the background in the RD images. The background is the superposition of external electromagnetic noise and cosmic noise exhibiting less intensity in RD images, while the ship targets are strong

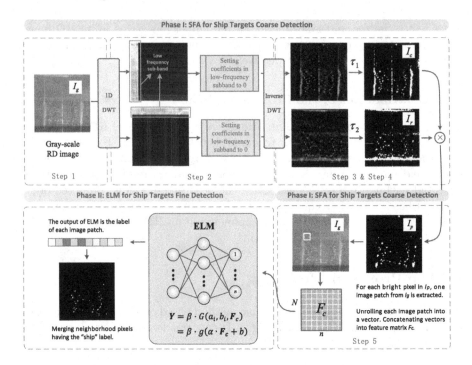

Fig. 2. The learning diagram of the proposed ship target detection algorithm. It subsumes two phases. Phase I is the SFA for ships coarse detection; Phase II is the ELM for ships fine detection. Here, $\otimes$ refers to element-wise product.

peak points in the RD images. Therefore, the ship targets are always present as isolated points in the spatial domain, whereas the background noise exhibits as relatively flatter areas, i.e., there will be no dramatic change in the gray level intensity map (I_g). Thus, this paper uses gray-level values to differentiate ship targets from the background.

The steps of the proposed ship target coarse detection method are shown as follows.

Step 1 - Converting the raw RD image into the gray-scale image I_g.

Step 2 - Performing row-wise/column-wise one-dimensional DWT on the gray-scale image I_g, respectively. For each transformation, one can get one low-frequency sub-band $[L_1]$, and n high-frequency sub-bands denoted as $[H_1, H_2, ..., H_n]$. Setting all the coefficients in the low-frequency sub-band to 0.

Step 3 - Performing the one-dimensional inverse DWT on the frequency domain images. Reconstructing the image only with the low- and high-frequency sub-bands: $[L_1, H_1, H_2, ..., H_n]$ in row-wise and column-wise directions, respectively. Here, the reconstructed row-wise and column-wise images are denoted as I_r and I_c, respectively. Performing the dynamic threshold optimization strategy [12] to find the optimal threshold τ in I_r and I_c:

$$\mathrm{Tw}(c) = \begin{cases} 1, & c \geq \tau \\ 0, & c < \tau \end{cases} \tag{1}$$

where c is the coefficients of the RD image after reconstruction, and Tw is the generated binary image.

Step 4 - Aggregating the reconstructed Images I_r and I_c to generate the potential region of interest:

$$I_p = I_r \otimes I_c, \tag{2}$$

where $\otimes$ denotes the element-wise product. By doing so, only the regions that are kept in both directions (I_r and I_c) will be denoted as bright pixels in the image I_p.

Step 5 - For each pixel in the potential region of interest (bright pixels), one 15×15 image patch from I_g is extracted. Unrolling each image patch and generating an 255-dimensional vector (c.f. $n = 225$ in Fig. 2). Concatenating the vectors into a matrix $\boldsymbol{F}_c$.

2.2   Phase II: ELM for Ship Target Fine Detection

Phase I removes most of the ionospheric and sea clutters and the background from the raw RD image. However, the detected results still contain different interferences. The second phase of the proposed method is to detect ship targets precisely. In this paper, we apply the ELM [14] to assign each pixel with one of the following labels, ship targets, ionospheric clutter, sea clutter, and the others. The ELM is a single-layer neural network, having the advantages of fast training speed and high classification accuracy. Essentially, the hidden later parameters, both hidden layer weights α and biases b are randomly assigned to avoid overfitting. Oppositely, the output layer weights β are optimized by the least-squares learning method, such as the Moore-Penrose inverse.

$$\beta = (\Psi^T \Psi + \frac{I}{C})^{-1} \Psi^T \boldsymbol{T}, \tag{3}$$

where I is the identity matrix, C is the offset term, $\Psi = \sigma(\boldsymbol{F}_c \cdot \alpha + b)$ is the hidden layer feature, and $\sigma(\cdot)$ is the sigmoid function. Then, in the testing stage when detecting ship targets from a new RD image, the following equation is used to calculate the label of each bright pixel in I_p.

$$\boldsymbol{Y} = \sigma(\boldsymbol{F}_c^{test} \cdot \alpha + b) \cdot \beta. \tag{4}$$

where $\boldsymbol{F}_c^{test}$ is the input feature of Phase II in the testing stage. After generating the label for each pixel in I_p, we need to show the detected ship targets on the original RD images. Here, the ship targets are obtained by merging neighbourhood pixels having the "ship target" class. In this paper, we use the bright region to show the detected ship targets in one RD image (c.f. Fig. 3 (d)).

3   Experimental Results

3.1   Experimental Setup

Dataset. In this paper, the HFSWR-RD dataset [12] is utilized to evaluate the performance of the proposed method. The specifications of HFSWR of this dataset are as follows. The carrier frequency is 4.7 MHz, the pulse repetition frequency is 2.3 kHz, the frequency modulation bandwidth is 30kHz, the coherent integration time is 144 ms, the pulse repetition period is 0.42 ms, the frequency modulation period is 128 m, the number of receive channels is 8, and the physical spacing between the channels is 14 m. Specifically, the HFSWR-RD dataset is composed of 200 raw RD images, with 100 RD images used for training and the rest for testing. For the choice of the threshold τ_1 and τ_2, 12,451 image segments are extracted from the gray-scale RD images. Here, each segment is labelled with one of the following two categories, potential region of interest or background. For the learning of the ELM classifier, we follow the work [12] and use 6,958 image patches to learn the network. The image patches have the patch size of 15 ×15, which comprise four classes: the ship target, sea clutter, ionospheric clutter, and the others.

Rival Methods. This paper verifies the effectiveness of the proposed strategy with the following algorithms. The CFAR-based method [15], the wavelet-based algorithm [2], the regression-based model [12], the multilayer subnet-based network [16] and width-growth model [17].

Evaluation Metric. The following four evaluation metrics are used in this paper, including the true positive rate (TPR), false discovery rate (FDR), false negative rate (FNR), and error ratio (ERO).

$$TPR = \frac{TP}{(TP + FN)}, \quad FDR = \frac{FP}{(FP + TP)},$$
$$FNR = 1 - V_d, \text{ and } ERO = FDR + FNR, \tag{5}$$

where TP, FN, and FP refer to the number of true positives, false negatives, and false positives, respectively.

3.2   Quantitative Analysis

Effectiveness of Spatio-Frequency Analysis. This paper utilizes a spatio-frequency strategy to detect the potential region of interest, i.e., I_p in Fig. 2. Here, experiments are conducted with various potential domains ranging from purely reconstructed row/column-wise RD images (I_r or I_c) and the combined domain (I_p). It is worth noting that in these experiments, the learning steps of Phase II and the number of training patterns are the same for a fair comparison. Table 1 compares the FDR and the inference time of the above-mentioned

Table 1. Performance of the ship target detection with various feature sources

Combination	FDR (%)	Inference time (s)
I_r	7.5	5.9
I_c	6.8	5.7
$I_r \otimes I_c$ (Ours)	**4.6**	**3.8**

Table 2. Performance of the proposed algorithm with the existing ship target detection methods

Method	TPR (%)	FDR (%)	FNR (%)	ERO (%)
Existing methods				
CFAR-based method [15]	85.4	13.4	14.6	28.0
Wavelet-based method [2]	90.3	8.3	9.7	18.0
Regression-based method [12]	91.3	5.9	8.7	14.6
Subnet-based method [16]	92.5	5.1	7.5	12.6
Width-growth method [17]	92.9	5.0	7.1	12.1
The proposed method				
Ours	**93.9**	**4.6**	**6.1**	**10.7**

domains. Here, the best performance among all the strategies is highlighted in boldface. Through analysis, we found that the proposed spatio-frequency analysis obtains the lowest FDR score and inference time, which are 2.9% and 2.1 s lower or longer than the analysis on I_r, respectively. Hence, the effectiveness of the proposed spatio-frequency strategy is verified.

Effectiveness of the Proposed Method. This section evaluates the performance of the proposed method with the existing ship target detection method on the HFSWR-RD dataset. Table 2 tabulates the overall performance of these methods with respect to the TPR, FDR, FNR and ERO. Experimental results show that the proposed ship target detection algorithm provides superior performance to the comparison algorithms from 1% to 8.5% of improvements in terms of the TPR. Additionally, the proposed method scores 4.6% in FDR, which achieves the lowest value when compared to the existing ship target detection strategies. However, the other methods in the literature record a worse FDR rate, ranging from 5.0% to 13.4%.

The average inference time of the proposed strategy and the other state-of-the-art HFSWR ship detection methods, including wavelet-based [2], regression-based [12] and subnet-based [16] algorithms are tabulated in Table 3. We conclude that our proposed strategy requires about half of the inference time as that of the subnet-based method [16], which is a similar processing time compared to the regression-based model [12]. The reason for this is that the proposed algorithm utilizes a spatio-frequency analysis to search for the potential region of

Table 3. Inference time comparison of the proposed method with rival methods

Methods	Inference time (s)
Wavelet-based method [2]	4.6
Regression-based method [12]	3.9
Subnet-based method [16]	8.7
Ours	**3.8**

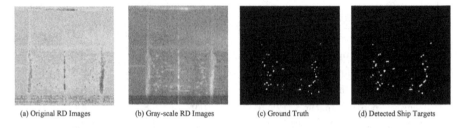

(a) Original RD Images   (b) Gray-scale RD Images   (c) Ground Truth   (d) Detected Ship Targets

Fig. 3. Visualization of samples of detection results by the proposed method

interest, which removes most of the background regions before the ELM-based fine detection. Hence, the proposed algorithm can efficiently detect ship targets from the RD images.

3.3 Qualitative Analysis

Figure 3 visualizes ship target detection results of the proposed strategy. Specifically, Fig. 3 (a) shows one original RD image, which contains ship targets, sea clutter, ionospheric clutter and background. Figure 3 (b) and (c) show the grayscale RD image and the ground truth of this RD image, while Fig. 3 (d) visualizes the detected ship targets (bright pixels in these two images) by the proposed algorithm. These figures' results indicate that the proposed method can detect the ship target very tightly to the human-annotated ground truth.

4   Conclusion

This paper proposes a novel ship target detection pipeline for high-frequency surface wave radar range-Doppler image. In particular, it is a coarse-to-fine detection framework based on spatio-frequency feature analysis and the extreme learning machine. The proposed method contains two phases: Phase I - ship target coarse detection using a discrete wavelet transform-based spatio-frequency analysis method, and Phase II - ship target fine detection based on an extreme learning machine. Exhaustive experimental results indicate the superiority of the proposed model over the existing ship target detection methods.

References

1. Zhang, W., Wu, Q.J., Yang, Y., Akilan, T., Li, M.: HKPM: a hierarchical key-area perception model for HFSWR maritime surveillance. IEEE Trans. Geosci. Remote Sens. **60**, 1–13 (2021)
2. Li, Q., Zhang, W., Li, M., Niu, J., Wu, Q.J.: Automatic detection of ship targets based on wavelet transform for HF surface wavelet radar. IEEE Geosci. Remote Sens. Lett. **14**(5), 714–718 (2017)
3. Tajdini, M.M., Gonzalez-Valdes, B., Martinez-Lorenzo, J.A., Morgenthaler, A.W., Rappaport, C.M.: Real-time modeling of forward-looking synthetic aperture ground penetrating radar scattering from rough terrain. IEEE Trans. Geosci. Remote Sens. **57**(5), 2754–2765 (2018)
4. Zhang, W., et al.: Fast ship detection with spatial-frequency analysis and ANOVA-based feature fusion. IEEE Geosci. Remote Sens. Lett. **19**, 1–5 (2021)
5. Sun, H.: Conceptual study on bistatic shipborne high frequency surface wave radar. IEEE Aerosp. Electron. Syst. Mag. **33**(3), 4–13 (2018)
6. Jalil, A., Yousaf, H., Baig, M.I.: Analysis of CFAR techniques. In: 2016 13th International Bhurban Conference on Applied Sciences and Technology (IBCAST), pp. 654–659. IEEE (2016)
7. Finn, H.: Adaptive detection mode with threshold control as a function of spatially sampled clutter-level estimates. RCA Rev. **29**, 414–465 (1968)
8. Liang, J.: Target CFAR detection method and software implementation with two-dimension data for HFSWR Qingdao: Ocean University of China (2014)
9. Gui, R.: Detecting target located in nonstationary background based on two-dimensions constant false alarm rate. Geomat. Inf. Sci. Wuhan Univ. **37**(3), 354–357 (2012)
10. Chen, Z., He, C., Zhao, C., Xie, F.: Enhanced target detection for HFSWR by 2-D music based on sparse recovery. IEEE Geosci. Remote Sens. Lett. **14**(11), 1983–1987 (2017)
11. Grosdidier, S., Baussard, A.: Ship detection based on morphological component analysis of high-frequency surface wave radar images. IET Radar Sonar Navig. **6**(9), 813–821 (2012)
12. Zhang, W., Li, Q., Wu, Q.J., Yang, Y., Li, M.: A novel ship target detection algorithm based on error self-adjustment extreme learning machine and cascade classifier. Cognit. Comput. **11**(1), 110–124 (2019)
13. Zhang, L., Li, Q., Wu, Q.J.: Target detection for HFSWR based on an S3D algorithm. IEEE Access **8**, 224825–224836 (2020)
14. Huang, G.-B., Zhu, Q.-Y., Siew, C.-K.: Extreme learning machine: theory and applications. Neurocomputing **70**(1–3), 489–501 (2006)
15. Hinz, J.O., Holters, M., Zölzer, U., Gupta, A., Fickenscher, T.: Presegmentation-based adaptive CFAR detection for HFSWR. In: 2012 IEEE Radar Conference, pp. 0665–0670. IEEE (2012)
16. Yang, Y., Wu, Q.J.: Multilayer extreme learning machine with subnetwork nodes for representation learning. IEEE Trans. Cybern. **46**(11), 2570–2583 (2015)
17. Zhang, W., Wu, Q.J., Yang, Y., Akilan, T., Zhang, H.: A width-growth model with subnetwork nodes and refinement structure for representation learning and image classification. IEEE Trans. Ind. Inform. **17**, 1562–1572 (2020)

Correction to: SimFormer: Real-to-Sim Transfer with Recurrent Restoration

Yingnan Ma, Fan Yang, Xudong Li, Chen Jiang, and Anup Basu

Correction to:
Chapter "SimFormer: Real-to-Sim Transfer with Recurrent
Restoration" in: S. Berretti and G.-M. Su (Eds.):
Smart Multimedia, LNCS 13497,
https://doi.org/10.1007/978-3-031-22061-6_33

In the originally published version of chapter 33, erroneously, a second, incorrect affiliation had been added for all of the authors. This has been corrected.

The updated original version of this chapter can be found at
https://doi.org/10.1007/978-3-031-22061-6_33

Author Index

Printed in the United States
by Baker & Taylor Publisher Services